International Symposium on Semiconductor and Electronic Technology (ISSET 2022)

Journal of Physics: Conference Series Volume 2370

Fuzhou, China
12-14 August 2022

ISBN: 978-1-7138-6478-3
ISSN: 1742-6588

Printed from e-media with permission by:

Curran Associates, Inc.
57 Morehouse Lane
Red Hook, NY 12571

Some format issues inherent in the e-media version may also appear in this print version.

Institute of Physics
Dirac House, Temple Back
Bristol BS1 6BE UK

Phone: 44 1 17 929 7481
Fax: 44 1 17 920 0979

techtracking@iop.org

Additional copies of this publication are available from:

Curran Associates, Inc.
57 Morehouse Lane
Red Hook, NY 12571 USA
Phone: 845-758-0400
Fax: 845-758-2633
Email: curran@proceedings.com
Web: www.proceedings.com

TABLE OF CONTENTS

Preface

The 2022 International Symposium on Semiconductor and Electronic Technology (ISSET 2022), scheduled to be held on August 12-14, 2022 in Fuzhou, China, was changed to be held as a virtual conference. ISSET 2022 promoted the development and application of theories and technologies in the fields of semiconductor and electronic technology in universities and enterprises, established business or research contact and found global partners for future careers for all the participants.

During the conference, we had the honor of having invited both Prof. Zhigang Zang from Chongqing University, China and Prof. Jiangzhao Chen from Chongqing University, China to serve as our Conference General Chairs. This online conference was composed of three parts: keynote speeches, oral presentations and online Q&A discussion, attracting 60 delegates from all over the world. Firstly, keynote speakers were each allocated 30-45 minutes to hold their speeches. Then in the oral presentations, the excellent papers we had selected were presented by their authors one by one.

At the beginning of the conference, five distinguished keynote speakers were invited to address their speeches. Among them, Prof. Wenli Zhou from Huazhong University of Science and Technology, China performed a thought-provoking speech on: Reservoir Computing Based on Carbon Nanotube Network. Due to the great interest on the implementation of Reservoir Computing (RC) on novel hardware, her team fabricated a physical reservoir consisting of single-walled carbon nanotubes (SWCNTs) network with phosphomolybdic acid modification and investigated its computing architecture and characteristic dynamics, demonstrating high performances in several test benches. Moreover, Prof. Junjie Qi from University of Science and Technology Beijing, China shared with us his study: In-situ Atomic Scale Studies of Structural Transitions for 2D TMDs. His team studied the phase transition and interlayer structure transition of bilayer PtSe2 under electron beam irradiation and combined atomic resolution STEM and DFT calculation to study the behavior of interlayer stacking orders and interlayer rotations in bilayer PtSe2, revealing the phase transition and interlayer structure transition mechanisms of PtSe2 at atomic scale. Their dramatic speeches had triggered heated discussion in the third part of the conference. Besides, every participant praised this conference for disseminating useful and insightful knowledge.

After months of hard work and preparation, we have selected and compiled a bunch of high-quality papers into the proceedings after rigorously reviewing them. These papers feature but are not limited to the following topics: Semiconductor Physics, Magnetic Semiconductor, Electronic Packaging Technology, Advanced Electromagnetics, etc. All the papers have been checked through rigorous review and processes to meet the requirements of publication.

On behalf of the conference organizing committee, we would like to extend our sincere acknowledgement to all the keynote speakers, peer reviewers, and all the participants. Especially, our thanks go to the Journal of Physics: Conference Series, for the endeavor of all its colleagues in publishing this paper volume. We truly believe that all the attendees have had fruitful discussions and gained valuable knowledge, and will enjoy the opportunity for future collaborations.

Committee of ISSET 2022

Journal of Physics: Conference Series **2370** (2022) 011001 doi:10.1088/1742-6596/2370/1/011001

Committee member

Conference General Chairs
Prof. Zhigang Zang, Chongqing University, China
Prof. Jiangzhao Chen, Chongqing University, China

Technical Program Committee Chair
Prof. Yuan Yang, Xi 'an University of Technology, China
Prof. Wenming Wu, University of Chinese Academy of Sciences, China

Local Committees
A. Prof Dezhong Cao, Xi'an Polytechnic University, China
A. Prof. Guoxing Huang, Fujian Normal University, China
Dr. Qiang Wu, Southwest Jiaotong University, China
Dr. Jianhui Bu, University of Chinese Academy of Sciences, China

Technical Program Committees
Prof. Xiaoyuan Wang, Hangzhou Dianzi University, China
Prof. Yunbin He, Hubei University, China
Prof. Xinbo Zou, ShanghaiTech University, China
Prof. Na Ren, Zhejiang University, China
Prof. Wei Feng, Northeast Forestry University, China
Prof. Jinzhong Wang, Harbin Institute of Technology, China
Prof. Weihua Tang, Xiamen University, China
Prof. Zhongjie Guo, Xi 'an University of Technology, China
Prof. Fengjuan Wang, Xi 'an University of Technology, China
Prof. Sichao Du, Zhejiang University, China
Prof. Xingzhan Wei, Chinese Academy of Sciences, China
Prof. Peng Hu, Northwest University, China
Prof. Song Ye, Chengdu University of Information Technology, China
Prof. Wenli Zhou, Huazhong University of Science and Technology, China
Prof. Yiying Gu, Dalian University of Technology, China
Prof. Zhenyang Zhong, Fudan University, China
Prof. Junjie Qi, University of Science and Technology Beijing, China
A. Prof. Hua Ye, University of Chinese Academy of Sciences, China
A. Prof. Huaxin Wang，Chongqing University, China
A. Prof. Qijin Cheng, Xiamen University, China
A. Prof. Jingkai Qin, Harbin Institute of Technology(Shenzhen), China
A. Prof. Ying Yang, Xi 'an University of Technology, China
A. Research fellow Yanfang Wang, Shanghai Jiao Tong University, China
A. Prof. Yunxin Wang, Beijing University of Technology, China
A. Prof. Liu Chu, Nantong University, China
Dr. Tingwei Zhou, Southwest University, China

Journal of Physics: Conference Series **2370** (2022) 011002 doi:10.1088/1742-6596/2370/1/011002

Peer Review Statement

All papers published in this volume have been reviewed through processes administered by the Editors. Reviews were conducted by expert referees to the professional and scientific standards expected of a proceedings journal published by IOP Publishing Publishing.

- **Type of peer review:** Double Anonymous
- **Conference submission management system:** Morressier
- **Number of submissions received:** 61
- **Number of submissions sent for review:** 60
- **Number of submissions accepted:** 34
- **Acceptance Rate (Submissions Accepted / Submissions Received × 100):** 55.7
- **Average number of reviews per paper:** 2
- **Total number of reviewers involved:** 15
- **Contact person for queries:**
  Name: Fengxin Cen
  Email: alan.cen@gsrassn.org
  Affiliation: Global Scientific Research Association

Journal of Physics: Conference Series **2370** (2022) 012001 doi:10.1088/1742-6596/2370/1/012001

Design of Temperature Measuring Device in Student Dormitory Based on MSP430

Author: Henghao Cheng[1], Xuan Pang[1], Jiaxin Deng[1], Yan Cui[1]

[1]China University of Petroleum (East China), Qingdao, Shandong, 266580, China

Corresponding author: Henghao Cheng

Email: 1319258069@qq.com

Abstract: Aiming at the phenomenon that the temperature of each dormitory is different in winter due to many factors in the dormitory of northern colleges and universities, this paper puts forward a simple and convenient dormitory temperature measurement device. The device takes an MSP430 single chip microcomputer as the control core, collects ambient temperature data through temperature sensor DS18B20, collects battery voltage with AD, and displays the collected battery voltage and ambient temperature through an OLED display screen. At the same time, according to the temperature measured by the temperature sensor, the device can output PWM to control the speed of the motor and simulate the intelligent temperature control fan. The experimental data shows that the device has high accuracy, is easy to use, and is easy to implement.

1. Introduction

With the continuous development of modern sensors in the field of electronic applications, temperature measurement and monitoring have become more and more important in people's lives. In daily life, different temperatures have a great impact on our college students' learning status [1].

The winter temperature in the north is low, so most of the student dormitories in Colleges and universities in the north will be equipped with heating. However, due to the geographical location of the dormitory, the quality of heating equipment, and other factors, the temperature of each dormitory is different, and some students in the dormitory will feel that the temperature of the dormitory is very low. Therefore, to facilitate these students to measure the temperature of the dormitory, some scholars designed a high-precision temperature measurement system based on FPGA [2], while others designed an embedded temperature measurement system based on DSP [3]. However, the design process of these systems is cumbersome and difficult for students to implement. This paper designs a temperature measuring instrument based on MSP430 single-chip microcomputer. Just turn on the power switch, and the measured temperature value can be directly displayed through the OLED display screen, which is very convenient and simple. Moreover, after inserting the small fan, the device can output PWM to control the speed of the DC motor according to the temperature measured by the temperature sensor and simulate the intelligent temperature control fan. Finally, through the experimental test, it can be concluded that the device has high accuracy and can be used.

2. The overall structure of the system

The block diagram of the overall structure of the dormitory temperature measuring device system is shown in Figure 1. It can be seen that the system can be divided into seven modules: main control chip,

RS232 serial communication module, power supply voltage stabilization module, AD voltage acquisition module, temperature acquisition module, OLED display module, and motor drive module. In the system, MSP430G2553 single-chip microcomputer is used as the main controller, AMS1117-3.3 voltage stabilizing module is used to supply power to the single chip microcomputer, temperature sensor DS18B20 is used to collect ambient temperature data, and AD is used to collect battery voltage. The collected battery voltage and ambient temperature are displayed on the OLED display screen, and PWM is output to control the rotation speed of the motor.

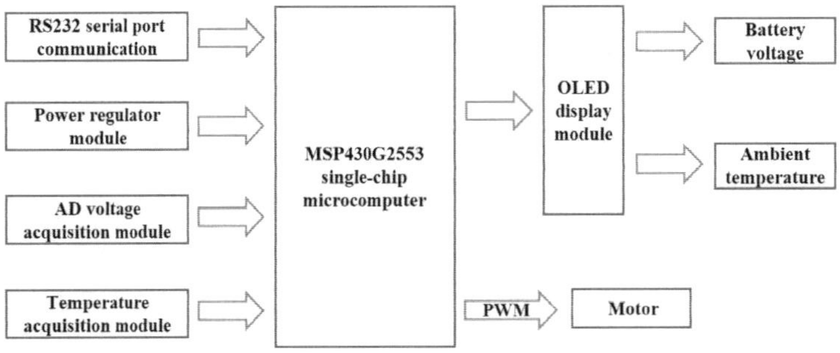

Figure 1 Block diagram of the overall system structure

3. Hardware circuit design of the system

3.1 Main controller

The main control chip is the core of data processing, which can connect different modules and coordinate the reasonable operation of each module simultaneously. Therefore, the choice of the main control chip is important.

The main control chip of this device adopts MSP430G2553 single-chip microcomputer. The single-chip microcomputer has many characteristics, such as ultra-low power consumption, strong processing power, fast operation speed, and stable system work [4].

The minimum system circuit of the MSP430G2553 single-chip microcomputer is shown in Figure 2. In the figure, the No.1 and No.28 pins of the single chip microcomputer are respectively connected with the 3.3V power supply and the ground to supply power for the single chip microcomputer. Pin No.5 (P1.3) is connected to the power supply module to collect the battery voltage, so that the single chip microcomputer can read the voltage value in the power supply module. Pin No.1 (P2.1) is connected to the temperature measurement module. Pin No.3 (P1.1) and pin No.4 (P1.2) are respectively connected with RXD_232 and TXD_232 serial ports of MAX3232 chip of serial communication module to realize serial communication. Pin No.19 (P3.5) and Pin No.20 (P3.6) are respectively connected with the corresponding pins of the L9110S chip of the motor drive module and control the motor to output corresponding PWM waves. Pin No.16 (P2.3) is connected to the CS# pin of the OLED chip, which is used to output the chip selection signal of the OLED and control the chip selection of the OLED. Pin No.17 (P2.4) is connected to the RES# pin of the OLED, which is used to output the reset signal of the OLED, and the OLED can be reset by outputting a low level. Pin 18 (P2.5) is connected to the D/C# pin on the OLED, which is used to output data and commands to control the OLED display content. Pin No.2 (P1.6) is connected to pin D0 of OLED, which is used to transmit the clock signal to OLED and control the timing of OLED. Pin No.23 (P1.7) is connected to pin D1 of OLED and is used to send the data to be displayed to OLED, such as prompt text, voltage, temperature, etc.

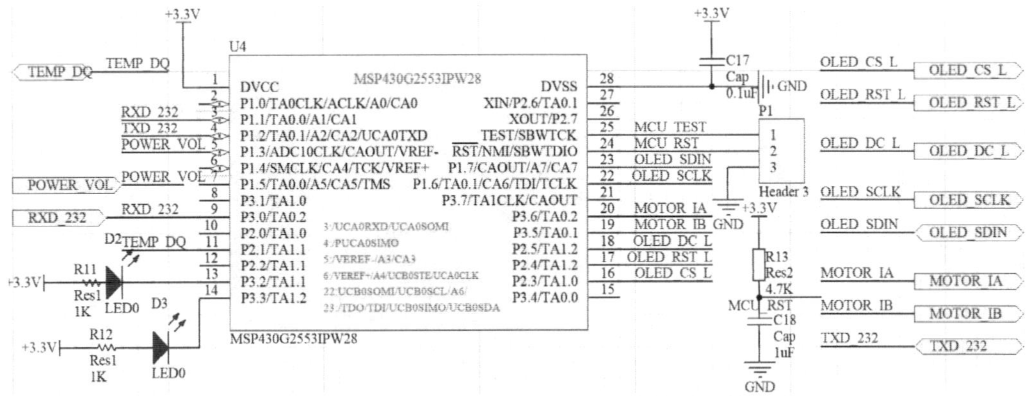

Figure 2 Circuit diagram of the minimum system of single chip microcomputer

3.2 Power supply module

The power supply voltage of the MSP430 single-chip microcomputer is generally between 1.8 and 3.6V. Therefore, in order to make MSP430 single-chip microcomputer work normally, the power supply module of this device takes the AMS1117-3.3V chip as the core, and the lithium battery with a rated voltage of 3.7V will finally output a stable voltage of 3.3V to supply power to the single-chip microcomputer.

AMS1117 is a three-terminal voltage regulator. Its working principle is to sample the output voltage and then feed it back to the regulating circuit to adjust the impedance of the regulating tube of the output stage. By judging whether the output voltage is high or low, it can adjust the impedance of the output stage to change the voltage drop of the regulating tube and, finally, maintain the stability of the output voltage [5].

The specific circuit of this power module is shown in Figure 3. The module uses AMS1117 as the core chip. AMS1117 can reduce the voltage of lithium batteries to the working voltage range of single-chip microcomputers. At the same time, it is matched with two polar capacitors and two common patch capacitors to form a filter circuit, which plays a role in stabilizing voltage and filtering, and can continuously and stably output 3.3V voltage. In addition, a line is connected to the OLED between the positive electrode of the battery and the ground in order to collect the battery voltage with AD.

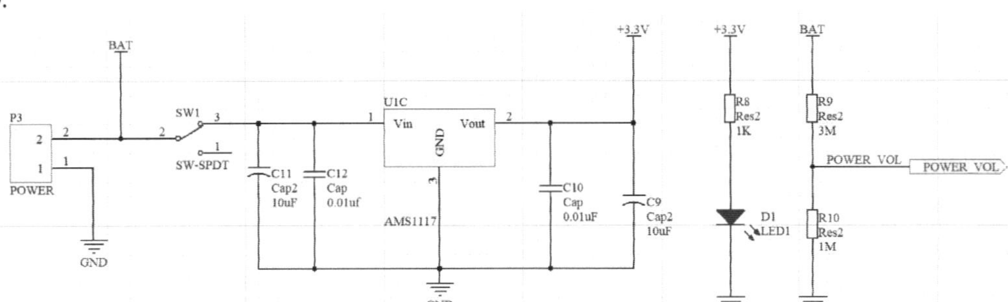

Figure 3 Power module-specific circuit diagram

3.3 Temperature acquisition module

In order to measure the external environmental temperature, it is necessary to adopt a temperature sensor, which converts the analog quantity of the sensed temperature into a digital signal that can be used for output, and the data is received by a single-chip microcomputer.

This device adopts DS18B20, which is common in the market, as the temperature sensor. This digital temperature sensor is common, with its small size, high precision, convenient use, and low price, and is suitable for various non-extreme temperature occasions. The step of reading the temperature is very complicated. First, the temperature conversion should be realized, and the delay time is 750 ms. Then, different pulse signals are generated by high and low-temperature coefficient crystal oscillators and sent to the corresponding subtraction counters as pulse inputs. Finally, DS18B20 counts the generated clock pulses to complete the temperature measurement [6].

The specific circuit of the temperature acquisition circuit module is shown in Figure 4. Pin 1 and Pin 3 of the DS18B20 chip are respectively grounded and powered by the power supply, and Pin 2 converts the collected temperature analog signal into a digital signal and sends it to the single-chip microcomputer's main control chip.

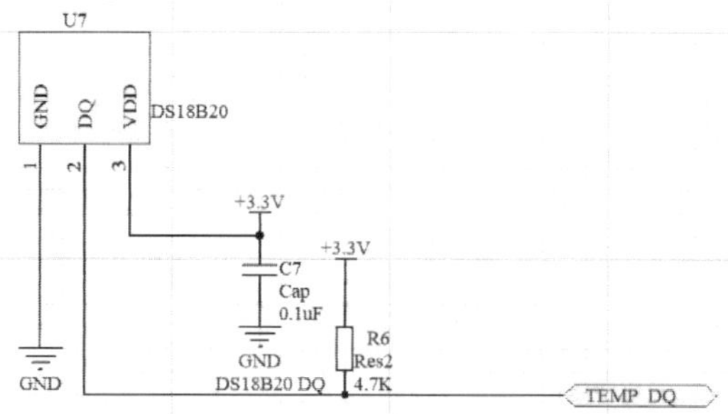

Figure 4 Circuit diagram of the temperature acquisition module

3.4 Serial communication module

Because it is to realize the communication between single-chip microcomputer and PC, MAX232 is used to convert TTL level to RS232 type. The MAX232 chip is internally provided with an electrostatic protection circuit to protect the device from personal electrostatic damage. It contains two groups of independent RS232 level conversion channels, which can convert the TTL level output by the single-chip microcomputer into 232 level or 232 level output into TTL level, which is convenient for the single-chip microcomputer to receive [7].

The specific circuit of the serial communication module is shown in Figure 5. The whole circuit is very simple, and there are only a few capacitors around the circuit. Capacitors C14 and C15 provide the internal voltage doubling charge pump of MAX3232. Pin 2 and pin 6 of MAX3232 are two times positive and negative voltage output terminals, respectively. C13 is the filter capacitor of positive output voltage, and C16 is the filter capacitor of negative output voltage. The communication function of the serial port is completed by three wires: ground wire, sending wire, and receiving wire. This chip shares the ground with the GND of another single-chip microcomputer, and two interfaces, RXD_232 and TXD_232, are cross-connected to realize TTL level communication. Because serial communication is asynchronous, the port can send data on one line and receive data on another line.

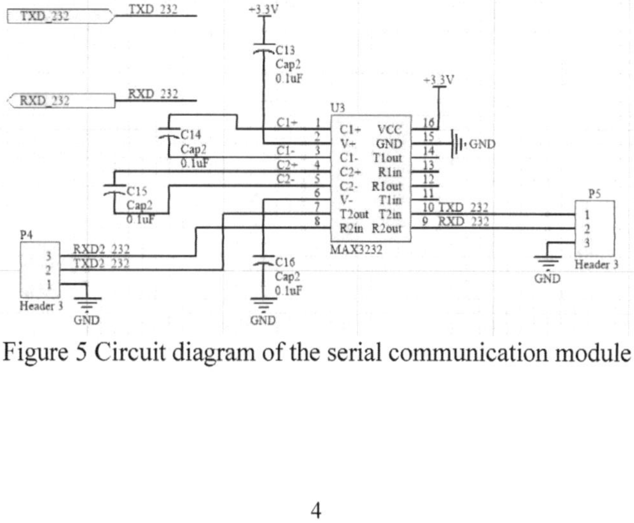

Figure 5 Circuit diagram of the serial communication module

3.5 OLED display module

In order to display the measured temperature and battery voltage, an OLED display screen is needed. This device adopts the electronic OLED display produced by Zhongjingyuan, and its size is 0.96 inches. The OLED can choose white, blue, and yellow blue colors: white is pure white, that is, the screen is black, and the font is white; Blue is pure blue, that is, the screen is black, and the font is blue [8]; The blue screen can be divided into two parts, the upper part is yellow, and the lower part is blue. This OLED has a high resolution and a variety of interface modes, and this device adopts the SPI interface mode.

The specific circuit of the OLED display module is shown in Figure 6. The D0 pin is the clock pin, the D1 pin is the data pin, the D/C#E pin is the data and command control pin, and the CS# pin is the chip select pin. C4 and C5 capacitors are used for internal boosting, and the resistance value of the resistor connected with the 26 pins should be about 560K. If it is too small, it may affect the life of the display screen; if it is too large, the brightness of the screen will be dark.

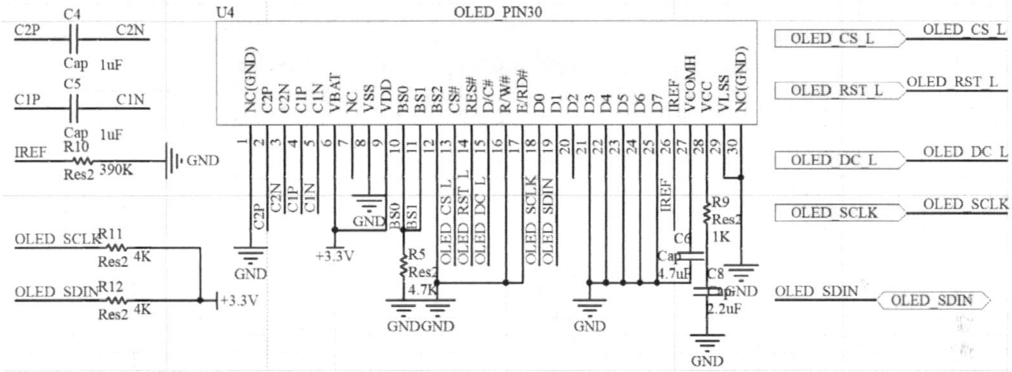

Figure 6 Circuit diagram of the OLED display module

3.6 Motor drive module

As the amplitude of the voltage directly output by the single-chip microcomputer is limited, its value can't achieve the purpose of driving the DC motor, so it is necessary to amplify the current to drive the DC motor to rotate and simulate the intelligent temperature control fan [9].

The motor drive module of this device mainly adopts the L9110S motor drive chip. The L9110S driver chip is widely used. It has a built-in H-bridge driver circuit designed with a MOS transistor, and its low on-resistance (only 0.56Ω) ensures that the chip has extremely low power loss, which makes the L9110S have a longer working time. SOP8 /DFN8 package is adopted. Although there are only eight pins, the control logic is rich and can meet almost all the requirements of motor speed control. It can drive 800ma continuous current and has a built-in over-temperature shutdown protection circuit. Once the motor is locked, the L9110S can automatically shut down.

The specific circuit of the motor drive module is shown in Figure 7. The 2-pin and 3-pin connection can handle 800mA of 3.3V DC external power supply, and the GND pin is connected to GND, in which the capacitance can increase the reliability of the chip. Pin 6 and 7 are connected with the single chip microcomputer; Pin 1 and 4 are connected to the motor as output terminals, which have great current driving ability and can directly drive the motor to move in the forward and backward directions.

Among them, according to the temperature measured by the temperature sensor, the single-chip microcomputer can output PWM waves with different duty ratios to control the speed of the DC motor so as to analog the intelligent temperature control fan.

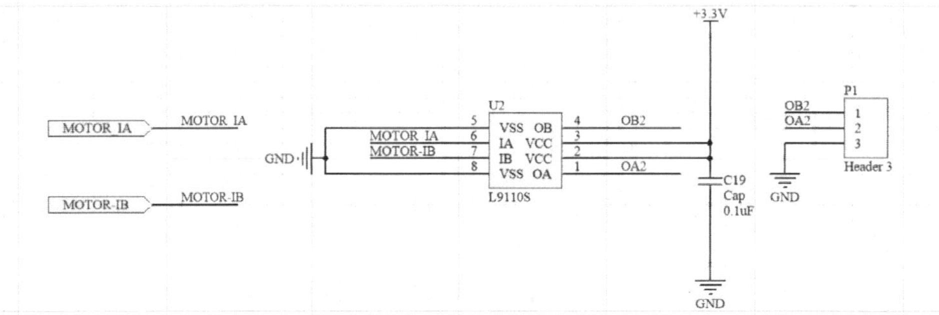

Figure 7 Circuit diagram of the motor drive module

4. Software design of the system

The software design of the system is programmed by CCS, and the program is divided into different modules, and the program is modularized, which is convenient to write, modify and burn each module. The software module can be divided into AD voltage acquisition module, DS18B20 temperature acquisition module, OLED display module, and PWM output module. The software flow chart is shown in Figure 8.

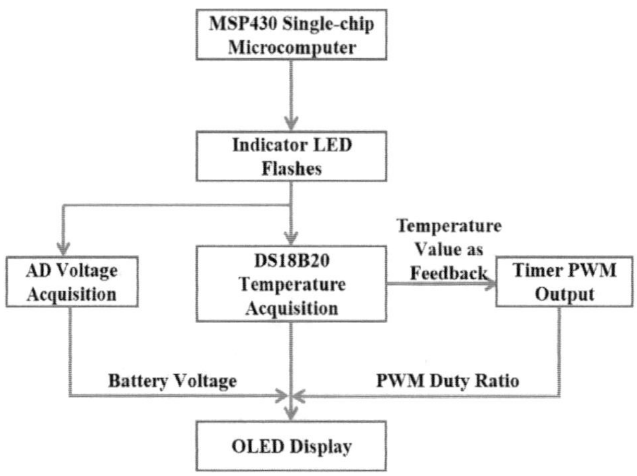

Figure 8 Flow chart of software design

4.1 AD voltage module program

The module program collects the battery voltage after resistance partial voltage through channel A3, and the last step uses the corresponding formula to calculate according to the selected internal reference voltage. For example, if the resistance division is 1:3 and the measured voltage is 1/4 of the battery voltage, it is multiplied by 4 to obtain the battery voltage.

4.2 Temperature acquisition program

According to the DS18B20 manual, the module initializes and wakes it up, then writes the instruction, finally reads the 16-bit data sent by it, and gets the final ambient temperature by weighting.

4.3 OLED display program

This device mainly uses a seven-pin SPI interface OLED, in which D0 is the clock pin for clock input, D1 is the data pin, and the DC pin is used for command control. Different signals are written to different pins by the program to realize the OLED display.

4.4 PWM output program
The module program not only defines the PWM output of the pin, but also takes the read temperature value as the feedback value, and corresponds to the temperature from 24 to 34 degrees to ten PWM output levels, thus achieving the function of making the PWM output change with the real-time changes of the environment.

5. Experimental data

5.1 Temperature measurement
The temperature measuring device designed in this paper is shown in Figure 9. It is placed in several different dormitories to measure the temperature. After the data is stable, it is compared with the data measured by the temperature gun. The specific data are shown in Table 1. According to the comparison, the temperature measurement device designed in this paper is more accurate.

Table 1 Test results of temperature measurement data

Temperature measured by this device	Temperature measured by the temperature measuring gun	Voltage value
25.1°C	25.1°C	3.7V
25.7°C	25.6°C	3.7V
23.2°C	23.4°C	3.7V
26.9°C	26.8°C	3.7V
24.4°C	24.5°C	3.7V

Figure 9 Physical diagram of temperature measuring device

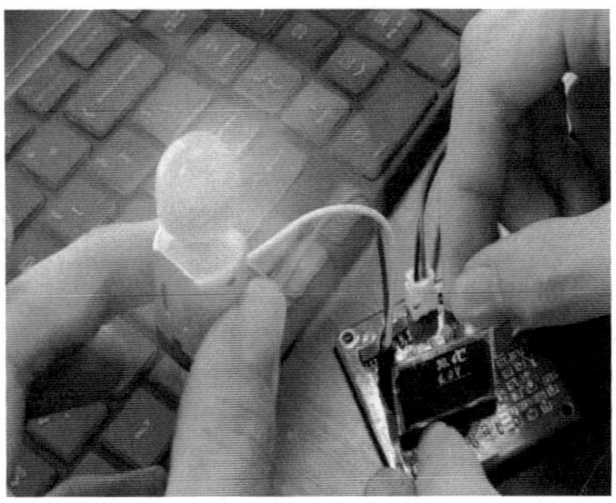

Figure 10 Working diagram of intelligent temperature control fan

5.2 Intelligent temperature control fan
Connect the small fan controlled by the motor to the device, and its operation is shown in Figure 10. It is obvious that the speed of the fan changes with the temperature, and the intelligent temperature control fan is successfully simulated.

6. Summary
This paper designs a simple temperature measuring device based on MSP430 single-chip microcomputer. The device can measure the ambient temperature of the dormitory and simulate the intelligent temperature control fan. The rotation speed of the fan will change with the measured temperature. After many and repeated tests, it can be determined that the device system operates stably, and the deviation between the collected ambient temperature data and the real value is small, which has high practicability and popularization.

Reference

[1] Meteorological knowledge in life [j] Guizhou Meteorology, 2008 (02): 25.
[2] Wang L Q, Wen F and Zhang K H Design of high precision S-type thermocouple temperature measurement system based on FPGA [j] Application of Electronic Technology, 2022,48 (02): 41-45.
[3] Liu Y J Design and implementation of an embedded temperature measurement system based on DSP [j] Foreign Electronic Measurement Technology, 2014,33 (01): 78-81.
[4] Chen S C, Li X B, Cui M and Su L W Design of wireless temperature and humidity alarm system based on MSP430 [j] Electronic Design Engineering, 2019,27 (10): 183-187.
[5] Liao Z Z A dynamic seat with music rhythm [p] Guangxi Zhuang Autonomous Region: CN212546264U, 2021-02-19
[6] Si S W Design of combustible gas detection device with wireless transmission function [d] Hebei: Hebei University of Science and Technology, 2011.
[7] Chen J Design and implementation of an input turbidimeter based on near-infrared light [d] Chongqing: Chongqing University, 2019.
[8] LV L H Design of electronic control system for interactive brushing guidance model [d] Shaanxi: Xi'an University of Electronic Science and Technology, 2015.
[9] Nantong University An AGV trolley with omni-directional drive and steering: CN201610815943.9 [P] 2017-05-10.

A CMOS down-conversion mixer for microwave radar application

Shaopan Lin[a], Boming Su[a], Sikai Chen[a], Yi Wu[a], Guochi Huang[a*]

[a] College of Photo and Electronic Engineering, Fujian Normal University, Fuzhou City, Fujian Province, 350007, China

[*]Corresponding author: guochi.huang@fjnu.edu.cn

Abstract: This paper introduces a CMOS down-conversion mixer for microwave radar application. The mixer is based on the classical Gilbert cell architecture, using PMOS current bleeding technology and cross-coupling as a transconductance stage to further elevate the conversion gain and decrease the noise factor, while introducing source negative feedback technology to enhance the linearity performance of the mixer. Using the SMIC 0.13 um CMOS process, the simulation results prove that the frequency conversion gain of the mixer is 12.31 dB, the noise factor is 21.16 dB, and the -1dB compression point of the output is -12.66 dBm when the input RF signal is 10 GHz and the output IF signal is 10 MHz. The third-order intermodulation cut-off point is -1.70 dBm, and the power dissipation is 3.16 mW under a 1.2 V power supply.

1. INTRODUCTION

In recent years, the fifth generation mobile communication(5G) has become a hot spot for global research and development. The rapid development of mobile communication technology is inseparable from the exploration and research of RF front-end integrated circuit chips. Meanwhile, since the advantages of high integration and low power dissipation of the CMOS process have been continuously improved, designing RF integrated circuits using the CMOS process has become a hot research topic [1]. The front end of the transceiver plays a major role in the wireless communication system, while the mixer is the core module in the whole RF transceiver system, so its performance is very important. For a receiver system, a good mixer can reduce the front-end LNA gain module on the gain of pressure, at the same time its linearity affects the dynamic range of the receiver system, and noise also determines the sensitivity of system performance. Therefore, it is of great significance to design a mixer to ensure the excellent performance of gain, noise coefficient, linearity, and power dissipation.

Mixers can be divided into passive mixers and active mixers. Passive mixers have a large noise factor and generally need to provide large local vibration power, so their application is limited. Compared with passive mixers, active mixers has the advantages of high gain, low noise, and the double-balanced active mixer based on the Gilbert multiplier not only has the advantages of high isolation, high conversion gain, and low noise figure, but also can easily optimize and improve the circuit structure to achieve the compromise adjustment of its various performance indicators. Therefore, this structure is widely used in the current mixer design.

The architecture of a double-balanced active mixer based on the Gilbert multiplier is shown in Figure 1, where the conversion gain is determined by the transconductance(GM) and load impedance(RL) of the transconductance stage transistor. The transconductance(GM) of the transconductance stage transistor is limited by the bias current and overdrive voltage. At present, various performance-enhancing optimization technologies[2][3][4][5] have emerged based on this architecture. The mixer transconductance stage with the current bleeding technique[7] has become a common and effective method to improve the conversion gain. It can increase the current flowing through the transconductance stage while keeping the current flowing through the switching transistor and the load unchanged, so it will not lead to the reduction of load impedance (RL) and increase of overdrive voltage.

In addition, in order to have enough linearity in the whole RF transceiver system, the design of the mixer must also have good RF input matching. The main work of this paper is to optimize the double-balanced Gilbert mixer using current bleeding technology, by introducing cross-coupling in the transconductance stage to further improve the transconductance of the mixer, by introducing negative feedback technology in the source terminals to ensure the linearity of the mixer.

The rest of this work is organized as follows: The second part describes the proposed optimization method for mixers. The third section describes the measurement results, and the last is the summary.

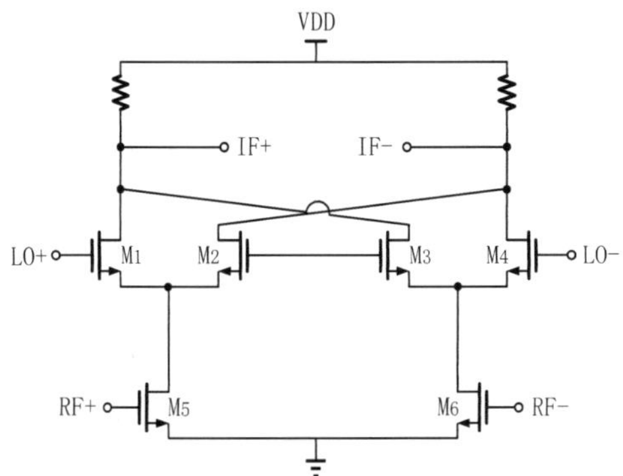

Figure 1. A double-balanced active mixer based on Gilbert multiplier.

2. CIRCUIT DESIGN

A double-balanced mixer structure using current bleeding technology is shown in Figure 2. This technology introduces a pair of PMOS transistors into the transconductance stage of the mixer together with the original NMOS transistor as the transconductance stage. In this structure, the DC current through the transconductance stage NMOS transistors is the sum of the DC current through the LO switching stage and the DC current introduced through the transconductance stage PMOS transistors. Therefore, under the condition that the DC current flowing through the transconductance stage NMOS transistors remains unchanged, the current flowing through the LO switching stage can be actively reduced, which means that the size of the switching stage transistors can be reduced to reduce the gate-source parasitic capacitance of the switching stage to improve the switching efficiency. In addition, with the LO switching current decreasing, the load resistance can be appropriately increased to improve the frequency conversion gain of the mixer. Under this structure, the total transconductance of the circuit is the sum of the transconductance of NMOS transistor and PMOS transistor, which can be expressed as:

$$\mathrm{Gm} = \mathrm{gm}_{7,8} + gm_{5,6} \tag{1}$$

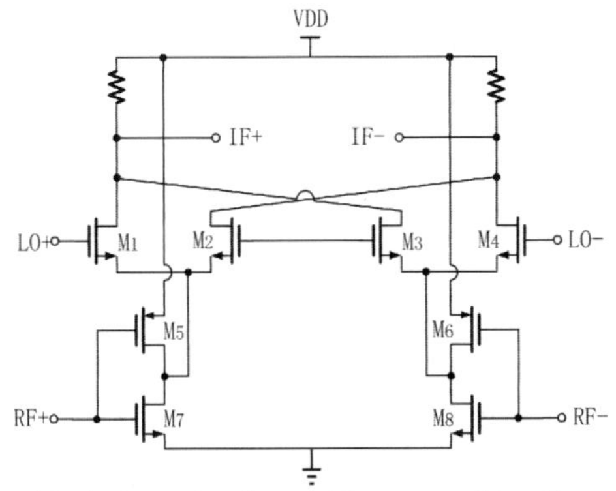

Figure 2. Double-balanced mixer with current bleeding technology.

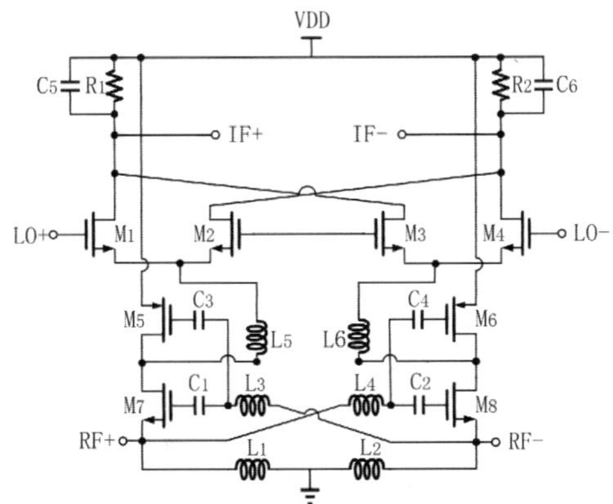

Figure 3. The proposed mixer.

The mixer proposed in this paper is shown in Figure 3. The proposed mixer is optimized for the load stage and the transconductance stage. The load level consists of R_1 and C_5 to form the RC load network. The RC load network can ameliorate the gain of the circuit, and can also play a role in filtering to reduce the influence of the load level noise on the circuit. The equivalent load impedance RL can be expressed as:

$$\mathrm{RL} = \frac{R_1}{1 + j\omega R_1 C_5} \tag{2}$$

Usually, mixer gain, noise factor, linearity, and other performance are determined by the transconductance stage circuit, so the design of the transconductance stage circuit is very important. The proposed mixer transconductance stage transistors consist of M_5, M_7, and M_6, M_8, which construct two PMOS-NMOS complementary transistors. Capacitors C_1, C_2, C_3, and C_4 are used to isolate DC and are connected in series with inductors L_3 and L_4 to increase frequency conversion gain and reduce noise. Inductors L_1 and L_2, on the one hand, are used to form a DC path and do not consume DC voltage drop; on the other hand, they can also achieve resonance with the capacitance of the node, including M_1, M_2 gate-source parasitic capacitance and solder pad parasitic capacitance, so as to reduce the noise coefficient and improve the third-order intermodulation cut-off point (IIP3). Inductors L_5 and L_6 are connected in series to the common source terminals of switch stage pairs on both sides respectively. By properly adjusting the value of inductors L_5 and L_6, the parasitic capacitance of the common source terminal can be resonated to suppress the charging and discharging effect of the parasitic capacitance of the common source terminal, thus reducing the 1/ f noise of the switch transistors.

Compared with Figure 2 traditional current bleeding technology, two levels of the transconductance of NMOS transistor gate respectively with capacitor C_1, C_2, C_3, C_4, and inductance L_3 and L_4 series and coupling to the other side of the RF input. In this way, the RF signal input from the source of M_7 and M_8 can be cross-coupled to the gate of two pairs of transconductance transistors composed of current bleeding technology for a second amplification. Therefore, the transconductance of the circuit can be expressed as:

$$\mathrm{Gm} = \left(\frac{2C_{1,2}}{Cgs_{7,8}+C_{1,2}}gm_{7,8}\right)+gm_{5,6} \tag{3}$$

In this formula, $Cgs_{7,8}$ is the gate-source parasitic capacitance of transistors M_7, M_8, $GM_{5,6}$, and $GM_{7,8}$ are the transconductance of four transconductance transistors. It can be seen that when $C_{1,2} >> Cgs_{7,8}$, the equivalent transconductance Gm can be expressed as $2gm_{7,8}+GM_{5,6}$. Therefore, the total frequency conversion gain of the mixer can be expressed as:

$$\mathrm{Av} = \frac{2}{\pi}\mathrm{Gm}RL = \frac{2R_1(2\mathrm{gm}_{7,8}+gm_{5,6})}{\pi(1+j\omega R_1C_5)} \tag{4}$$

In addition, reasonable adjustment of capacitor C_1, C_2, C_3, and C_4 values can easily match the rf input port with inductors L_1, L_2, L_3, L_4, L_5, and L_6 to 50 Ω. Figure 4 shows the layout of the mixers proposed in this article. The total area of the layout is 636×602.674 um^2.

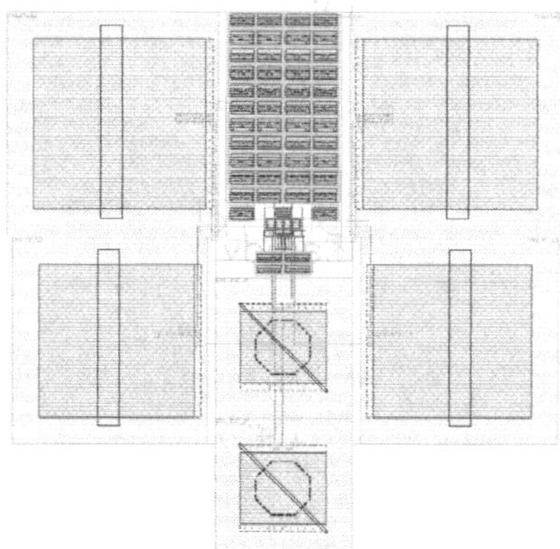

Figure 4. The layout of the proposed mixer.

3. SIMULATION RESULTS

With the Cadence Virtuoso tool, representative different process corners(TT, SS, and FF) are selected to conduct simulation and validation of the mixer, where "TT" refers to "Typical NFET and Typical PFET", "SS" refers to "Slow NFET and Slow PFET", and "FF" refers to "Fast NFET and Fast PFET". For the RF input port source, the sine wave is defined with a frequency specified at 10 GHz and an amplitude specified at -30 dBm. The LO port source is defined as a sine wave with a frequency specified at 10.01 GHz and an amplitude specified at 0 dBm. The output IF port is defined as a dc source. Supply voltage specified 1.2 V. Figure 5 shows the results of the proposed mixer's gain and noise factor and RF port matching S11 at different process corners, where the green curve represents typical "TT", the red curve is "FF", and the yellow curve is "SS".

Figure 5(a) shows that the gain of the proposed mixer is 12.31 dB at "TT", 10.14 dB at "FF", and 12.20 dB at "SS". The difference in gain results remains within 2.17 dB under different processing corners.

Figure 5(b) shows that the NF of the proposed mixer is 21.16dB at "TT", 21.97dB at "FF" and 21.53dB at "SS", showing a small fluctuation.

Figure 5(c) shows the result of S11 matching in the RF port.When S11<10dB, the bandwidth is 9.37 GHz-10.42 GHz under "TT" and 9.91 GHz-10.81 GHz under "FF". The bandwidth under "SS" is 9.04 GHz to 10.33 GHz. When the RF frequency is 10 GHz at different process corners, the bandwidth is within the range, the ports match well, and the bandwidth is about 1 GHz.

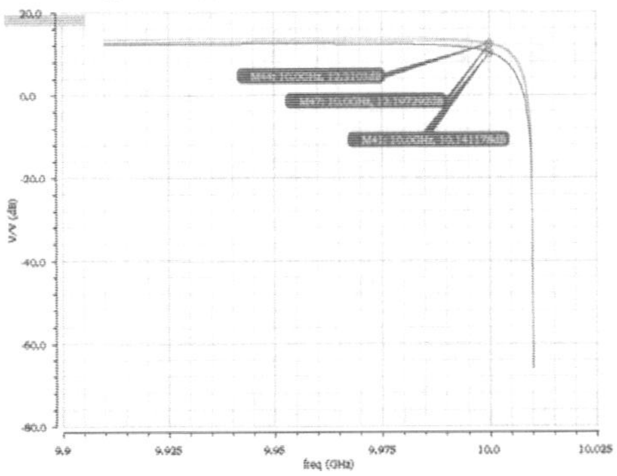

(a)Gain simulation results

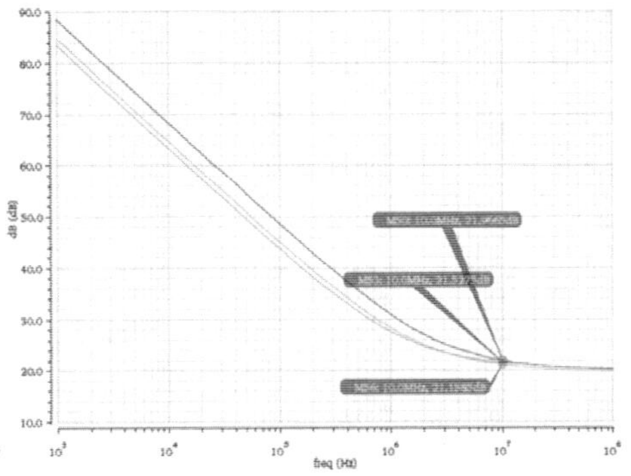

(b)NF simulaion results

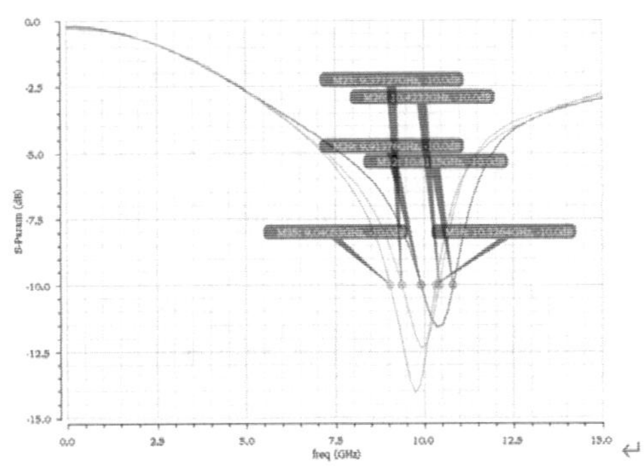

(c)S11 simulation results

Figure 5. Simulation results of gain, noise factor and S11 at different process corners.

HB was used to analyze the 1dB compression point when the intermediate frequency output was 10 MHz, and the result was -12.66 dBm, as shown in Figure 6.

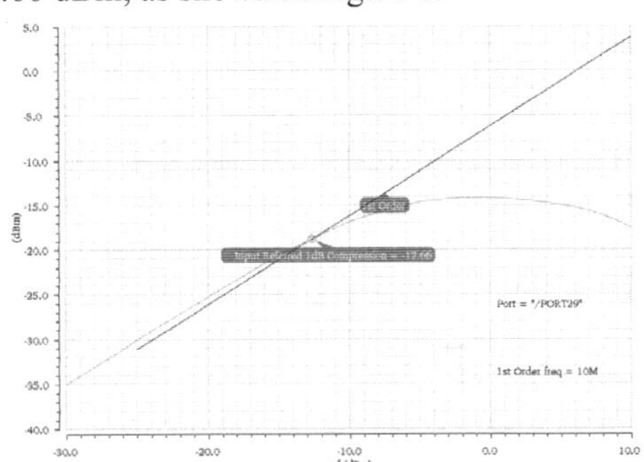

Figure 6. -1dB compression point obtained at 10MHz by HB analysis.

PSS and PAC were used for IIP3 analysis. For the fundamental wave of 10 MHz, RF dual-tone signals were selected as 10 GHz and 10.001 GHz, so the first harmonic was 9 MHz and the third harmonic was 11 MHz. The simulation result of IIP3 was -1.70 dBm, as shown in Figure 7.

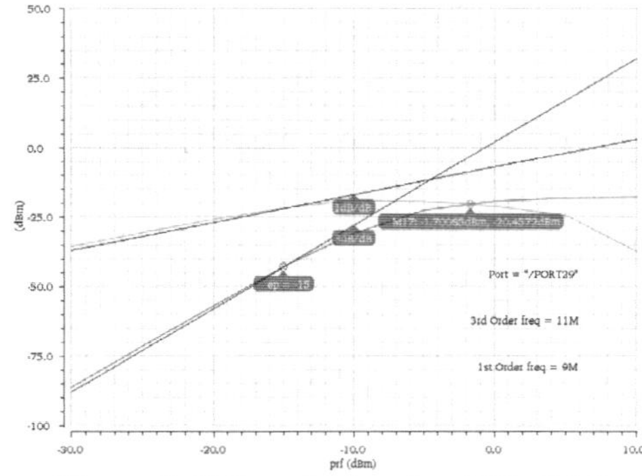

Figure 7. IIP3 analysis using PSS+PAC.

Table 1 shows the performance of the mixers proposed in this article, and also lists the performance of the mixers in several other published articles for comparison.

Table 1. Performance summary compared to other mixers.

reference	This work	[4]	[2]	[3]	[8]
RF(GHz)	10	2.401	2.4	5.2	9.9
CMOS tech(nm)	130	130	65	180	180
Supply Voltage(V)	1.2	1.2	1.5	1.8	1.8
DC Power (mW)	3.16	5.82	3.82	——	9
Gain(dB)	12.31	13.61	14.64	9.48	7.913
P_{1dB}(dBm)	-12.66	-13	——	4	-15.688
IIP3 point(dBm)	-1.7	-4.46	-15.89	——	-5.907

4. CONCLUSION

In this work, SMIC 0.13 um CMOS process is used to design a Gilbert double-balanced mixer for a 10 GHz microwave radar application. In this paper, the common current bleeding technology is used to improve the gain of the mixer, and cross-coupling technology and source negative feedback technology are added to the transconductance stage to further optimize the performance. The simulation results show that the gain of the mixer is 12.31 dB, the noise factor is 21.16 dB, the 1dB compression point is -12.66 dBm, the third-order intermodulation cut-off point is -1.70 dBm, and the power dissipation is 3.16 mW. Moreover, the RF port is well matched after the introduction of cross-coupling technology.

REFERENCES

[1] PH Woerlee, M.J.Knitel, R.V.Langevelde, D.B.M.Klaassen, L.F.Tiemeijer, A.J.Scholten, et al., "RF-CMOS performance trends", IEEE Trans Electron Devices, vol. 48, no. 8, pp. 1776-1782, Aug. 2001.

[2] Sung-Hyun Pang, Youngwoon Kim and Tae-Yeoul Yun, "High-Gain, Low-Power, and Low-Noise CMOS Mixer Using Current-Reused Bleeding Amplification", IEEE Microwave and Wireless Components Letters, pp. 418-421,2021

[3] H.Saoudi, M.Dhieb, H.Ghariani and M.Laniani, "A 5.2 GHz High Gain High Isolation Double Balance Gilbert Cell Mixer With Linearity Improvement In TSMC 180nm CMOS Technology For WLAN Application", 2017 18th International Conference on Sciences and Techniques of Automatic Control and Computer Engineering(STA), pp. 133-138, 2017.

[4] W. K. Chong, et al, "A 0.12-mm^2 2.4-GHz CMOS Inductorless High Isolation Subharmonic Mixer With Effective Current-Reuse Transconductance", IEEE Trans. Microw. Theory Techn. 63(2015)2427, July 2015.

[5] Seong-Kyun Kim, Chenglin Cui, Guochi Huang, SoYoung Kim and Byung-Sung Kim, "A 77 GHz Low LO Power Mixer With a Split Self-Driven Switching Cell in 65 nm CMOS Technology", IEEE Microwave and Wireless Components Letters, vol. 22, pp. 480-482, Sept. 2012.
[6] Arif A. Siddiqi, "Design Methodology And Investigation Of GHz Range RF CMOS Mixers", M.S.thesis Department of Electronics, Sept. 2000.
[7] S.G.Lee and J.K.Choi, "Current-Reuse Bleeding Mixer", Electron.Lett, vol.36, no.8,pp.696-697, Apr. 2000.
[8] Anju Katarmal, Akash Mecwan and Manish Patel, "RF CMOS Double Balanced Gilbert Cell Mixer for 5G Application", 2021 3rd International Conference on Signal Processing and Communication(ICPSC), May. 2021.

A spatiotemporal registration method between roadside lidar and camera

Jianguo Zhu[1], Yanyan Dong[2,*], Xiaobin Xu[1], Jiewen Wang[1], Jianqing Wu[3]

[1] Shandong High-Speed Dongying Development Co.LTD, Dongying, China, 25700

[2] School of Traffic and Transportation, Lanzhou Jiaotong University, Lanzhou, China, 730070

[3] School of Qilu Transportation, Shandong University. Jinan 250061, China;

*Corresponding author: 11200004@stu.lzjtu.edu.cn

Abstract. With the rapid development of intelligent transportation, it is of great significance to improve road traffic safety and traffic efficiency by tracking road users in real time. In this paper, a roadside multi-source data acquisition platform is designed based on two main sensors, lidar and camera, to perform time synchronization between lidar data and camera data. The point cloud data and image data of the same frame are the research objects. Based on Zhang Zhengyou's calibration method and the plane target model theory, the camera internal parameters and the joint external parameters of the lidar and camera are respectively calibrated to obtain the internal and external parameter matrix. The error is 0.09 pixels, and the external parameter calibration error is 0.22 pixels. The registration test of lidar and camera under different roadside layout conditions is designed to verify the adaptability of the registration method. When the camera is arranged and installed, the distance between the horizontal and vertical directions should not exceed 90 cm.

1. Introduction

With the continuous increase of car ownership, the complex traffic environment leads to increasing traffic pressure, and problems such as traffic congestion and traffic safety are becoming more and more significant. Therefore, obtaining reliable and accurate road target trajectory information plays a crucial role in judging the traffic situation, reducing the risk of traffic conflict, and ensuring the efficiency of road traffic. The current common multi-target trajectory tracking methods mostly rely on a single sensor, which is prone to problems such as low accuracy and poor robustness [1]. Since the current multi-sensor registration methods are mostly based on vehicle-mounted objects [2-4], and roadside sensors are arranged in different poses, choosing a registration method suitable for roadside sensors is helpful to obtain reliable and accurate registration results. In view of this, The roadside multi-source data acquisition platform is designed based on two main sensors, lidar and camera. The calibration board is used as the sensor data registration medium to collect the lidar and camera data of the calibration board under different attitudes. Based on the lidar data collection frame and the timestamp relationship between the two kinds of data, time synchronization between the lidar data and the camera data is carried out; the point cloud data and image data of the same frame after time synchronization are taken as the research objects; based on calibration method proposed by Zhang et al. [5,6] and plane target model theory, registration calibration was performed on the camera internal parameters and the combined external parameters of the lidar and the camera to the internal and external parameter matrix; the registration test of the lidar and the camera under different roadside layout conditions were designed to verify the adaptability of the registration method. The obtained internal and external parameter matrix lay the foundation for the subsequent trajectory information fusion.

2. Roadside lidar and camera spatiotemporal registration

2.1. Roadside multi-source data collection platform

In this paper, a roadside multi-source data acquisition platform is designed, which is mainly composed of HY1080 camera produced by Jerry Microcom, a lidar produced by Leishen, a portable power supply, an inverter, a camera and a notebook computer. The portable power supply is a 12V lead-acid battery, and the voltage is raised to 220V through the inverter for the laptop and lidar to work normally. The camera is a USB interface, which can be directly plugged into the laptop for use, as shown in Figure 1.

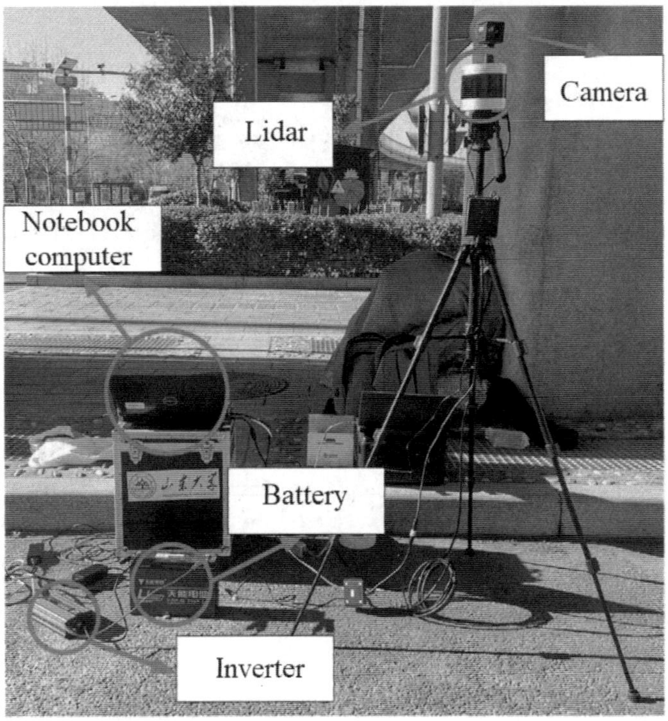

Figure 1. Roadside multi-source data collection platform

2.2. Time registration of roadside lidar and camera data

The frequency of data collected by the lidar is 10 Hz, and the frequency of data collected by the camera is 30 Hz. Based on the timestamp of the point cloud data, the article finds the image data with a difference of 0.1 s from the timestamp, and then completes the data synchronization and time synchronization algorithm. Specific steps are as follows:

- The collected bag data package is entered.
- According to the Ros topic, the images and point cloud topics in the data package are extacted.
- The timestamps of the two topics are extrated, named Ts 1, Ts 2.
- If Ts 1-Ts 2<0.1s, the corresponding point cloud and image data are saved.
- All the data are traversed, and the above point cloud and image data are saved to different folders.

2.3. Spatial registration of roadside lidar and camera data

2.3.1. Camera internal parameter calibration.

In this paper, the calibration method of Zhang et al. [5,6] is used to calibrate the internal parameters of the camera. During the calibration test of the internal parameters of the camera, to obtain better registration and calibration results, the calibration plate should appear as far as possible in each position in the camera field of view. To this end, the calibration board is placed at different distances, such as 3 m, 4 m, and 5 m, and various postures such as left, right, forward, and backward leaning are made at the same distance. The image is input to the monocular camera calibration toolbox (camera calibrator) in Matlab to calibrate the camera internal parameters. The number of inner corner points of the calibration plate used in the camera internal parameter calibration test is "6 × 8", and the size of a single grid is 80 m, as shown in Figure 2.

Figure 2. Calibration process of camera internal parameters

After completing the calibration of the internal parameters of the camera, the positional relationship between the calibration plate and the camera can be obtained by calculation, as shown in Figure 3 and Figure 4. Taking the calibration error as the judging standard, pictures with higher calibration errors are deleted in turn until the remaining 20 pictures are completed. The camera intrinsic parameter matrix is then obtained. The average error of the calibration results of the camera's internal reference standard is 0.09 pixels.

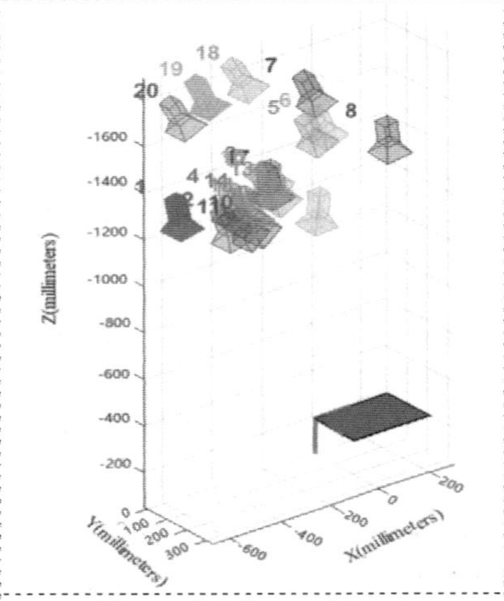

Figure 3. The positional relationship (the calibration board is the reference)

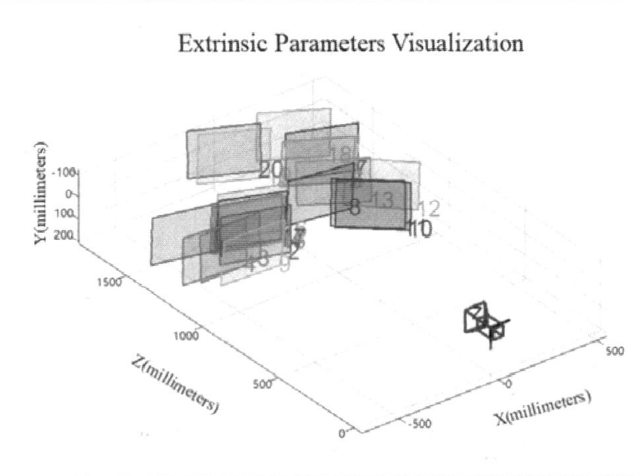

Figure 4. The positional relationship between the camera and the calibration board (the camera is the reference)

2.3.2. External parameter calibration of roadside lidar and camera

In this paper, the plane target model theory based on lidar and camera is used, and Autoware software is used to carry out the joint calibration test of roadside lidar and camera. The calibration test process consists of three parts, which are multi-source data acquisition, multi-source data time registration and multi-source data space registration. In the multi-source data acquisition part, it is necessary to ensure that the calibration plate can completely appear in the field of view of the lidar during the acquisition process. It can be seen from many experiments that the calibration plate with a smaller lattice cannot be used to complete the external parameter calibration test. Therefore, the external parameter calibration test was carried out using a calibration plate with a number of inner corner points "6×8" and a single grid size of 108 mm. In terms of multi-source data time registration, because Autoware software performs calibration by playing bag files, to ensure that the lidar and camera data participating in the registration and calibration are in the same frame, the playback of bag files should be paused and then the data should be captured by Autoware software. The point cloud reprojection error means that a point in the lidar coordinate system is projected to the camera coordinate system through the obtained calibration matrix, and the pixel difference between the projected point of the camera coordinate system and the real point is compared. The success of the external parameter calibration test can be judged by the re-projection of the point cloud on the image and the re-projection error value, as shown in Figure 5. The final camera and lidar are jointly calibrated with external parameters, and the point cloud reprojection error is 0.22 pixels.

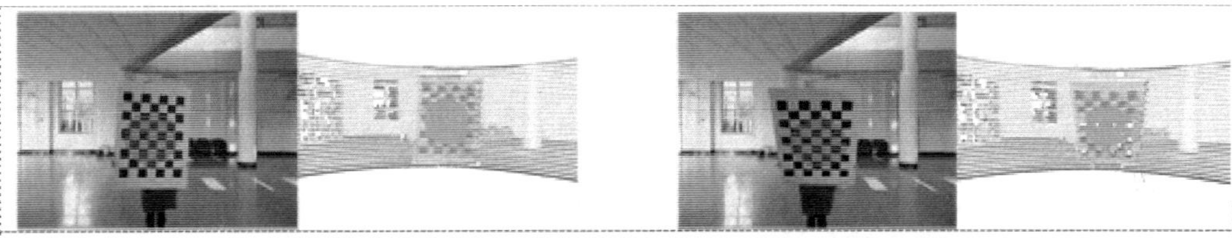

Figure 5. Reprojection of the point cloud of the calibration board.

After the calibration is completed, the Sensing module of Autoware is used to realize the fusion of image data and point cloud data in the same time and space. The relevant fusion effect is shown in Figure 6 and Figure 7. The fusion effect from the calibration plate registration and the actual scene registration fusion effect. Two aspects prove that this spatial registration experiment is successful.

Figure 6. The effect of registration and fusion of the calibration board.

Figure 7. The effect of registration and fusion of the calibration board

2.3.3. Installation attitude

It is found that the distance between the lidar and the camera is different in the roadside facilities of the vehicle-road coordination system currently constructed in various places, and the installation posture is different. Based on this, this chapter also explores the relationship between the roadside lidar and the camera. This test considered two layout modes, vertical and horizontal, as shown in Figure 8 and Figure 9, and the distances were 30 cm, 60 cm, 90 cm, 120 cm, and 150 cm.

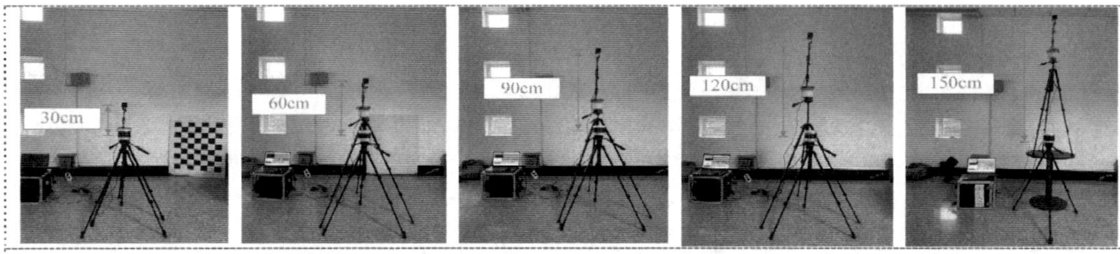

Figure 8. Equipment layout diagram under different vertical spacing.

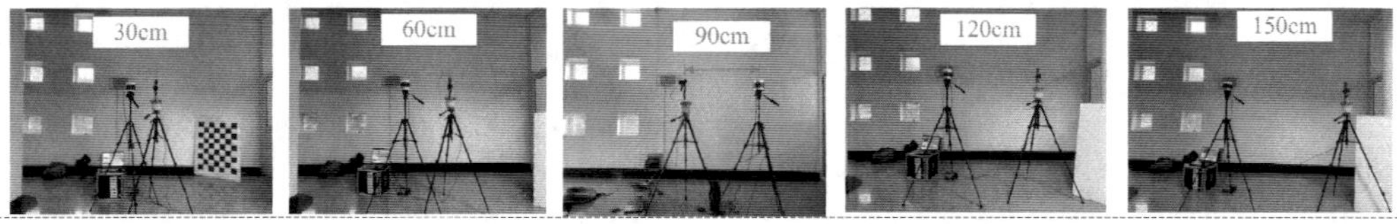

Figure 9. Equipment layout diagram under different horizontal spacing.

Under different layout distances, better point cloud reprojection results can be obtained after the data is registered and calculated. Correspondingly, the calculation results of the reprojection error under each attitude are shown in Table 1.

Table 1. Statistical table of reprojection error of lidar and camera in different posture.

Reprojection error	Installation attitude
0.546	The horizontal distance is 30 cm, and the vertical distance is 0 cm
0.689	Horizontal distance 60 cm, and the vertical distance is 0 cm
0.763	The horizontal distance is 90 cm, and the vertical distance is 0 cm
0.849	The horizontal distance is 120 cm, and the vertical distance is 0 cm
0.987	The horizontal distance is 150 cm, and the vertical distance is 0 cm
0.363	Horizontal distance 0 m, and the vertical distance is 30 cm
0.550	Horizontal distance 0 m, and the vertical distance is 60 cm
0.570	Horizontal distance is 0 m, and the vertical distance is 90 cm
0.761	Horizontal distance is 0 m, and the vertical distance is 120 cm
0.858	Horizontal distance is 0 m, and the vertical distance is 150 cm

Table 1 shows that as the distance between the two sensors increases, the reprojection error tends to increase. But both are less than 1 pixel. At different calibration distances, point cloud and image fusion can be achieved, but as the distance increases, the quality of the fusion effect gradually decreases. The specific performance is: in the horizontal direction, the point cloud projection of the front object overlaps with the background projection. When the horizontal distance is 30 cm and 60 cm, the overlapping phenomenon is not significant. When the distance is 90cm, the overlap between the two point clouds is more serious. It can be seen that when the overlapping area of the two projections reaches at least 80%, it is difficult to distinguish the object point cloud and the background point cloud, and the fusion effect is poor; in the vertical direction, the specific performance is that as the distance increases, the object point cloud It cannot be correctly projected on the object, the projection has a certain deviation, and as the distance increases, this phenomenon becomes more and more obvious. When the vertical distance between the two sensors is 30 cm, 60 cm and 90 cm, the point cloud can still be correctly projected on the calibration plate, but when the vertical distance is 120 cm, the point cloud cannot be correctly projected on the object. Therefore, considering both the reprojection error and the fusion effect, it is recommended that the horizontal and vertical distances between the roadside lidar and the camera should not be greater than 90 cm.

3. Result

This paper mainly researches the adaptability of the data registration method in roadside lidar and camera and explores the influence of different layout attitudes of roadside sensors on the registration and calibration results. The main conclusions are as follows:

- Based on lidar data, the time registration between lidar and camera data can be better accomplished by limiting the time difference between the time stamps of each data in the same frame, and the time difference between the two data in the same frame is not greater than 0.1s ;
- Based on Zhang et al.'s plane calibration method and the plane target model, the spatial registration between lidar and camera data can be well accomplished. The average error of camera internal parameter calibration is 0.09 pixels, and the external parameter calibration error is 0.22 pixels;
- By carrying out the registration calibration test of multiple sets of lidar and cameras at different layout distances, from the perspective of re-projection error and the fusion effect after registration and calibration, it is recommended to install roadside lidar and cameras. The distance between the horizontal and vertical directions does not exceed 90 cm.

The research results provide a practical reference for the layout and installation of roadside sensors and also accumulate technology and experience for the development of intelligent vehicle-road coordination systems.

Acknowledgement

This research was funded part by part by the National Natural Science Foundation of Jiangsu Province, grant number BK20200226, part by the Program of Science and Technology of Suzhou, grant number SYG202033.

References

[1] Lee S, Kim H and Sim S H 2022 Nontarget-based displacement measurement using LiDAR and camera, *Automation in Construction* , Volume 142: 104493.

[2] Choi J D and Kim M Y 2022 A sensor fusion system with thermal infrared camera and LiDAR for autonomous vehicles and deep learning based object detection, *ICT Express,* Volume 214, pp 361-365.

[3] An P, Liang J, Yu K, Fang B and Ma J 2022 Deep structural information fusion for 3D object detection on LiDAR–camera system, *Computer Vision and Image Understanding,* Volume 214 : 103295.

[4] Zhang X, Li Z, Gao X, Jin D and Li J 2021 Channel Attention in LiDAR-camera Fusion for Lane Line Segmentation, *Pattern Recognition,* Volume 118 : 108020.

[5] Zhang Z. 1999 Flexible camera calibration by viewing a plane from unknown orientations, *Conf on the seventh ieee international conference on computer vision. Ieee* , vol 1, pp 666-673.

[6] Tan Z, Zhao B, and Wang W 2022 Metal surface 3D imaging based on convolutional autoencoder and light plane calibration, *J, Optik,* Volume 259: 169007.

Advancement and Challenges of Field Effect Transistors based on Multi-gate Transistor

Jinhan Lin

Fuzhou University, 2 Wulongjiang Dadao, Minhou County, Fuzhou City, Fujian Province, China

Jinhan.lin.2020@mumail.ie

Abstract. The advancement and challenges of field effect transistors are based on multi-gate transistors from the perspective of structure and material. Multi-gate field-effect transistors (Multi-gate FET) have steeper sub-threshold slopes, which can reduce the short channel effect and improve mobility and drive current. A fin field-effect transistor (FinFET) and gate-all-around field-effect transistor (GAAFET) are attractive multi-gate structures most compatible with today's standard machining technologies. As the future moves towards smaller processes, FinFET and GAAFET processes limit the spacing between n-to-p devices. In order to increase the possibility of transistor miniaturization, innovative structures such as Forksheet FET and Complementary-FET (CFET) have been proposed.

1. Introduction

In recent years, transistors have been getting smaller and smaller, and therefore there are more and more of them in a single microchip. Until 2005, the semiconductor industry followed Moore's Law. But the traditional Metal-Oxide-Semiconductor Field-Effect Transistor(MOSFET) is facing physical limitations. In order to improve the performance of FET, some new technologies must be applied.

A classic MOSFET is shown in Figure 1. This is an N-channel MOSFET. The device consists of two N-type semiconductor regions, called source and drain, separated by a P-type semiconductor region called the substrate. Normally, the semiconductor is silicon. A thin layer of insulating material, such as silica, covers the area between the source and drain, and at the top of this layer is a metal electrode called the gate. Insulators are called gate oxides. Since the gate (G) is insulated from the drain (D) and the source (S), there are two PN junctions between D and S. In general, the substrate and the source are connected internally, equivalent to a PN junction between D and S. When the voltage is not added to G, there are few electrons in the channel between S and D, so the resistance is relatively large, and S and D cannot conduct. When a positive voltage is applied to the gate, electrons gather in the channel between S and D due to the attraction of the electric field, so the resistance decreases, and S and D turn on [1]. In this way, MOSFETs become switches in logic microcircuits. Different MOSFETs can be combined to achieve different logic functions.

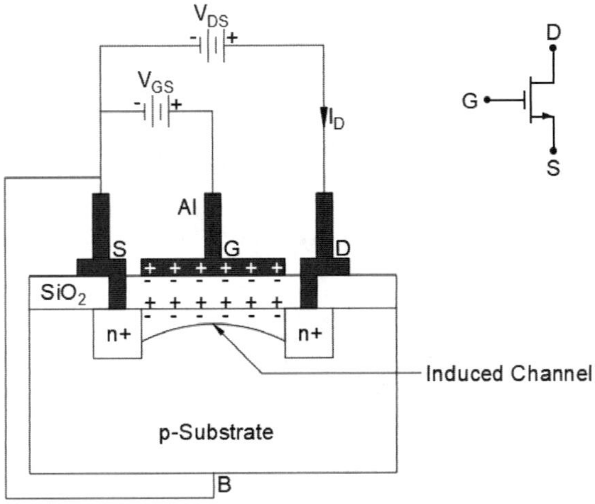

Figure 1. N-Channel MOSFET [2]

On the one hand, many transistors are needed to achieve complex functions. Therefore, transistors must be made small enough. On the other hand, the switching speed of the transistor limits the performance of the chip. According to the capacitor charging principle, switching speed is related to capacitor size [3]. The larger the capacitance, the longer the charging time, and the slower the switch conduction speed. Therefore, we need to reduce the capacitance to increase the speed of operation. According to this formula,

$$c = \frac{\varepsilon A}{4\pi K d} \quad (1)$$

the capacitance can be reduced in three ways increasing the dielectric layer thickness, changing the dielectric constant, and reducing the area. However, if the dielectric layer thickness is too large, the electric field in the channel is not strong enough to conduct. Changing the dielectric constant requires changing the dielectric material, which is very limited for a long time. Therefore, the only feasible method is to reduce the area, that is, reduce the length and width of the channel. So transistors follow this strategy and keep shrinking. But as transistors get smaller and the distance between source and drain becomes very small, short-channel effects can occur. As a result, MOSFET is difficult to continue miniaturization.

With the advent of a multi-gate transistor, the performance of FET can be improved. In this review, we discuss why current FET models encounter a fundamental barrier to further miniaturization, how multi-gate architectures can overcome this barrier, and how multi-gate architectures should be developed in the future.

2. Advancement

2.1. Pursue the perfect switch

A perfect switch must satisfy three conditions. The first is that the current flowing through the switch is zero when the switch is open. The second is that the switch resistance is zero when the switch is closed. The last one is that the switch can change the state from on to off Instantaneously and smoothly, and vice versa. MOSFET is not a perfect switch for logic microcircuits. Its off current is not zero, on current is limited, and switching states need some time. In addition, MOSFET as a switch occurs gradually over a range of gate voltages. It does not instantly change state at a precise gate voltage. As transistors shrink, their switching performance worsens [3].

The solution is to build the FET structure vertically, shifting it from horizontal to vertical, and design a gate electrode wrapped around several sides of the conductive channel to improve the electrostatic control of the channel. This multi-gate structure can further reduce the size of the transistor without degrading its performance. Therefore, after 16/14 nm, the structure of FET changes from the traditional planar structure to FinFET (Figure 2) structure.

Figure 2. FinFET

The FinFET process uses epitaxial growth technology to insert SiGe and SiC strain materials and conducts source and drain doping, which makes the source and drain active region protrusion increase the thickness and surface area of the active region, and reduces the contact resistance of the source and drain. At the same time, the light doping of FinFET in the active region greatly reduces the scattering effect of particles, improves the mobility of carriers, and increases the switching speed. However, with the further microfabrication of the chip, FinFET will also reach its limit at the 3/2nm-Node after the 5nm-Node. The FinFET transistor cannot survive due to the close fin distance, the re-emergence of leakage, and the limits of physical materials. Therefore, GAAFET is equivalent to an improvement based on FinFET, and the transistor structure under this technology has changed. The gate and drain are no longer fins but become "small sticks," vertical through the gate, so that the gate can realize the source-drain wrapped in four sides, from three contact surfaces to four contact surfaces, and divided into several four contact surfaces. Apparently, control of the current is further improved. GAAFET has two structures (Figure3); one is the common GAAFET which uses Nanowire as the electronic transistor fin; The other is MBCFET with thick fins in the form of nanosheets. Both of these two methods can achieve 3nm-Node.

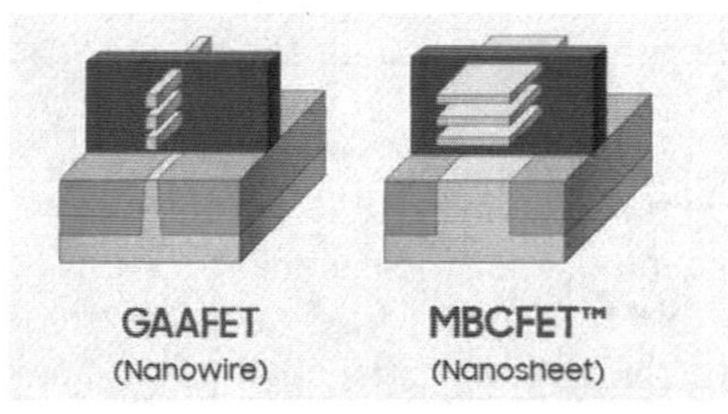

Figure 3. GAAFET MBCFET

As future moves towards smaller processes continue, smaller spacing between nFET and pFET devices within standard cells will be required. When the Complementary FET(CFET) node reaches 1nm, the IMEC uses a Complementary FET(CFET) structure shown in Figure 4 to stack n-shaped FETS on a P-shaped FET. Specifically, it is composed of two independent nanosheet FETs (P-type and N-type). P-type nanowires are superimposed on n-type nanowires. The bottleneck of N-P separation is eliminated by stacking transistors with different conductive types through stacking, and the area of standard cells is greatly reduced [4].

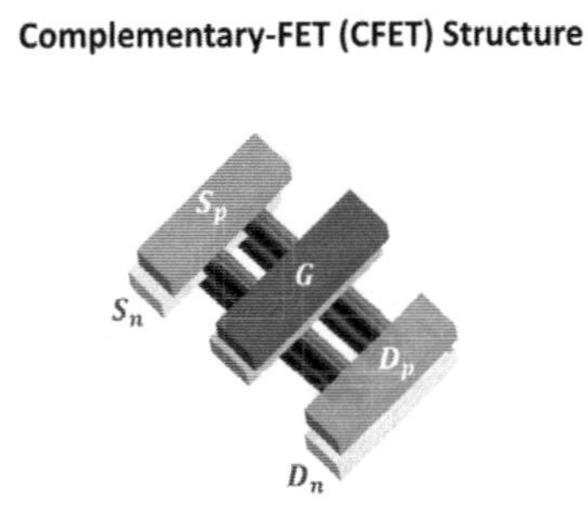

Figure 4. Complementary FET(CFET) Structure [4]

In general, with the miniaturization of FET and some new materials and structures, the new generation of FET becomes a perfect switch.

2.2. Reduction of the short channel effect

The short channel effect refers to a series of phenomena caused by the sharing of charge in the channel region between the gate and the source and drains of MOSFET [5]. Specifically, MOSFET can be viewed as a three port of switch, its working principle is to put the electrons from the source to the drain, the process is done through the channel, and another one is the role of the grid, decide the channel is open, or shut down these operations are done through a port with specific voltage The transistor as a similar switch device, open no problem, but not tight, the reason is that the size is too small, there are a lot of internal electric field interference, in reality can not be ignored, this will lead to the grid end of the electric field can not play a full role, so the switch is not tight Consequence is a leakage current, transistor in the closed state, but will cause energy loss and current This is also the current transistor design and circuit design of one of the major problems In addition, they lead to a series of problems, including the threshold voltage roll drain induced barrier lowering (DIBL) speed low saturation increased mobility of reverse leakage current Hot carrier effect and other problems [6].

As transistors get smaller in size, the leakage current becomes more and more severe. The energy loss becomes more and more severe because of the short channel effect, and most of the leakage current flows through the region below the channel, which is a very thin layer on the surface of the silicon wafer below the oxide insulator (Figure 5). The region below the channel is called the depletion layer, but the electrons travel in the channel and in the physical model we need the depletion layer to balance the charge, but in the short channel device we don't have to put the depletion layer and the channel together and then the leakage current flows through So IBM came up with the idea of taking out this silicon and replacing it with an insulating layer, where the rest of the silicon is underneath the insulating layer, so that the channel is separated from the depletion layer, because the electrons are coming from the poles, but the barrier between the poles and the depletion layer is separated by the insulating layer, so that in addition to the channel, you have leakage currents Intel built on that and then came up with the idea of putting a gate directly under the oxide layer, with channels on either side or insulating silicon, and then putting a gate on the outside and you had FinFET and this multi-gate structure of FinFET reduced the leakage current significantly

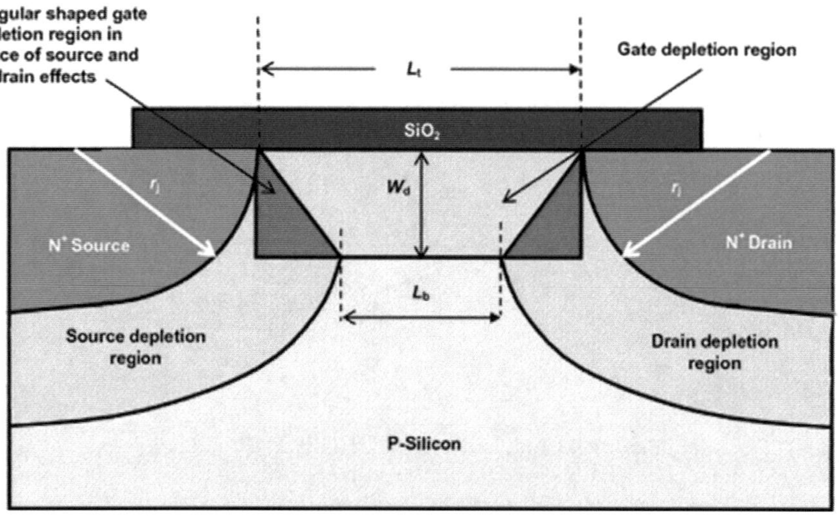

Figure 5. The short channel effect [6]

For FINFETs with 5 nm-Nodes, channel control capability is insufficient. Therefore, people tend to develop GAA FET structure to improve the interaction between gate and source/drain. In 2006, Lee et al. successfully produced a full-gate FINFET with a fin width of 3 nm for the first time, as shown in Figure 6, Singh et al. demonstrated the excellent electrical properties of GAA N-FET and P-FET structures based on Si nanowires, using new channel materials [7]. It can further reduce the gate length [8].

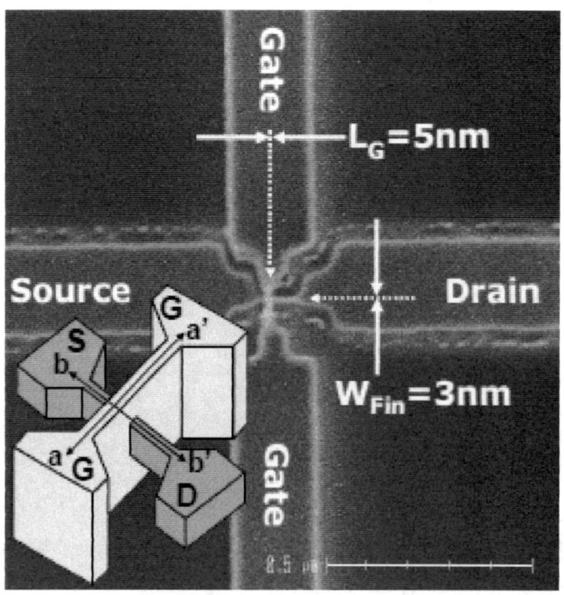

Figure 6. SEM top-view of 3 nm silicon-fin and sub-5 nm gate [7].

3. Challenge

3.1. FinFET

FinFET technology has been put into practical application, but there are still some challenges and difficulties in the manufacturing process. First, the FinFET structure has very high access resistance, which is caused by the very thin structure and the difficulty of forming a three-dimensional junction [9]. Second, the FinFET structure will affect the voltage and current characteristics of the device. Metal gate granularity (MGG) and fin edge roughness (FER) can cause changes in threshold voltage, while line edge roughness (LER) can cause changes in maximum drive current [10]. These possible manufacturing difficulties can affect FinFET performance. However, slight doping in the channel can change the threshold voltage of some devices or better control the leakage current under the fin. High temperature (300-400 K) implantation or plasma doping can reduce this damage [11]. Compared with the classical planar FET, the structure of the transistor is also changed from the traditional 2D structure to the 3D structure, which satisfies both mechanical and electrical properties.

3.2. GAAFET
GAAFET achieves better transistor structure, but it also brings many challenges to the processing. Because FinFET technology has been relatively mature, chip manufacturers need to change the production of GAAFET at a large cost. The introduction of some new materials makes the process more complicated and makes the requirements for environmental conditions higher. In addition, the back end line (BEOL) is where the copper interconnects in the chip. The interconnections at each node become more and more compact, resulting in severe resistive-capacitance delays [12]. Another challenge of GAAFET is that as the density of the FET increases, changes in the internal structure become increasingly difficult to detect, and even small changes can cause the system to slow down [13]. For MBCFET, measuring the replacement of a single nanosheet is the most difficult, and the worst-performing nanosheet determines the quality of the entire transistor.

4. Conclusion
At present, as transistors shrink further, GAAFET is proving to be the successor to FINFET for very large-scale process nodes. In the future, as transistors become smaller in size, full-gate structures such as CFET and Forksheet-FET may be needed. In short, the right technological path for scaling up FET depends on the market, which is something to look forward to.

Reference
[1] Oh K S. MOSFET basics[J]. Fairchild Technical Manual, AN9010, 2000.
[2] Samak, Chinmay Vilas, and Tanmay Vilas Samak. "Novel Design of a Magnetically Switchable MOSFET Using Magnetoresistive Elements." Telkomnika 16.5 (2018): 2233-238. Web.
[3] Steven H Voldman (1998). The impact of MOSFET technology evolution and scaling on electrostatic discharge protection. , 38(11), 1649–1668. doi:10.1016/s0026-2714(98)00169-3
[4] Cheng, C.-K., Ho, C.-T., Lee, D., Lin, B., & Park, D. (2021). Complementary-FET (CFET) Standard Cell Synthesis Framework for Design and System Technology Co-Optimization Using SMT. IEEE Transactions on Very Large Scale Integration (VLSI) Systems, 29(6), 1178–1191. doi:10.1109/tvlsi.2021.3065639
[5] Ferain, Isabelle; Colinge, Cynthia A.; Colinge, Jean-Pierre (2011). Multi-gate transistors as the future of classical metal–oxide–semiconductor field-effect transistors. , 479(7373), 310–316. doi:10.1038/nature10676
[6] Khanna, V.K. (2016). Short-Channel Effects in MOSFETs. In: Integrated Nanoelectronics. NanoScience and Technology. Springer, New Delhi. https://doi.org/10.1007/978-81-322-3625-2_5
[7] Singh, N. et al., Balasubramanian, N. and Kwong, D.L., 2006. High-performance fully depleted silicon nanowire (diameter/spl les/5 nm) gate-all-around CMOS devices. IEEE Electron Device Letters, 27(5), pp.383-386.
[8] Zhang S. Review of modern field effect transistor technologies for scaling[C]//Journal of Physics: Conference Series. IOP Publishing, 2020, 1617(1): 012054.
[9] Jurczak, M., Collaert, N., Veloso, A., Hoffmann, T. and Biesemans, S., 2009, October. Review of FINFET technology. In 2009 IEEE international SOI conference (pp. 1-4). IEEE.

[10] Wang, X., Brown, A.R., Cheng, B. and Asenov, A., 2011, December. Statistical variability and reliability in nanoscale FinFETs. In 2011 International Electron Devices Meeting (pp. 5-4). IEEE.
[11] Maszara, W.P. and Lin, M.R., 2013, September. FinFETs—Technology and circuit design challenges. In 2013 Proceedings of the European Solid-State Device Research Conference (ESSDERC) (pp. 3-8). IEEE.
[12] Lapedus, M. (2020) 5/3nm wars begin. Available at:https://semiengineering.com/5-3nm-warsbegin/(Accessed 23 January 2020).
[13] Sperling, E. and Rambo, S. (2020) Reliability Challenges Grow For 5/3nm Available at: https://semiengineering.com/reliability-challenges-grow-for-5-3nm/ (Accessed 7 April 2020)

Study on the Characteristics of FSRD with a Corrugated Anode and p$^+$ Adjustment Region at Cathode

Cailin Wang*, Li Yu, Wuhua Yang, Lei Guan

Xi'an University of Technology

e-mail：wangcailin8511@xaut.edu.cn

Abstract. The reverse recovery characteristic of the power diode will directly affect the reliability of the inverter application. In order to improve the reverse recovery characteristics of the fast and soft recovery diode (FSRD), a new FSRD structure with Corrugated Anode and P$^+$ adjusting Cathode (CA-PC-FSRD) is studied by the Sentaurus-TCAD simulator and compared with the characteristics of conventional FSRD. The influence of key parameters, such as the height and width of anode corrugation and the width of the p$^+$ adjusting region at the cathode, on the characteristics are discussed. Simulation results show that the CA-PC-FSRD has excellent soft recovery characteristics compared with the conventional FSRD at the same voltage level.

1. Introduction

With the wide application of high-power devices such as GTO, IGCT, and IGBT, their anti-parallel diodes are required to have more excellent features and higher reliability [1-2]. In order to improve the softness factor and shorten the reverse recovery time, the carrier lifetime control technology combined with the optimal design of the device structure is proposed [3-5]. However, the proposed complicated anode and cathode structures of the device have increased manufacturing costs significantly. Therefore, this paper proposes a new FSRD with Corrugated Anode and P+ adjustment region at Cathode (CA-PC-FSRD) to improve the device features and reliability while having a simpler manufacturing process.

2. Device structure and process

Figure 1 shows the conventional FSRD and the CA-PC-FSRD. Conventional FSRD adopts the pn$^-$nn$^+$ structure, in which p$^+$ anode and n$^+$ cathode regions are shallow, and the n buffer layer is thicker. CA-PC-FSRD adopts a corrugated anode region which is formed by the selective boron ion implantation and the driven-in of boron impurities at high temperature, while the cathode side adds a p$^+$ adjustment region through boron ion implantation.

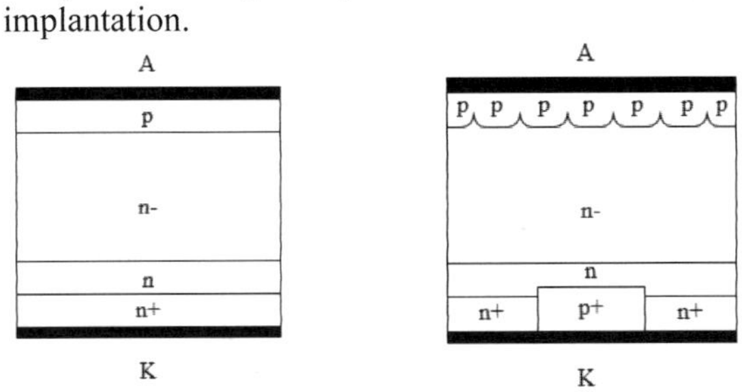

(a) Conventional FSRD (CD) (b) CA-PC-FSRD

Figure 1. Comparison of the two FSRD structures

3. Simulation and discussion

In order to study the characteristics of CA-PC-FSRD, this paper takes the 3.3kV voltage level as an example, and the device structural model is set up by the Sentaurus-TCAD simulator. For comparison, the structural parameters of the two structural models are the same except for the corrugated anode and the p^+ adjustment region at the cathode, as shown in Table 1.

Table 1. Structural model parameters.

Regions	**concentration（cm^{-3}）**	**thickness（μm）**
p anode region	1e18	9
n^- base region	2.5e13	314
n buffer layer	8e15	14
p^+ adjustment region	1e19	6
n^+ cathode region	1e19	4

3.1. Mechanism and characteristics

The dependence of the anode hole injection efficiency and cathode electron injection efficiency on current density is firstly analyzed in Figure 2 (J_{rated} represents the rated current density of 125A/cm^2). It is obvious that both the anode hole injection efficiency and the cathode electron injection efficiency of CA-PC-FSRD are lower than those of conventional FSRD. This is because the vertical and horizontal diffusion of the corrugated anode region causes a low p+ doping concentration near the peak, so that the injected electrons from the cathode side can easily pass through it, as shown in Figure 3, while the lower cathode electron injection efficiency is due to the p^+ adjustment region on the cathode side. The lower anode hole injection efficiency is helpful to improve reverse recovery time, but the on-state voltage drop will increase; the lower cathode electron injection efficiency is helpful to improve the softness and eliminate the voltage oscillations during the reverse recovery, thus improving the reliability of the device.

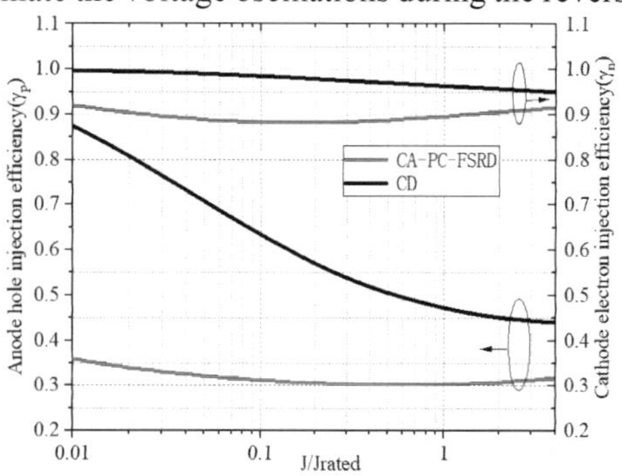

Figure 2. Dependence of injection efficiency on current density

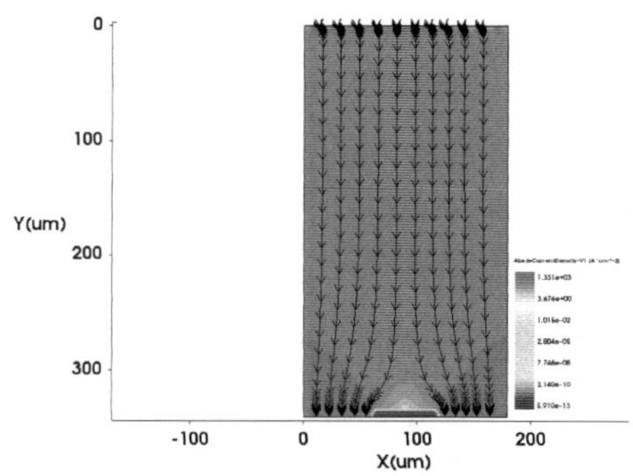

Figure 3. Electron current density distribution inside the device

The reverse blocking, forward conducting, and reverse recovery characteristics curves of two diodes are shown in Figures 4, 5, and 6, respectively. The reverse blocking characteristics of the two diodes are almost the same, but the forward conducting and reverse blocking recovery characteristics have obvious differences. The forward voltage drop of conventional FSRD is lower, but its reverse recovery characteristic is poorer. The reverse peak current of CA-PC-FSRD is much lower than that of conventional FSRD, and the softness increases significantly, eliminating the voltage oscillation, but the forward voltage drop and reverse recovery time are increased. This is not only because the special corrugated anode structure changes the on-state carrier concentration distribution, but also because the p^+ adjustment region improves the carrier concentration distribution during the end of the reverse recovery. The current density distribution during the reverse recovery is shown in Figure 7. At t=0.66μs, the diode begins to support the reverse voltage, and the electrons and holes in the n^- region are extracted into the cathode and anode sides, respectively. At t=0.74μs, the pnp transistor is turned on, the holes are injected into the n^- region from the p^+ adjustment region at the cathode, and the current density through the anode and cathode sides is enhanced. At t=1.03μs, the current decreases. At t=1.08μs, the reverse recovery finishes, and the reverse current reaches the leakage current level.

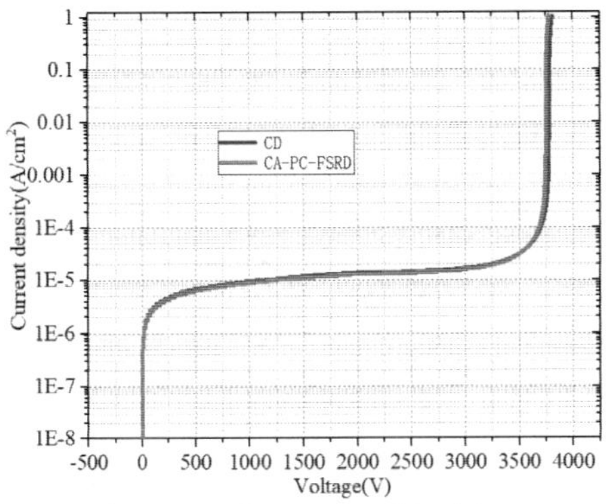

Figure 4. The reverse blocking characteristics curve

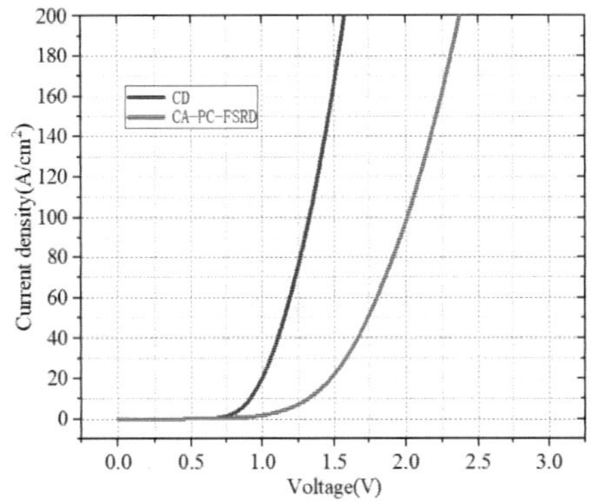

Figure 5. The forward conducting characteristics curve

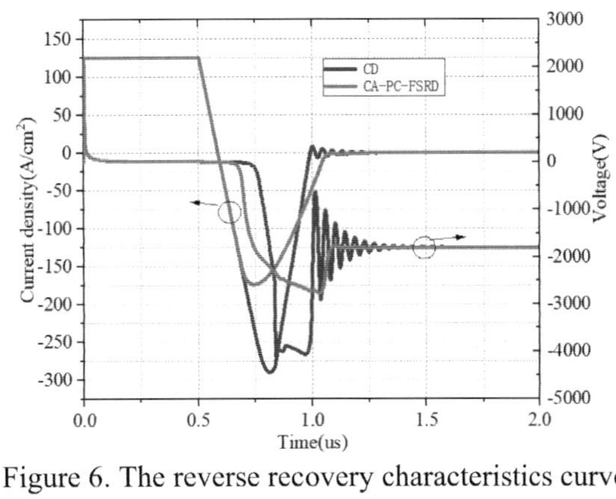

Figure 6. The reverse recovery characteristics curve

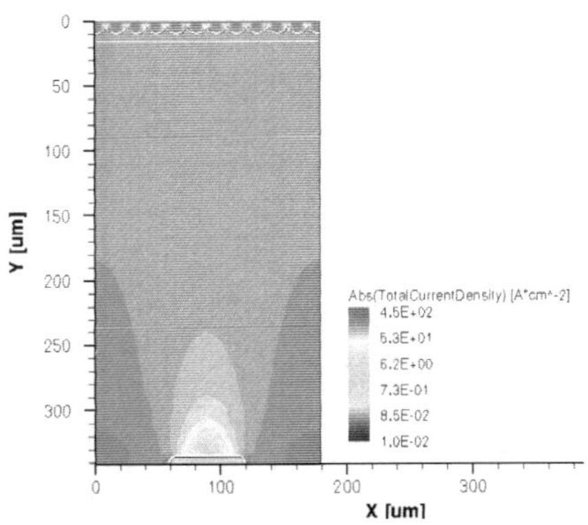

(a)t=0.66μs

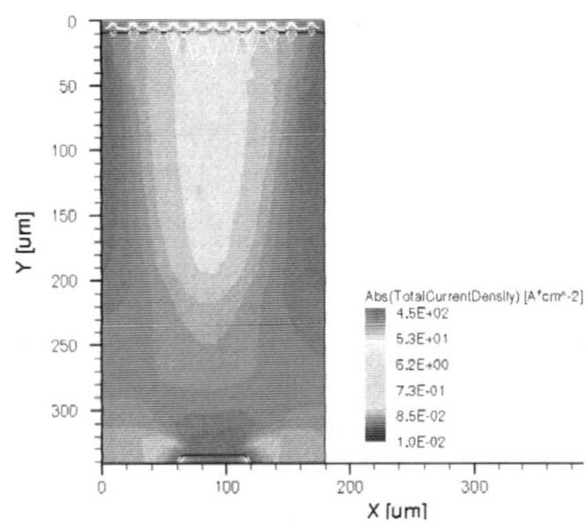

(b) t=0.74μs

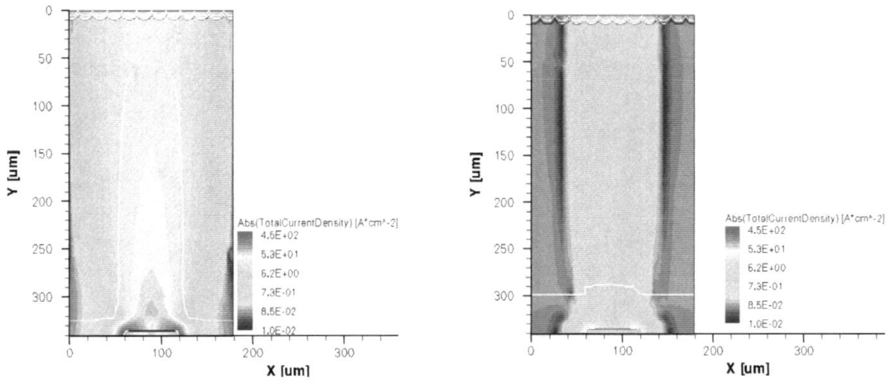

(c) t=1.03μs (d)t=1.08μs

Figure 7. Current density distribution during reverse recovery

3.2. *Influence of the structural parameter on the characteristics*

The influences of the key parameters, such as the corrugation height and width and the width of the p^+ adjustment region at the cathode, on the forward conducting and the reverse recovery characteristics are analyzed below.

A. Corrugation height. The width of the corrugated window is 5μm, the number of corrugations is 11, and other structural parameters are unchanged. When the mask width (which determines the corrugation height) changes from 3.4μm to 5.8μm, the forward conducting characteristics and reverse recovery characteristics under different corrugation heights are shown in Figure 8. It can be seen that with the mask width increasing (i.e., the higher the corrugation height), the forward voltage drop becomes larger, the reverse peak current becomes smaller, and the recovery time becomes shorter.

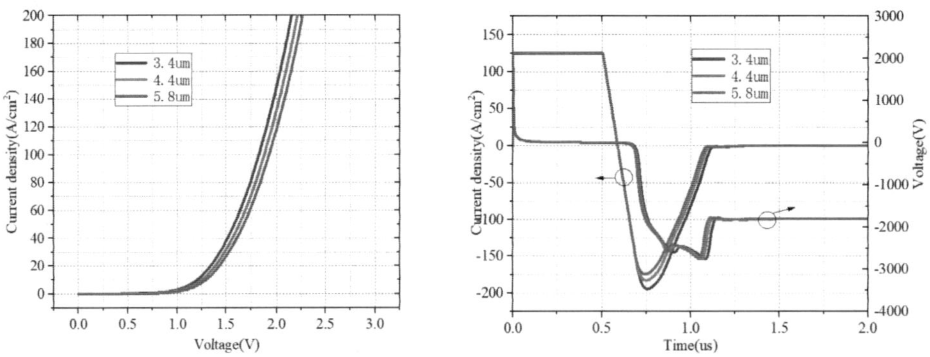

(a)The forward conducting characteristics (b) The reverse recovery characteristics

Figure 8. Influence of the corrugation height on forward conducting and reverse recovery characteristics

B. Window width. The width of the mask is 13μm, the number of corrugations is 11, and other parameters are unchanged. When the doping window width (which determines the corrugation width) changes from 3μm to 5μm, the variation of the forward conducting and reverse recovery characteristics with the doping window width are shown in Figure 9. The wider the window width (corrugation width), the lower the forward voltage drop and the greater the reverse peak current; when the window width is 5μm, the reverse recovery time is slightly longer, and the softness is best.

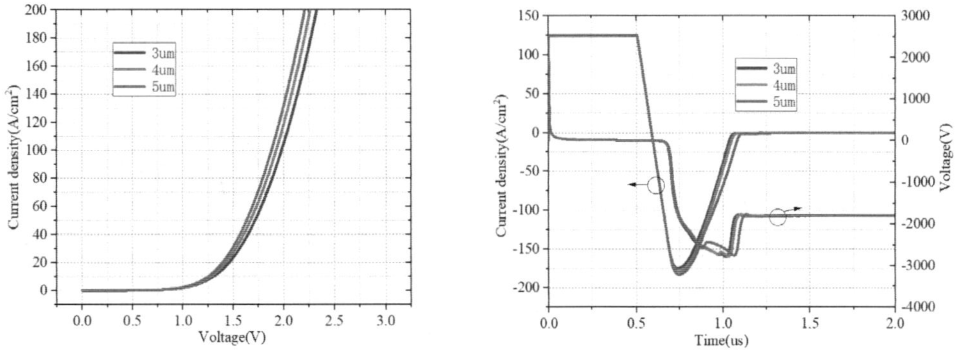

(a)The forward conducting characteristics (b) The reverse recovery characteristics

Figure 9. Influence of the corrugation width on forward conducting and reverse recovery characteristics

C. Width of the p⁺ adjustment region. The parameters of the anode and other regions are unchanged. When the width of the p^+ adjustment region at the cathode changes from 50 μm to 70μm, the variation of the forward conducting and reverse recovery characteristics with the p^+ adjustment region width are shown in Figure 10. It can be seen that, with the p^+ adjustment region width increasing, the forward voltage drop slightly increases, the reverse recovery time becomes slightly longer, and the reverse peak current is unchanged.

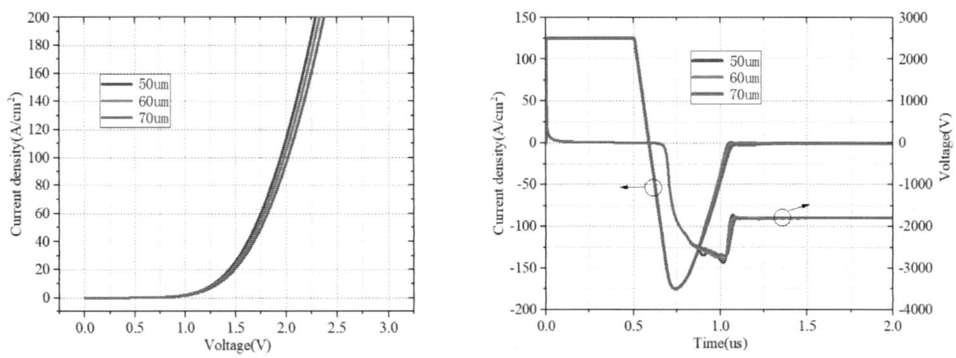

(a)The forward conducting characteristics (b) The reverse recovery characteristics

Figure 10. Influence of p^+ adjustment region on forward conducting and reverse recovery characteristics

4. Conclusion

The comparison of the characteristics of the two FSRD shows that the CA-PC-FSRD has a lower reverse peak current and a little increased forward voltage drop. And the structural parameters have little influence on the characteristics. This implies the CA-PC-FSRD can not only obtain good dynamic characteristics, but also improve the freedom of the manufacturing process.

Acknowledgments

The authors wish to acknowledge financial support from the Key R&D Program of Shaanxi Province (Two Chains Integration Key Project) (No. 2021LLRH-02) and the Ph.D. Start-up Foundation of Xi'an University of Technology (No. 103-451121007).

References

[1] Yi R, Zhao Zm, Yuan Lq. 2008 *Transactions of China Electrotechnical Society*, p63-80.
[2] Li Fz, Xu Qf, Lai Jj. 2011 *Transactions of China Electrotechnical Society*, p 9172-9176.

[3] Zhang L., Wang CL. 2017 *Microelectronics Reliability* p75-79.
[4] Ling C., Ge Bm, Bi Dq. 2012 *Power System Protection and Control* p 35-39.
[5] Baburske R, Heinze B, Lutz J. 2008. *IEEE Transaction on Electron Device* p2164-2172.
[6] Liu Yy, Jin R, Wu D. 2017 *International Conf. on Circuits, Devices, and Systems (ICCDS)*, p 25-28.

Thermal analysis and package optimization for the multi-wavelength laser bar module

Wanjun Sun[a], Lijiao Yang[a], Jianan Xie[a], Xiaxia Zhang[a], Yang Sha[a], Shuai Mi[a], Tao Lin*[a,b]

[a] College of Automation and Information Engineering, Xi'an University of Technology, Xi'an, Shaanxi 710048, China

[b] Shaanxi Key Laboratory of Complex System Control and Intelligent Information Processing, Xi'an University of Technology, Xi'an, Shaanxi 710048, China

* Corresponding author: lintao@xaut.edu.cn

ABSTRACT: To meet the demand of the laser application market for semiconductor laser function diversification, this paper proposes a handheld multi-wavelength laser stack packaging module. The 808 nm, 915 nm, and 980 nm semiconductor laser bars are selected for the integrated assembly of the multi-wavelength module. However, with the continuous increase of the optical power density of the integrated packaging module, thermal management has become a key factor affecting the performance and life of the device. In this paper, the finite element analysis method is used to analyze the influence of the arrangement of three laser stripe chips in the stacked packaging structure on the heat dissipation of the multi-wavelength module. The simulation results show that the heat dissipation efficiency of the multi-wavelength module can be improved by 13.74 % by changing the arrangement of chips. A back heat dissipation packaging structure with higher heat dissipation efficiency is proposed, and the influence of chip arrangement and working mode on the thermal characteristics of the device in the new packaging structure is simulated and analyzed. According to the simulation results, a chip arrangement that can make the heat dissipation effect of the multi-wavelength module achieve the best is determined.

1. INTRODUCTION

In recent years, due to the advantages of small size, high brightness, and high electro-optic conversion efficiency of semiconductor lasers, lasers have been widely used in material processing, medical, military, information, and other fields [1-3]. Especially in the field of medical cosmetology, with the improvement of laser performance and the decline in price, the market demand for semiconductor lasers in the medical field is increasing, and semiconductor lasers have become the mainstream of laser medical cosmetology. According to the selective photothermal effect theory, the biological effects of different lasers are also different due to the differences in wavelength and type, and the application fields will also be different. Therefore, the laser modules integrated and packaged by different band laser chips have a broad application prospect in the field of biomedicine, which is a hot topic of research and exploration. At present, the working wavelength of laser therapeutic apparatus on the market is single, and multi-wavelength laser therapeutic apparatus can make up for this deficiency. Therefore, multi-wavelength laser therapy will become the mainstream of the laser medical market [4-6].

As the semiconductor laser chip is more tightly packaged and the spacing between the adjacent laser emitting units of the device decreases, the heat dissipation of the laser becomes more and more difficult.

Although with the development of semiconductor lasers, the electro-optical conversion efficiency can be as high as 50%-70%, but still 30%-50% of the electric energy converted to heat [7]. The small size of the semiconductor laser makes it produce huge heat flux. A semiconductor laser can produce higher heat flux than rocket ballistic reentry into the atmosphere. High chip temperature will directly lead to the decline of semiconductor laser lifetime and reliability, even damaging the chip. Therefore, the thermal characteristics become a key factor restricting the further improvement of the performance of semiconductor multi-wavelength modules [8-10]. At present, the research focus of semiconductor laser thermal management at home and abroad mainly focuses on the form of a single-wavelength single tube, bar array, etc., while the lack of comprehensive analysis of thermal characteristics of multi-wavelength modules with different wavelengths. To make up for this deficiency, the finite element analysis method is used in this paper. Firstly, the finite element model of the multi-wavelength module is established. The arrangement of different output wavelength chips in the multi-wavelength module and the influence of different working modes on the temperature of the device are mainly studied. A back heat dissipation packaging structure with better heat dissipation capacity is proposed, which is of certain significance to optimize the packaging structure of multi-chips and improve the performance and application scope of multi-wavelength modules.

2. MODELING AND BOUNDARY CONDITIONS

Figure 1 shows the structure diagram of the multi-wavelength module packaged by conduction cooling [11]. The multi-wavelength module in the figure is composed of the main heat sink, laser bar, solder layer, and sub-heat sink CuW. The size of the main heat sink is 25000×15000×7500μm. The size of the secondary heat sink is 10000×5000×1000μm. The multi-wavelength module is composed of 3Pcs laser chips with a central wavelength of 808nm, 915nm, and 980nm. The actual packaging parameters of three laser chips are: 808nm laser bar consists of 19 light emitting units with a width of 150μm, period of 500μm, and cavity length of 1mm. 915nm laser bar consists of 24 light emitting units with 190μm width, 400μm period, and 4.3mm cavity length. The number, width, and period of the 980nm laser stripe are the same as the 915nm laser stripe, and the cavity length is 4mm.

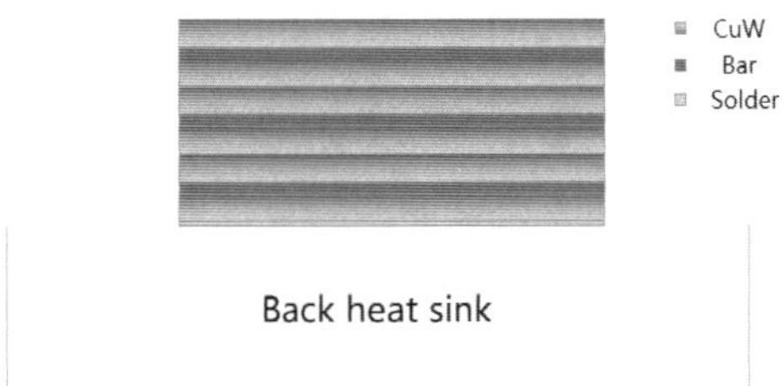

Figure 1 Schematic diagram of the structure of the conduction cooling packaged multi-wavelength module [11]

The length of three wavelength laser chips is 10 mm. In this paper, the multi-wavelength module uses the conduction cooling method, the simulation process assumes that the material parameters are isotropic. Table 1 shows the 808nm, 915nm, and 980nm laser bar layer structure. In this paper, the multi-wavelength stack module has three laser bars. Under the same output current, the output powers of the three devices are 50W, 70W, and 70W.

Table 1 Structure of each layer of 808nm, 915nm, and 980nm laser bar

	808nm	**915nm**	**980nm**
Layer	**Material**	**Material**	**Material**
n-contact	Au	Au	Au
substrate	GaAs	GaAs	GaAs
Lower cladding	$Al_{0.5}GaAs$	$Al_{0.3}Ga_{0.7}As$	$Al_{0.38}GaAs$
Lower waveguide	$Al_{0.33}GaAs$	$Al_{0.225}Ga_{0.745}As$	$Al_{0.22}GaAs$

Active	$Al_{0.12}Ga_{0.795}In_{0.085}As$	AlGaInAs	$In_{0.175}Ga_{0.825}As$
Upper waveguide	$Al_{0.33}GaAs$	$Al_{0.225}Ga_{0.745}As$	$Al_{0.22}GaAs$
Upper cladding	$Al_{0.5}GaAs$	$Al_{0.3}Ga_{0.7}As$	$Al_{0.38}GaAs$
Cap	GaAs	GaAs	GaAs
Insulating	SiO_2	SiO_2	SiO_2
p-contact	Au	Au	Au

Thermal simulation of the multi-wavelength module in this paper, the accuracy of the simulation results of the temperature field mainly depends on whether the boundary conditions imposed on the model are closer to the actual situation. The constant surface temperature, internal active layer heat generation rate, and heat flow rate that exist in the operation of the multi-wavelength module require boundary conditions to be solved. Constraints are imposed on the model from the following three aspects according to the actual situation. At the same time, some assumptions are made for the simulation process for the convenience of calculation [12-15]:

(1) During actual operation, the device is mounted on the TEC, and the temperature of the bottom surface of the heat sink will remain constant. Therefore, the bottom surface temperature was set to 25°C during the simulation.

(2) Assuming that only the active layer generates waste heat during the working process of the semiconductor laser bar, the heat generation rate is only set on the active layer during the simulation process. When the multi-wavelength module works, assuming that the photoelectric conversion efficiency of the chip is not affected by temperature, 50%, 55%, and 60%, the thermal power of the device are 50W, 57W, and 48W. Therefore, the thermal powers loaded by the single luminescent unit of the three wavelength semiconductor lasers are $1\times10^{15}W/m^3$, $3.63\times10^{14}W/m^3$, and $3.68\times10^{14}W/m^3$.

(3) multi-wavelength module mainly adopts the cooling package structure. In the simulation process, the influence of thermal radiation on the device is negligible, and the thermal radiation of the device is ignored. However, the convection heat dissipation of the device in the air is considered. The natural convection coefficient of the model at room temperature is 6 W/m^2K.

3. SIMULATION RESULTS AND DISCUSSION

3.1 Influence of laser bar arrangement on the temperature of the multi-wavelength module

Due to the different center wavelengths and sizes of the three laser bars, the structural parameters are also different, and the threshold current required for the normal operation of different laser bars will also be different. Therefore, the arrangement of the three laser stripe chips in superposition to form a multi-wavelength two-dimensional array module will also have a certain impact on the waste heat generated by the multi-wavelength module. The optimal multi-wavelength module arrangement can be obtained by analyzing the influence of the packaging structure composed of three kinds of wavelength laser bars in the multi-wavelength module on the device's operating temperature. Figure 2 shows the temperature distribution of multi-wavelength packaging modules in different packaging structures under continuous working conditions. By comparing the simulation results, it can be concluded that when the multi-wavelength module works, and when the chip arrangement in the multi-wavelength laser packaging structure is 808-915-980 (Figure 2(f)), The top temperature of the device reaches the maximum 77.34°C in the six arrangements. When the chip arrangement is 915-980-808 (Figure 2(a)), the top temperature of the device reaches the minimum value of 66.71°C. The difference between the two temperatures was 10.63°C, and the heat dissipation efficiency of the module was increased by 13.74% by changing the arrangement of chips. The simulation results show that the maximum temperature of the multi-the wavelength module appears in the active layer of the laser bar. This is because the heat dissipation path in the middle of the multi-wavelength module packaging structure can only be dissipated through the upper and lower secondary heat sinks. However, the two heat dissipation paths are also interfered with by the heat generated by the other two laser bars, so that the waste heat generated by the middle laser bar cannot be dissipated, resulting in heat accumulation.

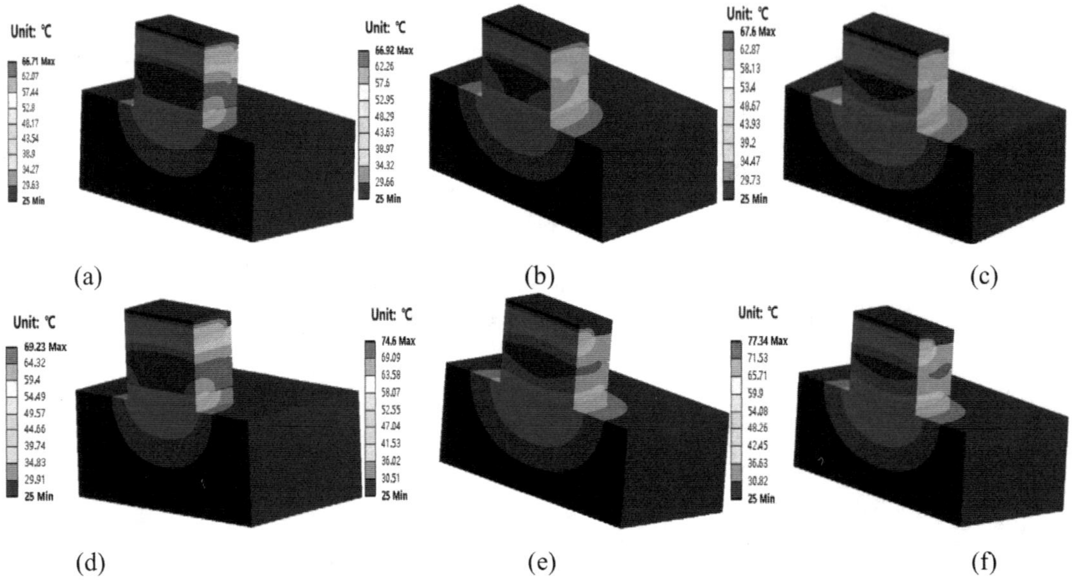

Figure 2 Temperature distribution of devices under different arrangement of chips, chip top-down order is (a)915-980-808 (b) 915-808-980 (c) 980-915-808 (d) 980-808-915 (e) 808-980-915 (f) 808-915-980

It is proved that the heat dissipation efficiency of multi-wavelength laser can be effectively improved by packaging multi-wavelength modules according to the optimal arrangement order of these three different wavelength laser bars. When the three laser bars work at the same time, the minimum highest temperature of the device in the packaging structures with different arrangements is 66.71°C, this far exceeds the maximum temperature of 50°C that the laser bar can support, which will cause damage to the device life and even complete damage [16]. In order to improve the heat dissipation efficiency of the multi-wavelength module, the conduction cooling package structure was optimized, and the back heat dissipation package multi-wavelength module was proposed.

3.2 Back heat dissipation package structure

In this paper above conductive cooling packaging, the structure is optimized by referring to the back heat dissipation packaging structure proposed by the same research group, and the back heat sink is increased. Figure 3 is a structural schematic and finite element model of the back heat dissipation packaging module. Figure 3(a) is a schematic of the back heat dissipation structure of a multi-wavelength packaging module, the material of the back heat sink is CuW, size is 1000×6000×5000μm. Figure 3(b) shows the finite element model of the back heat dissipation packaging structure of a multi-wavelength module.

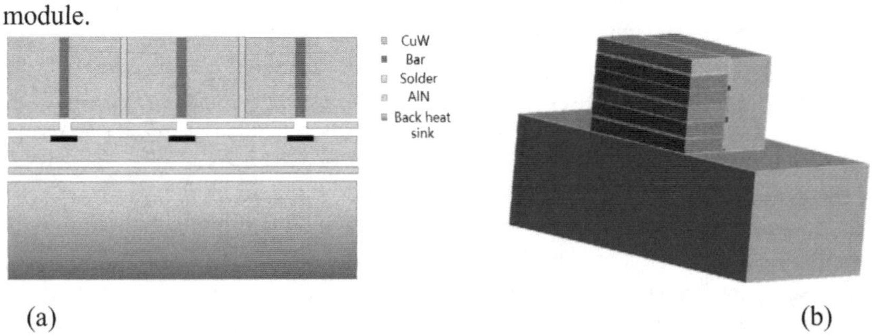

Figure 3 Schematic diagram and finite element model of the back heat dissipation package module

Figure 4 shows the temperature distribution of the multi-wavelength module when the back heat dissipation is adopted, results show that under continuous conditions, the top temperature of the multi-wavelength module reaches the maximum 59.42°C when the arrangement of laser bars is 808-915-980, the top temperature of the multi-wavelength module reaches the minimum value of 52.56°C when laser bar arrangement is 915-980-808. The difference in the temperature between the two arrangements is 6.86°C, and the heat dissipation efficiency of the device with different chip arrangements is increased by 11.54%. Compared with the conductive cooling package, the influence of the arrangement of laser stripe chips on the working temperature of the multi-wavelength module is reduced, this is because the back heat sink expands the cooling path of the multi-wavelength module, and a large amount of heat accumulated by laser bars in the middle layer of the module is dissipated, a heat dissipation path is added to the laser bar to reduce the thermal crosstalk between the laser bars. The simulation results show that the overall temperature of the multi-wavelength module is higher than that of other arrangements when the 808 nm laser bar is packaged at the top layer. When the 980nm laser chip is arranged in the narrow position of the heat dissipation path, the heat dissipation efficiency of the multi-wavelength module is optimized, showing that when the chip arrangement is 915-980-808, the back heat dissipation packaging structure is the best.

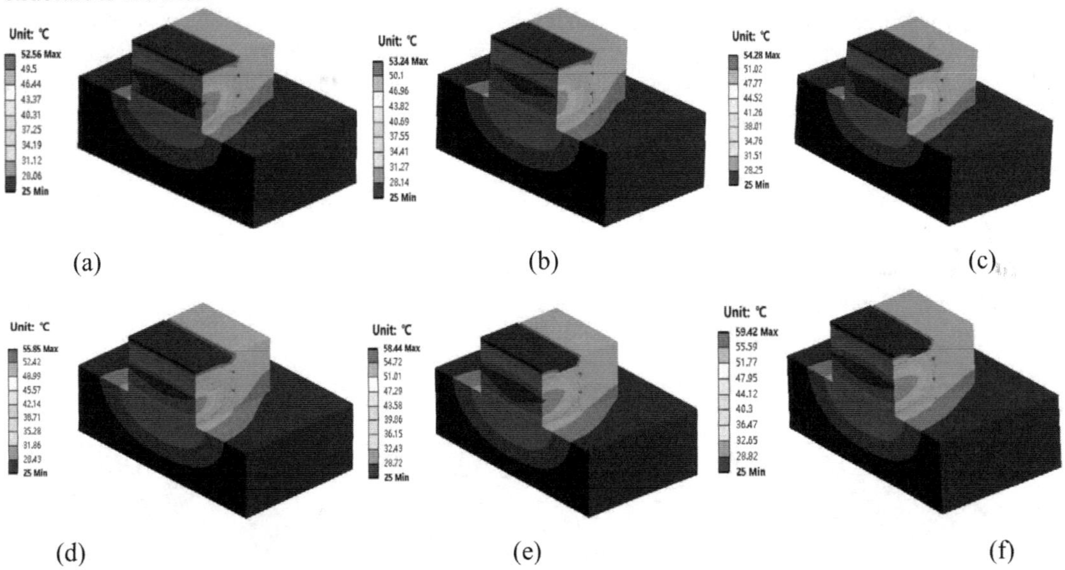

Figure 4 Temperature distribution of multi-wavelength module when the back heat dissipation is adopted, chip top-down order is (a)915-980-808 (b) 915-808-980 (c) 980-915-808 (d) 980-808-915 (e) 808-980-915 (f) 808-915-980

In this paper, we discuss the device a-f which is formed by six kinds of arrangement of three kinds of chips in a multi-wavelength packaging module. Figure 5 is a temperature histogram for device a-f with conduction cooling and backside cooling package structure. It can be seen that the temperature of the back heat dissipation packaging structure is greatly reduced compared with the cold conduction packaging structure. Compared with the conduction cooling packaging structure, the temperature of the device a-f with the back heat dissipation packaging structure decreased by 14.15°C, 13.68°C, 13.32°C, 13.41°C, 16.02°C, and 17.92°C. The back heat dissipation packaging structure effectively improves the heat dissipation rate of the device and achieves the effect of structural optimization.

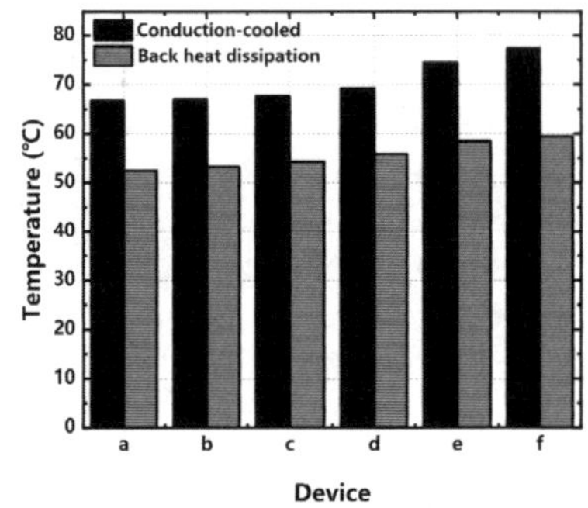

Figure 5 Temperature histogram for device a-f with conduction cooling and backside cooling package structure

3.3 Influence of different working modes on multi-wavelength modules

To determine the optimal working mode of the multi-wavelength module, transient thermal simulation of the multi-wavelength module packaged by back heat dissipation under pulse state is carried out. Figure 6 is the transient temperature distribution of multi-wavelength modules under pulse state. Similar to the above steady-state simulation results, when the arrangement of the three laser bars is 808-980-915, the maximum temperature of the device reaches 30.35°C. When the laser chip array is 915-980-808, the top temperature of the device reaches the minimum value of 29.55°C, and the heat dissipation efficiency of the device by changing the arrangement of laser bars increases by 2.6%. By comparison, the influence of the arrangement of laser stripe chips in multi-wavelength packaging module on the working temperature of the device becomes smaller under the state of the pulse, because when the laser bar is in pulse state, the chip does not work continuously, the waste heat generated by the laser bar can be dissipated in time, and there will be no large heat accumulation.

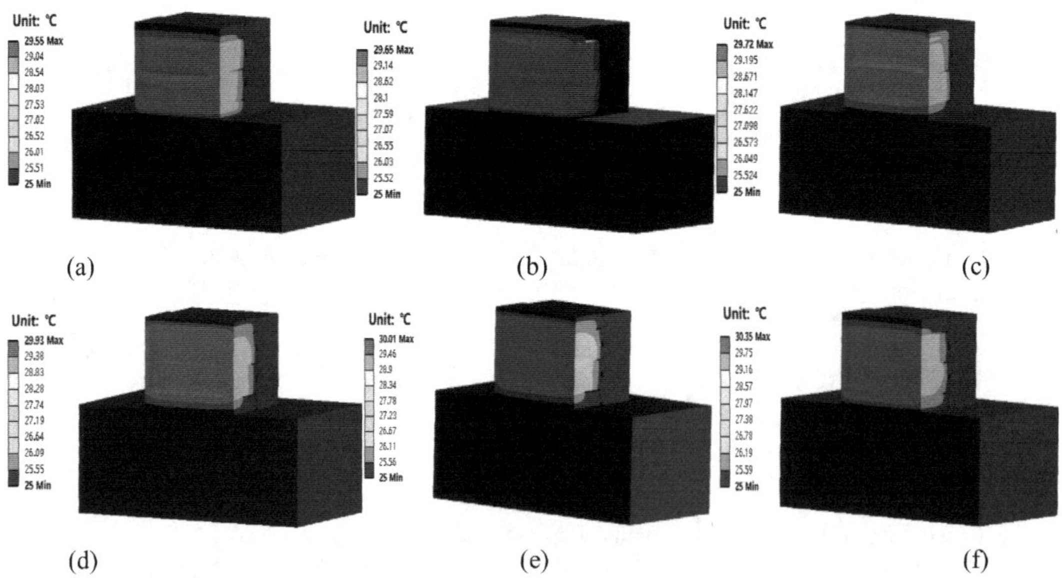

Figure 6 Transient temperature distribution of multi-wavelength modules under pulse state, chip top-down order is (a) 915-980-808 (b) 915-808-980 (c) 980-915-808 (d) 980-808-915 (e) 808-980-915 (f) 808-915-980

Figure 7 is a temperature histogram of a-f under different operating modes. It can be seen that the device temperature of the multi-wavelength module in the pulse state is much lower than that in the continuous state. In the pulse state, the effect of the multi-wavelength module chip arrangement on the device temperature is greatly reduced. By changing the chip arrangement, the temperature of the device is almost unchanged. Compared with the continuous operation mode, the temperature of the device a-f in pulse state is reduced by 23.01°C, 23.59°C, 24.56°C, 25.92°C, 28.43°C, and 29.07°C. In the pulse state, the top temperature of the device reaches a maximum of 30.35°C, which is within the temperature range that the semiconductor laser rod can withstand. According to the simulation results, the working mode of the multi-wavelength module should be a pulse, and the arrangement modes of the laser chips in the back heat dissipation package are 915 nm, 980 nm, and 808 nm.

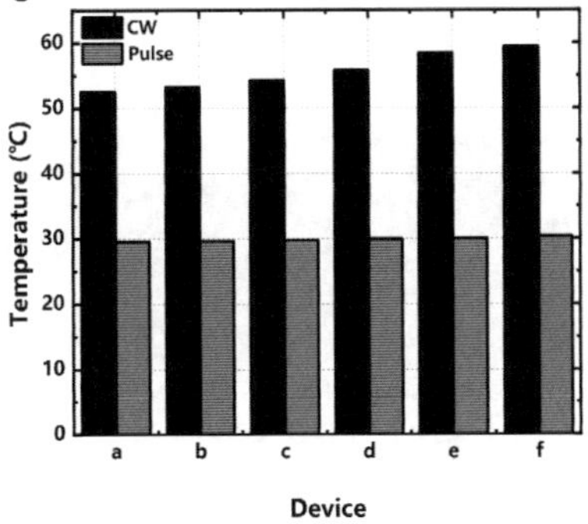

Figure 7 Temperature comparison curves of devices a-f under different operating modes

4. CONCLUSION

For a multi-wavelength stack module, the internal laser bar arrangement has an important influence on its performance. In this paper, the finite element method is used to simulate and analyze the influence of the arrangement of the bars on the temperature distribution of the multi-wavelength module. When the module adopts the conduction cooling package structure, the heat dissipation capacity of the device is improved by 13.74% by changing the laser bar arrangement. When the chip arrangement is 915-980-808, the heat dissipation capacity of the device is the best, and the temperature reaches 66.71°C. In order to make the multi-wavelength module have a better heat dissipation effect, the conduction cooling packaging structure is further optimized, and a back heat dissipation packaging structure is determined. In the back heat dissipation packaging structure, the heat dissipation capacity of the device is improved by 11.54% by changing the arrangement of laser bars. The lowest temperature is 52.56°C when the laser bar arrangement is 915-980-808. Compared with the conductive cooling package structure, the heat dissipation capacity of the device is increased by 23.17%, and the influence of the arrangement of laser stripe chips on the working temperature of the multi-wavelength module is reduced. Finally, the influence of different operating modes on the temperature of the multi-wavelength module is discussed. By comparing the operating temperature of the device under continuous and pulse states, it is found that the device temperature is reduced by 29.07°C under the pulse state. According to the simulation results, the multi-wavelength package module should work in pulse mode.

Acknowledgment
This work was supported by the key research and development projects in the Shaanxi innovation capability support project (2021TD-25), Shaanxi province (2020GY-044), and the Xi'an Science and technology plan project (2020KJRC0077).

REFERENCES

[1] Kim, D.S., Holloway, C., Bongtae, H., "Method for predicting junction temperature distribution in a high-power laser diode bar," Applied Optics. 55(27), 7487-7496 (2016).
[2] Extance, A., "Laser weapons get real," Nature. 521(7553), 408-410 (2015).

[3] Gu, B., "Current Situation and Application Development Trend of Laser Processing Technology and Industry," Metalworking. 10, 37-42 (2020).
[4] Ren, S.J., Chen, Z.B., Dai, Y.Y., "Development of control system for three-wavelength composite laser therapeutic instrument," Laser Journal. 40(7), 5 (2019).
[5] Jafarpour, T., Smith, F., "Design of multi-wavelength low-level laser therapy device for assisting bone reconstruction applications," Majlesi Journal of Electrical Engineering. 14(02), 117-125 (2020).
[6] Zhong, H., Jiang, D., Lan, H., "Low-cost multi-wavelength photoacoustic imaging based on portable continuous-wave laser diode module," IEEE Transactions on Biomedical Circuits and Systems. 99, 1-1 (2020).
[7] Liu, R.K., Wang, C.C., "Review of heat dissipation methods for high power semiconductor lasers," Electro-Optic Technology Application. 34(06), 1-6 (2019).
[8] Zhang, E.T., "Vertical stack of 808nm CW 2000W semiconductor laser," Infrared and Laser Engineering. 40(06), 1075-1080 (2011).
[9] Bull, S., Lim J.J., Amuzuvi, C.K., "Emulation of the operation and degradation of high power laser bars using simulation tools," Semiconductor Science and Technology. 27(9), 137-143 (2012).
[10] Fang, J.Y., Zhang, H., Zou, Y.G., "Thermal management of a semiconductor laser array based on a graphite heat sink," Applied Optics. 58(28), 7708-7715 (2019).
[11] Yuan, Q.H., Jing, H.Q., Zhong, L., "Thermal Stress in High-Power Semiconductor Laser Packaging," Chinese Journal of Lasers. 46(10), 1001009 (2019).
[12] Li, X.N., Zhang, Y.X., Wang, J.W., "Influence of Package Structure on the Performance of the Single Emitter Diode Laser," IEEE Transactions on Components Packaging & Manufacturing Technology. 2(10), 1592-1599 (2011).
[13] Wang, W., Gao, X., Zhou, Z.P., "Steady-state thermal analysis of hundred-watt semiconductor laser with multichip-packaging," Infrared Laser Eng. 43(5), 1438-4443 (2014).
[14] Tian, Z.H., Sun, C.L., Cao, J.S., "Junction temperature measurement of high power diode lasers," Optics and Precision Engineering. 19(6), 1244-1249 (2011).
[15] Jing, H.Q., Zhong, L., "Thermal analysis of laminated structure of high power density laser diode," Chinese Journal of Luminescence. 37(01), 81-87 (2016).
[16] Zhang, P., Wang J.W., Hou, D., "A 3000 W 808 nm QCW G-stack semiconductor laser array," International Symposium on High Power Laser Systems and Applications. International Society for Optics and Photonics. 92550W, 1-7 (2015).

Failure Analysis of the Solder Discoloration After Vacuum Sintering of Power Chips

Song Ling Li*, Ming Hua Zhang, Jing Ming Fei, Yu Yan Xiang, Xiao Feng Sun, Tao Chen, Qiong Wu, Cong Hui Peng, Qing Chen, Xue Ying Liu, Long Cheng and Ya Guang Zhang

Beijing Satellite Manufacturing Co., Ltd. Beijing 100094, China

Email: lisongling67@163.com

Abstract. In this paper, the discoloration of the solder after the vacuum sintering of the power chip is analyzed. The morphology and composition of the discoloration area were characterized and analyzed by the metallographic microscope, SEM, EDS and XPS, and the discoloration mechanism was studied. The analysis results show that the main components of the color-changing solder are C, O, Pb, Sn, Ni, Ag, and Au, which are the normal components of the solder after vacuum sintering. The reason for color-changing is the oxidation of the Sn element, and the different thickness of oxide shows different colors. The discoloration solder was treated with formic acid, and the solder returned to bright silver-white. The test further confirmed that the discoloration was caused by metal oxidation. The problem of solder discoloration is solved by optimizing the process to reduce the sintering time and vacuum defoaming time.

1. Introduction

Power devices are developing in the direction of high current, high voltage, high frequency, integration and intelligence [1]. Therefore, the thermal performance of the device becomes increasingly important. The quality of the chip solder layer is critical to the thermal and reliable performance of the device. The reliability of the chip and substrate interconnection is particularly important in the evaluation of the packaging process of electronic products as a key technical indicator [2]. At present, the commonly used chip interconnection technology mainly includes instantaneous liquid phase diffusion welding technology [3], nano-silver low-temperature sintering technology [4], and vacuum welding technology [5]. Among them, the vacuum welding technology represented by formic acid reduction vacuum defoaming welding technology is the most common, mature and reliable process method for the assembly of power device chips. Formic acid reduction vacuum sintering refers to the protective gas filled with formic acid and nitrogen mixture gas in a vacuum chamber, and the intermetallic compound is formed between the solder and the metal to be welded by heating [6]. The commonly used soldering materials include high-temperature solder such as tin-lead-silver, gold-tin, tin-silver-copper, etc. The solder layer plays a supporting and heat dissipation role in the package structure, which is the weak part of the package structure. Chip sintering failure mainly includes void rate exceeding the standard, insufficient shear strength, chip stress damage, etc., which is the main source of device packaging failure and seriously affects product reliability and package yield. Therefore, it is very necessary to analyze the failure problems generated during the sintering process of the power chip, find out the causes of failure, and formulate targeted solutions to improve product quality.

The internal visual inspection of the products involved in this paper found different degrees of discoloration of sintered chip solder (Figure 1) (×10). About 60% of the products showed blue-purple solder, while the rest showed yellow-brown. Tracing the production process, the solder after sintering was bright silver-white, and no obvious discoloration was found during bonding and bonding visual inspection for 3 days. After the internal visual inspection, the yellow-brown products gradually turned blue with the extension of storage time, indicating that the discoloration problem is worsening with time. In this paper, failure analysis and mechanism analysis are carried out to form corrective measures to avoid failure again.

Figure 1. Metallographic microscope photograph of the discoloration area

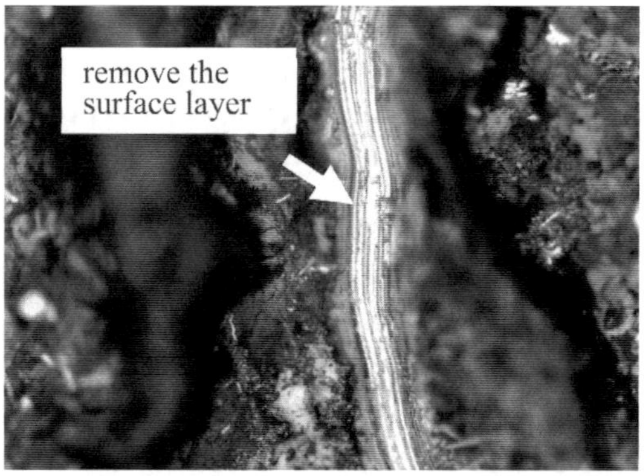

Figure 2. Metallographic microscope photograph of the discoloration area after removing the surface layer

2. Characterization of discoloration substances

The morphology was characterized by metallographic microscopy and SEM, and the composition was analyzed with EDX and XPS.

2.1 Metallographic

The solder discoloration area mainly appears in the solder overflow position. The surface of the overflow solder is not smooth, on which there are many folds and holes. The discoloration degree of solder near the chip position is lighter, and the discoloration is more obvious away from the chip position. The normal solder is observed after welding. The surface is bright silver-white and smooth. A 1000-time metallographic microscope is used to zoom in on the discoloration area (Figure 1) (×1000). Different areas have different discoloration. The main color is blue, while there are also purple, yellow, brown, etc. The discoloration material is similar to a film attached to the surface of the solder. After the mechanical force is applied to remove the surface of the solder, the interior shows the bright silver-white color of the fresh solder (Figure 2), indicating that the thickness of the discoloration layer is thin.

2.2 SEM and EDX

Using SEM and EDX micro examination and composition analysis for solder discolored areas (Figure 3 and Figure 4), It is found that there is no obvious excess adhesion on the surface of the solder, and the main components of the solder surface are C, O, Sn, Pb, Ag, Au and Ni. Sn, Pb and Ag are the material components of Sn5Pb92.5Ag2.5 solder, Au and Ni as the shell surface composition. C and O are caused by organic matter adhered to the surface of solder and oxidation of solder, in which the O element accounts for 11.41% and the C element accounts for 3.05%, indicating that there is significant oxidation on the surface of the solder.

After scraping off a layer of a blue spot on the surface of the solder, SEM and EDX analyses were conducted again (Figure 5 and Figure 6). The results of microscopic morphology and energy spectrum analysis showed that the main components of the solder were also C, Sn, Pb, Ag, Au and Ni. The content of the O element was too small to be detected, indicating that no obvious oxidation occurred in the internal solder. According to two EDX analysis results, the discoloration of the solder around the chip may be caused by solder oxidation.

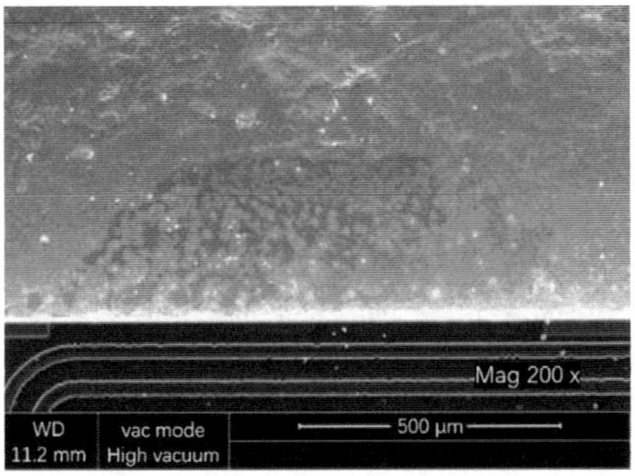

Figure 3. SEM result on the surface of the solder discoloration area

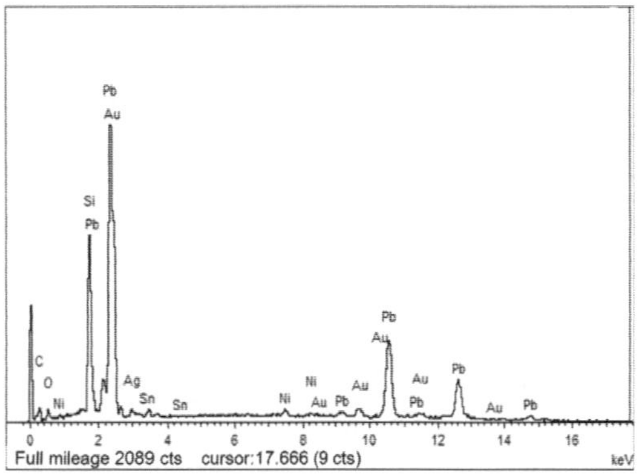

Figure 4. EDX result on the surface of the solder discoloration area

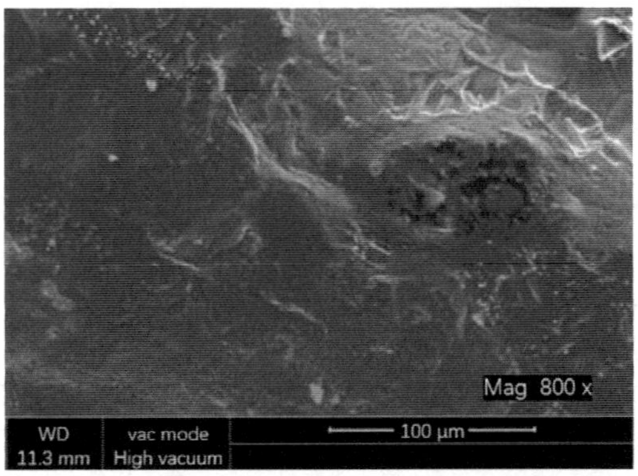

Figure 5. SEM result of discoloration zone after removing the discoloration layer

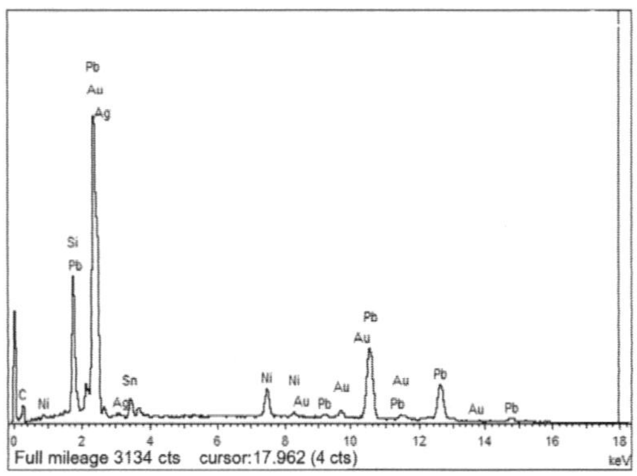

Figure 6. EDX result of discoloration zone after removing the discoloration layer

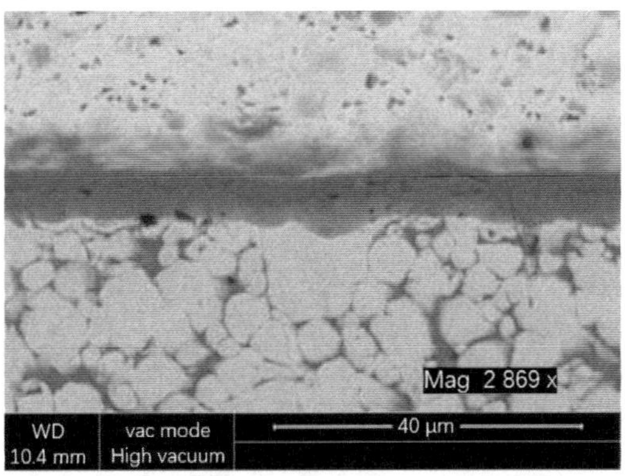

Figure 7. SEM photo of weld microstructure

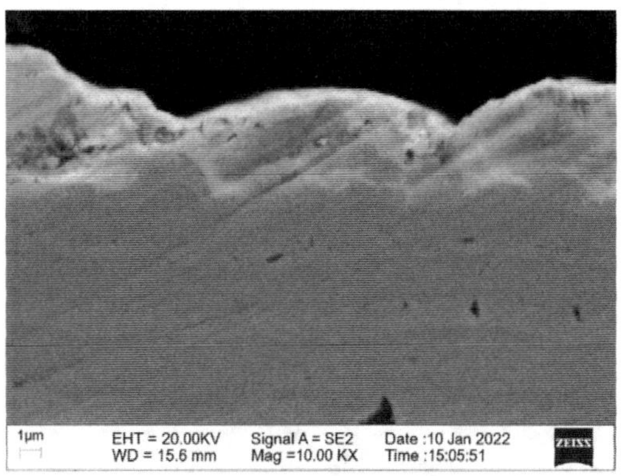

Figure 8. SEM photo of a section of solder overflow area of the device

The product was prepared into metallographic samples and the profile morphology was observed by SEM. Intermetallic compounds were formed between the chip and the solder, and between the solder and the nickel-plating layer on the bottom shell (Figure 7). Metallographic profile analysis was carried out on the solder overflow around the chip, and SEM was used to observe the profile. The thickness of the discoloration layer could not be seen from the profile picture of the solder in the discoloration area (Figure 8). The main components of the solder for the EDX analysis are Sn, Pb, Ag, Au, Ni, etc. The oxygen content of the solder surface accounts for 6%~9%, and the oxygen content measured internally is less than 2%. The results of the metallographic microscope, SEM and EDX showed that the discoloration and oxidation only occurred on the solder surface, and had no effect on the welding interface quality.

2.3 XPS

To further determine the phase of the solder surface, XPS is used to perform full-spectrum scanning and narrow-spectrum scanning of metallographic discoloration regions. From the full spectrum scanning data (Figure 9), it can be qualitatively analyzed that the elements in the discoloration area include C, O, Pb, Sn, Ni, Ag, etc.

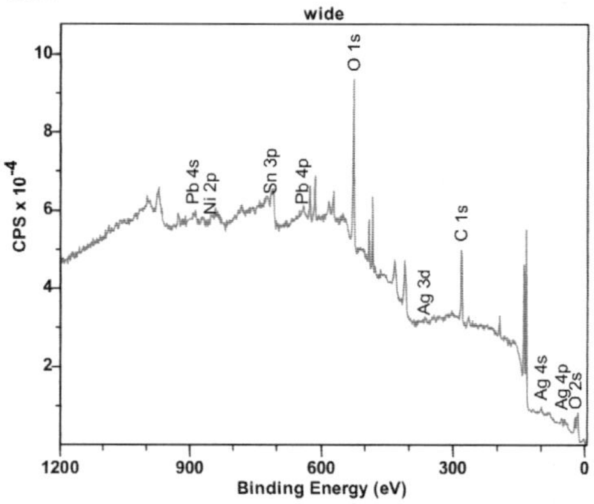

Figure 9. Full-screen scan results for discolored areas

According to the color and related literature, there are more likely oxides of Sn in the color material. The narrow spectrum scanning analysis was carried out for Sn and Pb that are focused on. From the narrow spectrum scanning analysis results of the Sn element (Figure 10), it can be seen that the Sn element mainly has two valence states, namely Sn 3d 3/2 and Sn 3d 5/2, respectively. By comparing the binding energy corresponding to the two main peaks of the Sn element and searching the standard database, it is found that the corresponding state is mainly SnO_2 and SnO and the oxides between the two.

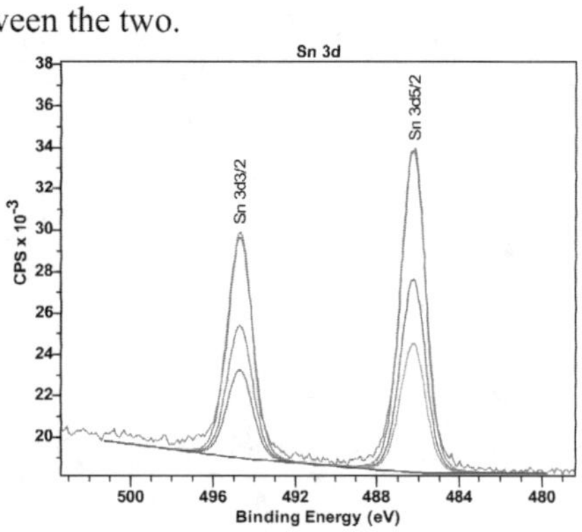

Figure 10. Peak fitting curve of Narrow spectrum of the Sn Element

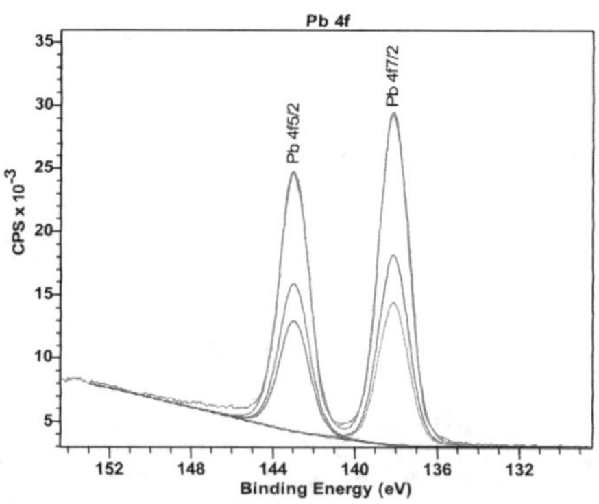

Figure 11. Peak fitting curve of Narrow spectrum of the Pb element

From the narrow spectrum scanning analysis results of the Pb element, it can be seen that there are two valence states (Figure 11), namely, Pb 4f7/2 and Pb 4f5/2. By searching the standard database, it is found that the corresponding valence states are mainly PbO_2 and Pb.

3. Mechanism Analysis

3.1 Oxidation analysis of Sn-Pb alloy

Sn-Pb is a low eutectic alloy. Both Sn and Pb are relatively active metals, which can react with oxygen in the air at room temperature on the surface. The surface of molten Sn-Pb alloy is easily oxidized, and the rate of oxidation and the state of oxidation are related to the temperature [7]. At 250 °C, the thickness of the oxide layer increases over time, and the color changes from silvery white to light yellow, yellow, brown, purple, gray, etc. As the temperature increases, the oxidation rate of the alloy surface increases. At 280 °C, the oxidation rate is slow, and at 350 °C~400 °C, the oxidation rate of tin accelerates, while that of lead decreases. At lower temperatures, tin and lead are in a high valence state in their oxides, but when the temperature increases, their valence state decreases, thus showing different surface states. The sintering temperature of the failed power device is between 335 °C and 345 °C. The total time of the high-temperature period is 8 min ~ 9 min. The visible solder surface after sintering is silver white, which indicates that the internal atmosphere of the sintering chamber has low oxygen content. The sintering time is short, and there is no solder oxidation discoloration caused by over-burning.

The oxides of both Sn and Pb have various valence states and different crystals [8]. The oxidation products of the pure metal Pb are mainly the quadrilateral phase and orthogonal phase. The former is stable at low temperatures, while the latter is stable at high temperatures with a transition temperature of 488 °C. A large number of experiments have proved that the oxidation reaction of solid Pb at room temperature generates a high-temperature stable orthogonal phase, which is mainly attributed to the epitaxial force of some specific crystal planes of the Pb metal substrate. Oxide of lead generated after oxidation of Pb easily interacts with trace amounts of H_2O and CO_2 in the air, and constantly penetrates the body phase, causing further corrosion of Pb. This oxide will form two main peaks with an interval of 2 eV in the XPS map of Pb. The interval between the main peaks in the XPS map of this batch of products is 4.9 eV, which can prove that further oxidation corrosion occurs. Studies have shown that the oxidation of Pb can make the solder lose its metallic luster. Even changes in gray-brown, blue, purple, etc. indicated by the solder of the failed device are not mainly caused by the Pb oxidation.

The oxidation products of Sn are not only related to temperature, but also related to oxidation time. In the air atmosphere, on the one hand, with the increase of oxidation time, the oxide layer of Sn gradually thickens. As the thickness increases, the oxide exhibits different appearance morphologies. In the relevant literature [9], it has also been found that with the thickness of the oxide layer, the color displayed is also inconsistent, appearing silver, gold, golden red, gray, and blue-violet related to the interference of light (Table 1). On the other hand, according to the bulk phase thermodynamic function, the stable oxidation product of Sn is SnO_2, followed by SnO [10]. As the oxidation time increases, more SnO_2 is formed.

Table 1. Relation between tin oxide film thickness and appearance color

Oxide layer of Sn thickness (μm)	< 0.7	0.7 ~ 1.5	1.5 ~2	2 ~ 3.5	> 3.5
The interference light	Ultraviolet light	Blu-ray	Greenlight	Yellow light	Red light
Color rendering	metallic	yellow	purple	blue	green

In Sn-Pb solder, although Sn and Pb are mixed at an atomic level, no new Sn-Pb-O compounds are formed at low or high temperatures, and they appear to be Sn-rich or Pb-rich phases [11]. The segregation of Pb in the Sn phase is due to the large strain energy given by Pb with a large atomic radius. The Sn segregation in the Pb-rich phase is due to the fact that Sn with a lower melting point gives less surface energy. When there is an aerobic atmosphere, due to the low standard oxide generation free energy of tin, more Sn tends to be enriched on the surface in the form of oxides, especially liquid oxidation. Even if Sn is rare, but due to the accelerated diffusion rate, Sn in the oxide film also accounts for more than 90%. So, the oxide of Sn is enriched on the surface, and the solder surface shows the oxide color of Sn. The blue and purple color exhibited by the increased thickness of the tin oxide layer matches the color of the failed device.

3.2 Influencing factors of Sn-Pb alloy oxidation

The main influencing factors of Sn-Pb alloy oxidation include temperature and humidity, internal impurity elements and alloy surface state. The devices with the same surface state of solder are placed in a dry nitrogen environment, an atmospheric environment, and an atmospheric environment after 10 minutes of high-temperature water vapor treatment, and then the surface state of solder is regularly observed. Among them, the surface of the sample solder after water vapor treatment changes significantly after one day. The sample in the atmospheric environment undergoes similar changes after 5 days, and the sample that has been in a dry nitrogen environment only slightly changes in color after 10 days. It indicates that humidity and a high heat environment will accelerate discoloration and deterioration.

Studies show that chalcogen, ferrous, and alkali metals have a worsening effect on the oxidation of Sn-Pb alloys [12] [13]. Other than that, organic matter is easy to oxidize and change in color. when there is more organic matter in the solder, under the guidance of water vapor, the organic molecules contained in the pores of the solder through capillary action will migrate to the surface, and accumulate on the surface and gradually oxidize, showing that the surface of the solder becomes yellow.

The degree of oxidation of solder is affected by the surface form of solder [14], when there are more microstructural defects on the surface of the solder. These pores, cracks and other defects make organic matter penetrate it, and cannot be cleaned and removed, so the organic matter on the surface of the solder oxidation discoloration accelerates the oxidation of tin and the lead elements.

In this paper, through the microscopic analysis of the surface of the solder of the failed product, it is found that the surface of the solder has more holes, folds and other defects, and the surface contains more C and O components. Sn-Pb alloys with rough surfaces are more likely to oxidize after exposure to large amounts of oxygen, thus leading to serious oxidation and discoloration of the solder surface. The analysis shows that after sintering, the diffusion of the solder is more obvious and the surface is uneven. It is related to the structure and composition of the bottom coating after sintering, which requires targeted process adjustment to improve the surface state of the solder.

4. Experimental verification

The formic acid reduction technology is used to reduce discolored products. Formic acid can react with metal oxides at high temperatures, and the reaction process is as follows Formula (1) ~ (3). Me stands for metal. Formic acid reacts with metal oxides at temperatures between 160 °C and 220 °C. After restoration, the product is visually inspected and the solder is restored to a silvery and shiny state. The test results further confirm that the discoloration substance on the surface of the solder is a metal oxide.

$$MeO+2(HCOOH)= Me\,(COOH)_2+H_2O\ (>150°C) \tag{1}$$

$$Me\,(COOH)_2= Me+2CO_2+H_2\ (>200°C) \tag{2}$$

$$MeO+H_2= Me+H_2O\ (>200°C) \tag{3}$$

The chip sintering process is optimized. The surface state of the solder can be effectively improved by reducing the sintering time and vacuum defoaming time. The surface of the solder after sintering is smooth, which has no obvious folds and holes and has a distinct metallic luster. After three days, five days, and ten days after observing the solder, there was no obvious change in the surface color, indicating that the problem of discoloration failure was solved.

5. Summary

In this paper, failure analysis was carried out on the discoloration of the solder after vacuum sintering of the power chip. The morphology and composition of the discoloration area were characterized and analyzed by a metallographic microscope, SEM, EDS and XPS, and the discoloration mechanism was studied. The EDS analysis results show that the main components of the color-changing solder are C, O, Pb, Sn, Ni, Ag, and Au, which are the normal components of the solder after vacuum sintering of the chip. The XPS analysis results show that the main compounds are Sn oxides and Pb oxides. The reason for color-changing is the oxides of Sn elements, and the main reason is that the oxides of Sn of different thicknesses show different colors. The discoloration solder was treated with formic acid, and the solder returned to bright silver-white. The test further confirmed that the discoloration was caused by metal oxidation. The problem of discoloration of solder was solved by reducing sintering time and vacuum defoaming time.

Acknowledgments

I am very grateful to the partners of the Power device project group (Project No. 2009ZYGQ0301) for playing an important role in my research direction. At the same time, I would like to thank my family for their support of my work.

References

[1] Baliga B J 2001 The future of power semiconductor device technology. *J. Proceedings of the IEEE*, 89(6): pp 822-832.
[2] Xu J L 2018 Cavity control of high-power chip brazing preformed strip. *J. Welding Technology*, 47(12): pp 89-91.
[3] Shirzadi and Wallach 2004 New method to diffusion bond superalloys. *J. Science and Technology of Welding and Joining*, 9(1): pp 37-40.
[4] Wang L, Yu J R, Dong X H, et al. 2015 Research Progress on Transient Liquid Phase Diffusion Welding Technology. *J. Machinery Manufacturing and Automation*, 44(04): pp 7-9.
[5] Wu Y b, Dai X P, Wang Y G, et al. 2015 Research progress on advanced interconnect technology in IGBT power module packaging. *J. Great work Rate Converter Technology*, 2015(02): 6-11
[6] Du C H, Chen F and Huang F X 2008 *Electronic micro-connection technology and materials* (Beijing: China Machine Press).
[7] Zhang C 1984 Anti-oxidation technology on the surface of molten tin-lead-filled brazing material. *J. Electronic Process Technology*, 1984(10): pp 2-8.
[8] Xiao B 2018 Study on mechanical behavior and microstructure evolution of tin-lead-silver brazing material and solder joints at very low temperatures. *D. Harbin Institute of Technology.*
[9] Dong C L 2017 Analysis of discoloration of tin layer on electronic connector surface during high-temperature welding. *J. Electroplating and Finishing*, 36(13): pp 696-700.
[10] Leonardo R G, Wislei R O, Leandro C P, et al. 2009 Wetting behavior and mechanical properties of Sn-Zn and Sn-Pb solder alloys. *J. Journal of Electronic Materials*, 38(11): pp 2405-14.
[11] Chu L X, Wu N Z, Gui L L, et al. 1988 Mechanism of oxidation and corrosion of Sn-Pb alloy, *J. Science China*, 18(08): pp 868-876.
[12] Stoneman A M, Mackay C A and Thwaites C J 1980 Oxidation and drossing of Potassium salts in Molten solders. *J. Metals Technology*, 7(01): pp 226-231.
[13] Tu K N and Zeng K 2001 Tin-lead (Sn Pb) solder reaction in flip chip technology. *J. Materials Science & Engineering R*, 34(1): pp 1-58.
[14] TAKASE Y, MAEDA S, OYAMA T, et al. 2010 Dissolution Mechanism of Lead from Tin-Lead Solder. *J. The Journal of Japan Institute for Interconnecting and Packaging Electronic Circuits*, 11(7): pp 525-528.

Non-Hermite Spinor Boltzmann Equation and Its Hermitization

Zheng-Chuan Wang

Department of Physics, University of Chinese Academy of Sciences, P. O. Box 4588, Beijing 100049, China.

wangzc@ucas.ac.cn

Abstract. It is known that the Wigner distribution function in the quantum Boltzmann equation is not positive elsewhere, which causes trouble for its application to the mesoscopic transport problem. Similarly, if we extend the quantum Boltzmann equation to the spinor Boltzmann equation, which includes the spin freedom in the usual Wigner distribution, and the spinor matrix distribution function therein is not Hermite, we must make this non-Hermite spinor Boltzmann equation Hermitization. So in this paper, we propose a spinor Boltzmann equation with Hermite distribution function and then obtain the equations of continuity satisfied by the charge density, charge current, spin accumulation, and spin current. The numerical results for these physical quantities are illustrated by an example of spin-polarized transport scattered by impurities in a system of spintronics. Their differences compared with the usual spinor Boltzmann equation are shown.

1. Introduction

Boltzmann equation, since its discovery, has been widely used in many kinds of classical systems to explore the dissipative and irreversible transport procedure therein, and the semi-classical Boltzmann equation has also been exploited to study the transport problem in the mesoscopic system, i.e., the spin-polarized transport in nanoscale spintronics devices[1,2,3,4,5]. However, in a system with a dimension smaller than De Broglie's wavelength, the semi-classical Boltzmann equation fails. One should treat its transport problem by the quantum Boltzmann equation. The quantum Boltzmann equation was originally proposed by Wigner in 1932[6,7]. Nowadays, it has been extended to include spin freedom and become the so-called spinor Boltzmann equation[8,9,10], and the latter is convenient for us to investigate the spin-polarized transport in spintronics[11,12].

As we know, the distribution function in the classical Boltzmann equation is always positive because it stands for the probability of a particle appearing in phase space. However, the distribution function in the quantum Boltzmann equation is not positive elsewhere[13] due to the uncertain relationship between the position and momentum in the quantum Wigner distribution function. In order to overcome the negative probability difficulty, Husimi made an average to the Wigner distribution function over a spatial region and obtained a positive so-called Husimi distribution function[14], but the spatial precision of this distribution function is lowered due to the average. Several other distribution functions were introduced by some authors, too[15,16]. In 2007, we expanded the quantum Wigner distribution function by the Planck constant $\hbar$[17], then divided it into the classical zero order, the first order, and higher order quantum correction terms, and got the equations satisfied by them. So we naturally obtained the positive zero distribution function from the Wigner distribution function, the negative part was separated into the quantum correction terms, and the spatial precision of the zero positive distribution function is as well as the original Wigner distribution function.

Despite the good spatial precision of zero order positive distribution function obtained by us[17], it is classical. When we want to investigate the quantum effect in the mesoscopic system, we should consider the first order and higher order correction terms in the distribution function. If we include the spin freedom in the distribution function, it will become the spinor distribution, a 2×2 matrix in spin space. However, this spinor distribution function is not Hermite because it is a Fourier transformation to the Hermite density matrix in spin space. As we know, the physical quantity should be Hermite in order to assure its real eigenvalues. Although the usual spinor distribution function contained the quantum effect, i.e., the off-diagonal elements of the 2×2 matrix in spin space, it is not Hermite and sometimes has no real eigenvalues, so we need to search for a Hermite spinor distribution function. In this paper, we introduce a Hermite spinor distribution function by Hermitization of the usual spinor distribution function, then derive the spinor Boltzmann equation for this new Hermite distribution function, and obtain the equation of continuity satisfied by the charge density, current, spin accumulation and spin current.

2. The Hermite Spinor Distribution Function and the Boltzmann Equation It Satisfying

Usually, the spinor distribution function is defined as the Fourier transformation on the density matrix in spin space. When we include the spin indices in the quantum Wigner distribution, their components are[18]:

$$f_{\sigma\sigma'}(p,X) = \frac{1}{2\pi\hbar}\int dx \exp\left(-\frac{ip}{\hbar}x\right)\rho_{\sigma\sigma'}\left(X+\frac{x}{2},X-\frac{x}{2}\right), \tag{1}$$

where σ and σ' stand for the spin indices, $\mathrm{p} = (\vec{p},\varepsilon)$ is the four-dimensional momentum, ε is the energy, $X = (R,T)$ where $\mathrm{R} = \frac{1}{2}(r_1 + r_2)$, $\mathrm{T} = \frac{1}{2}(t_1 + t_2)$ is the center of mass variable, the variable $\mathrm{x} = (\mathrm{r},\mathrm{t})$, which represents the differences between two position and time variables entering these functions, $\mathrm{r} = r_1 - r_2$ and $\mathrm{t} = t_1 - t_2$. Obviously, the spinor distribution function $f_W = \begin{pmatrix} f_{\uparrow\uparrow}(p,X) & f_{\uparrow\downarrow}(p,X) \\ f_{\downarrow\uparrow}(p,X) & f_{\downarrow\downarrow}(p,X)\end{pmatrix}$ is not Hermite because it is a Fourier transformation on a Hermite density matrix $\rho(\mathrm{p},\mathrm{X}) = \begin{pmatrix} \rho_{\uparrow\uparrow}(p,X) & \rho_{\uparrow\downarrow}(p,X) \\ \rho_{\downarrow\uparrow}(p,X) & \rho_{\downarrow\downarrow}(p,X)\end{pmatrix}$. The density matrix has real positive eigenvalues, but $f_W(p,X)$ has no such character. As a distribution function with corresponding physical meaning, the 2×2 matrix must be Hermitization so as to possess reasonable eigenvalues.

Due to the relation between density matrix and lesser Green function in non-equilibrium green function theory,

$$\rho_{\sigma\sigma'}(x,x') = -iG^{<}_{\sigma\sigma'}(x,x'), \tag{2}$$

the spinor distribution function can also be expressed by the lesser Green function[7,12]

$$f_{W\sigma\sigma'}(p,X) = \frac{1}{2\pi\hbar}\int dx \exp\left(-\frac{ip}{\hbar}x\right) G^{<}_{\sigma\sigma'}\left(X+\frac{x}{2},X-\frac{x}{2}\right) = iG^{<}_{\sigma\sigma'}(\mathrm{p},\mathrm{X}). \tag{3}$$

Starting from this definition, we can construct a Hermite spinor distribution function with the lesser green function or the Wigner distribution function, that is

$$\mathrm{f}(\mathrm{p},\mathrm{X}) = \frac{G^{<}(p,X)+G^{<\dagger}(p,X)}{2}, \tag{4}$$

so we have $\mathrm{f}^{\dagger} = f$, the Hermite spinor distribution function has real eigenvalues. It is just the physical distribution we are searching for. If we further define

$$\tilde{\mathrm{f}}(\mathrm{p},\mathrm{X}) = \frac{G^{<}(p,X)-G^{<\dagger}(p,X)}{2}, \tag{5}$$

we will find it is an anti-Hermite matrix. The lesser green function can be expressed by f and $\tilde{f}$, too, it is

$$G^<(p,X) = f(p,X) + \tilde{f}(p,X), G^{<\dagger}(p,X) = f(p,X) - \tilde{f}(p,X). \quad (6)$$

As we know, the lesser Green function $G^<(r,r')$ satisfies the Dyson equation in non-equilibrium Green function theory. If one makes a quantum Wigner transformation on both sides of the Dyson equation, one can obtain the equation satisfied by the lesser Green function $G^<(p,X)$ in phase space[7,12]

$$\left[\frac{\partial}{\partial T} + \frac{\vec{p}}{m}\cdot\vec{\nabla}_R + \frac{\partial U}{\partial X}\vec{\nabla}_p\right] G^<_{\sigma\sigma'}(p,X) = -2\pi i \sum_{\sigma''}(\Sigma^>_{\sigma\sigma''} G^<_{\sigma''\sigma'} - \Sigma^<_{\sigma\sigma''} G^>_{\sigma''\sigma'}). \quad (7)$$

where $G^>$ is the greater Green function, $\Sigma^<$ and $\Sigma^>$ are the self-energy. Substituting $G^<(p,X) = f(p,X) + \tilde{f}(p,X)$ into the above equation, and $G^{<\dagger}(p,X) = f(p,X) - \tilde{f}(p,X)$ into the Hermite equation of Eq.(7), we can get the equations satisfied by f and $\tilde{f}$, they are

$$\left[\frac{\partial}{\partial T} + \frac{\vec{p}}{m}\cdot\vec{\nabla}_R + \frac{\partial U}{\partial X}\vec{\nabla}_p\right] f(p,X) = -\pi i[\Sigma^>(f+\tilde{f}) - \Sigma^< G^>] + \pi i[G^{<\dagger}\Sigma^{>\dagger} - G^{>\dagger}\Sigma^{<\dagger}] \quad (8)$$

and

$$\left[\frac{\partial}{\partial T} + \frac{\vec{p}}{m}\cdot\vec{\nabla}_R + \frac{\partial U}{\partial X}\vec{\nabla}_p\right] \tilde{f}(p,X) = -\pi i[\Sigma^>(f+\tilde{f}) - \Sigma^< G^>] - \pi i[G^{<\dagger}\Sigma^{>\dagger} - G^{>\dagger}\Sigma^{<\dagger}]. \quad (9)$$

If we only consider the electron-impurity scattering at low concentration, the self-energy $\Sigma^{<,>}$ in the right side of Eq.(8) and (9) can be simplified as[7]

$$\Sigma^{<,>}_{\sigma\sigma'}(p,X) = n_i \int \frac{d^3p'}{(2\pi)^3}(T_{pp'})^2 G^{<,>}_{\sigma\sigma'}(p',X), \quad (10)$$

where

$$T_{pp'} = V_{pp'} + \int \frac{d^3p'}{(2\pi)^3}\frac{V_{pp'}T_{pp'}}{\omega-\varepsilon_p+i\delta}, \quad (11)$$

and n_i is the concentration of impurity, and $V_{pp'}$ is the Fourier transform of electron-impurity coulomb interaction, $G^>(p,X)$ is the distribution function of the hole. If we further define

$$f'(p,X) = \frac{G^>(p,X)+G^{>\dagger}(p,X)}{2}, \quad \tilde{f}'(p,X) = \frac{G^>(p,X)-G^{>\dagger}(p,X)}{2}, \quad (12)$$

we can see that f' is Hermite, $\tilde{f}'$ is anti-Hermite, f' is the Hermite spinor distribution function of the hole. Rewriting the right hand of Eq.(8) and (9) by means of expressions (10) and (11), we can obtain the final equations satisfied by f and $\tilde{f}$:

$$\left[\frac{\partial}{\partial T} + \frac{\vec{p}}{m}\cdot\vec{\nabla}_R + \frac{\partial U}{\partial X}\vec{\nabla}_p\right] f(p,X) = \pi n_i i \int \frac{d^3p'}{(2\pi)^3}(T_{pp'})^2[(f'(p)f(p) - f(p)f'(p')) -$$
$$\left(f'(p')\tilde{f}(p) + \tilde{f}(p)f'(p')\right) - \left(\tilde{f}'(p')f(p) + f(p)\tilde{f}'(p')\right) - (\tilde{f}'(p')\tilde{f}(p) - \tilde{f}(p)\tilde{f}'(p'))$$
$$+(f(p')f'(p) - f'(p)f(p')) + \left(f(p')\tilde{f}'(p) + \tilde{f}'(p)f(p')\right) + (\tilde{f}(p')f'(p) + f'(p)\tilde{f}(p'))$$
$$+(\tilde{f}(p')\tilde{f}'(p) - \tilde{f}'(p)\tilde{f}(p'))] \quad (13)$$

and

$$\left[\frac{\partial}{\partial T} + \frac{\vec{p}}{m}\cdot\vec{\nabla}_R + \frac{\partial U}{\partial X}\vec{\nabla}_p\right] \tilde{f}(p,X) = -\pi n_i i \int \frac{d^3p'}{(2\pi)^3}(T_{pp'})^2[(f'(p')f(p) + f(p)f'(p')) +$$

$$\left(f'(p)\tilde{f}(p)+\tilde{f}(p)f'(p')\right)+\left(\tilde{f}'(p')f(p)-f(p)\tilde{f}'(p')\right)+(\tilde{f}'(p)\tilde{f}(p)+\tilde{f}(p)\tilde{f}'(p))$$
$$-(f(p')f'(p)+f'(p)f(p'))-\left(f(p')\tilde{f}'(p)-\tilde{f}'(p)f(p')\right)-(\tilde{f}(p')f'(p)-f'(p)\tilde{f}(p'))$$
$$-(\tilde{f}(p')\tilde{f}'(p)+\tilde{f}'(p)\tilde{f}(p'))] \tag{14}$$

The above equations (13) and (14) are coupled together, which means we should solve both Eq.(13) and (14) to obtain the physical distribution function f, and the uninteresting distribution function $\tilde{f}$ which has an imaginary eigenvalue and affects f.

It should be pointed out that the distribution f and $\tilde{f}$ are the matrices, and Eq.(13) and (14) are the equations of matrices. In order to solve these matrix equations, we had better expand the matrices f and $\tilde{f}$ by the Pauli matrices $\vec{\sigma}$ and unit matrix. They are[8,11]

$$f(p,X)=f^0(p)+\left(-\frac{\partial f^0}{\partial\varepsilon}\right)[f_s(p,X)I+\vec{g}(p,X)\cdot\vec{\sigma}]$$

and

$$\tilde{f}(p,X)=f^0(p)+\left(-\frac{\partial f^0}{\partial\varepsilon}\right)[\tilde{f}_s(p,X)I+\vec{\tilde{g}}(p,X)\cdot\vec{\sigma}] \tag{15}$$

where $f^0(p)$ is the equilibrium distribution function, and $f_s(p,X)$ is the non-equilibrium scalar distribution function, while $\vec{g}(p,X)$ is the non-equilibrium vector distribution function. Substituting the decomposition (15) into Eq.(13), we can get the equations of scalar and vector distribution function $f_s(p,X)$ and $\vec{g}(p,X)$. They are

$$\left[\frac{\partial}{\partial T}+\frac{\vec{p}}{m}\cdot\vec{\nabla}_R+\frac{\partial U}{\partial X}\vec{\nabla}_p\right]f_s(p,X)=\pi n_i i\int\frac{d^3p'}{(2\pi)^3}(T_{pp'})^2[2\tilde{f}_s(p,X)f_s'(p',X)$$
$$+2\vec{\tilde{g}}'(p',X)\vec{g}(p,X)+2f_s(p,X)\tilde{f}_s'(p',X)+2\vec{g}(p,X)\cdot\vec{\tilde{g}}'(p',X)$$
$$+2\tilde{f}_s'(p,X)f_s(p',X)+2\vec{\tilde{g}}'(p,X)\cdot\vec{g}(p',X)+2f_s'(p,X)\tilde{f}_s(p',X)$$
$$+2\vec{g}'(p,X)\cdot\vec{\tilde{g}}(p',X)] \tag{16}$$

and

$$\left[\frac{\partial}{\partial T}+\frac{\vec{p}}{m}\cdot\vec{\nabla}_R+\frac{\partial U}{\partial X}\vec{\nabla}_p\right]\vec{g}(p,X)=\pi n_i i\int\frac{d^3p'}{(2\pi)^3}(T_{pp'})^2[-2\vec{\tilde{g}}'(p',X)\times\vec{g}(p,X)$$
$$+2f_s'(p',X)\vec{\tilde{g}}(p,X)+2\tilde{f}_s(p,X)\vec{g}'(p',X)+2\tilde{f}_s'(p',X)\vec{g}(p,X)$$
$$+2f_s(p,X)\vec{\tilde{g}}'(p',X)-2i\vec{\tilde{g}}(p,X)\times\vec{g}'(p',X)+2\vec{g}(p',X)\times\vec{g}'(p,X)$$
$$+2f_s'(p,X)\cdot\vec{\tilde{g}}'(p,X)]+2\tilde{f}_s'(p,X)\vec{g}(p',X)+2\tilde{f}_s(p',X)\vec{g}'(p,X)$$
$$+2f_s'(p,X)\vec{\tilde{g}}(p',X)+2i\vec{\tilde{g}}(p',X)\times\vec{\tilde{g}}'(p,X)] \tag{17}$$

Similarly, the equations satisfied by the scalar and vector distribution function $\tilde{f}_s(p,X)$ and $\vec{\tilde{g}}(p,X)$ can also be obtained:

$$\left[\frac{\partial}{\partial T}+\frac{\vec{p}}{m}\cdot\vec{\nabla}_R+\frac{\partial U}{\partial X}\vec{\nabla}_p\right]\tilde{f}_s(p,X)=-\pi n_i i\int\frac{d^3p'}{(2\pi)^3}(T_{pp'})^2[2f_s(p,X)f_s'(p',X)$$

$$+2\vec{g}'(p',X)\cdot\vec{g}(p,X)+2\tilde{f}_s(p,X)\tilde{f}_s'(p',X)+2\tilde{\vec{g}}(p,X)\cdot\tilde{\vec{g}}'(p',X)$$
$$+2f_s'(p,X)f_s(p',X)+2\vec{g}'(p,X)\cdot\vec{g}(p',X)+2\tilde{f}_s'(p,X)\tilde{f}_s(p',X)$$
$$+2\tilde{\vec{g}}'(p,X)\cdot\tilde{\vec{g}}(p',X)] \tag{18}$$

and

$$\left[\frac{\partial}{\partial T}+\frac{\vec{p}}{m}\cdot\vec{\nabla}_R+\frac{\partial U}{\partial X}\vec{\nabla}_p\right]\tilde{\vec{g}}(p,X)=-\pi n_i i\int\frac{d^3p'}{(2\pi)^3}(T_{pp'})^2[2i\tilde{\vec{g}}'(p',X)\times\tilde{\vec{g}}(p,X)$$
$$+2f_s'(p',X)\tilde{\vec{g}}(p,X)+2f_s(p,X)\tilde{\vec{g}}'(p',X)+2f_s'(p',X)\tilde{\vec{g}}(p,X)$$
$$+2\tilde{f}_s(p,X)\tilde{\vec{g}}'(p',X)+2i\vec{g}(p',X)\times\vec{g}'(p,X)+2i\tilde{\vec{g}}(p',X)\times\tilde{\vec{g}}'(p,X)$$
$$+2f_s(p',X)\vec{g}'(p,X)]+2f_s'(p,X)\vec{g}(p',X)+2\tilde{f}_s(p',X)\tilde{\vec{g}}'(p,X)$$
$$+2\tilde{f}_s'(p,X)\tilde{\vec{g}}(p',X)-2i\vec{g}(p,X)\times\vec{g}'(p,X)] \tag{19}$$

where $f_s'(p',X)$ and $\vec{g}'(p',X)$ are the scalar and vector distribution functions of the hole, and $\tilde{f}_s'(p',X)$ and $\tilde{\vec{g}}'(p',X)$ are its corresponding distribution function of the anti-Hermite part.

Eqs. (16), (17), (18), and (19) are coupled together. It is difficult to solve them simultaneously, and we had better simplify them further. Remember that $f_s(p,X)$ and $\tilde{f}_s(p,X)$ correspond to the real part and imaginary part of the same complex function, respectively. $\lim_{p\to\infty} f_s(p,X)\to 0$ and $\lim_{p\to\infty}\tilde{f}_s(p,X)\to 0$, so they meet the Kramers-Kronig relation,

$$\tilde{f}_s(p,X)=-\frac{1}{\pi}P\int_{-\infty}^{+\infty}\frac{f_s(p',X)}{p'-p}dp' \tag{20}$$

where P stands for the principle integral. Similarly, we can obtain the Kramers-Kronig relation for the vector distribution function $\vec{g}(p,X)$ and $\tilde{\vec{g}}(p,X)$,

$$\tilde{\vec{g}}(p,X)=-\frac{1}{\pi}P\int_{-\infty}^{+\infty}\frac{\vec{g}(p',X)}{p'-p}dp' . \tag{21}$$

Substituting Eq.(20) and (21) into Eqs.(16) and (17), we have

$$\left[\frac{\partial}{\partial T}+\frac{\vec{p}}{m}\cdot\vec{\nabla}_R+\frac{\partial U}{\partial X}\vec{\nabla}_p\right]f_s(p,X)=\pi n_i i\int\frac{d^3p'}{(2\pi)^3}(T_{pp'})^2[-\frac{2}{\pi}P\int_{-\infty}^{+\infty}\frac{f_s(p',X)}{p'-p}dp'\,f_s'(p',X)$$
$$+2\vec{g}'(p',X)\vec{g}(p,X)-\frac{2}{\pi}f_s(p,X)P\int_{-\infty}^{+\infty}\frac{f_s'(p'',X)}{p''-p'}dp''-\frac{2}{\pi}\vec{g}(p,X)\cdot P\int_{-\infty}^{+\infty}\frac{\vec{g}(p'',X)}{p''-p'}dp''$$
$$-\frac{2}{\pi}f_s(p',X)P\int_{-\infty}^{+\infty}\frac{f_s'(p',X)}{p'-p}dp'-\frac{2}{\pi}P\int_{-\infty}^{+\infty}\frac{\vec{g}'(p',X)}{p'-p'}dp'\cdot\vec{g}(p',X)$$
$$-\frac{2}{\pi}f_s'(p,X)P\int_{-\infty}^{+\infty}\frac{f_s'(p'',X)}{p''-p'}dp''-\frac{2}{\pi}\vec{g}'(p,X)\cdot P\int_{-\infty}^{+\infty}\frac{\vec{g}(p'',X)}{p''-p'}dp''] \tag{22}$$

and

$$\left[\frac{\partial}{\partial T}+\frac{\vec{p}}{m}\cdot\vec{\nabla}_R+\frac{\partial U}{\partial X}\vec{\nabla}_p\right]\vec{g}(p,X)=-\pi n_i i\int\frac{d^3p'}{(2\pi)^3}(T_{pp'})^2[-\frac{2}{\pi}P\int_{-\infty}^{+\infty}\frac{\vec{g}'(p'',X)}{p''-p'}dp''\times\vec{g}(p,X)$$

$$-\frac{2}{\pi}f_s'(p',X)P\int_{-\infty}^{+\infty}\frac{\vec{g}(p',X)}{p'-p}dp' + \frac{2}{\pi^2}P\int_{-\infty}^{+\infty}\frac{f_s(p',X)}{p'-p}dp'P\int_{-\infty}^{+\infty}\frac{\vec{g}'(p^{"},X)}{p^{"}-p'}dp^{"}$$
$$-\frac{2}{\pi}P\int_{-\infty}^{+\infty}\frac{f_s'(p^{"},X)}{p^{"}-p'}dp^{"}\vec{g}(p,X) - \frac{2}{\pi}f_s(p,X)P\int_{-\infty}^{+\infty}\frac{\vec{g}'(p^{"},X)}{p^{"}-p'}dp^{"}$$
$$-\frac{2}{\pi^2}iP\int_{-\infty}^{+\infty}\frac{\vec{g}(p',X)}{p'-p}dp' \times P\int_{-\infty}^{+\infty}\frac{\vec{g}'(p^{"},X)}{p^{"}-p'}dp^{"} + 2\vec{g}(p',X)\times\vec{g}'(p,X)$$
$$-\frac{2}{\pi}f_s'(p,X)\cdot P\int_{-\infty}^{+\infty}\frac{\vec{g}'(p^{"},X)}{p^{"}-p'}dp^{"}] - \frac{2}{\pi}P\int_{-\infty}^{+\infty}\frac{f_s'(p^{"},X)}{p^{"}-p'}dp^{"}\vec{g}(p',X)$$
$$-\frac{2}{\pi}P\int_{-\infty}^{+\infty}\frac{f_s(p',X)}{p'-p}dp'\vec{g}'(p,X) - \frac{2}{\pi}f_s'(p,X)P\int_{-\infty}^{+\infty}\frac{\vec{g}(p',X)}{p'-p}dp'$$
$$+\frac{2}{\pi^2}iP\int_{-\infty}^{+\infty}\frac{\vec{g}(p',X)}{p'-p}dp' \times P\int_{-\infty}^{+\infty}\frac{\vec{g}'(p^{"},X)}{p^{"}-p'}dp^{"}\,] \tag{23}$$

Eq.(22) and Eq.(23) are coupled together. They are closed equations, so we can solve them simultaneously to obtain the scalar and vector distribution function we want, which avoids solving the four Eqs. (16), (17), (18), and (19) as before. Eq.(22) and (23) are the decomposition of spinor Boltzmann equation with Hermite distribution function (8) by the scalar and vector distribution function. They are the starting point for our further exploration.

We can compare the differences between the spinor Boltzmann equation with the Hermite distribution function and the usual spinor Boltzmann equation with the non-Hermite distribution function. From Eq.(3) and (7), we can write the spinor Boltzmann equation with a non-Hermite distribution function as

$$\left[\frac{\partial}{\partial T} + \frac{\vec{p}}{m}\cdot\vec{\nabla}_R + \frac{\partial U}{\partial X}\vec{\nabla}_p\right] f_{non}(\mathrm{p},\mathrm{X}) = -2\pi i[\,\Sigma^{>} f_{non} - \Sigma^{<} G^{>}]. \tag{24}$$

It is different from the spinor Boltzmann equation with Hermite distribution function (8). The differences can be seen if we decompose the above equation (24) by the scalar and vector distribution function. Substituting Eq.(10) and Eq.(15) into Eq.(24), we can get the equations satisfied by the scalar and vector distribution function as

$$\left[\frac{\partial}{\partial T} + \frac{\vec{p}}{m}\cdot\vec{\nabla}_R + \frac{\partial U}{\partial X}\vec{\nabla}_p\right] f_{nons}(p,X) = -2\pi n_i i\int\frac{d^3p'}{(2\pi)^3}(T_{pp'})^2[f_{nons}'(p',X)f_{nons}(p,X)$$
$$+\vec{g}'_{nons}(p',X)\cdot\vec{g}(p,X) - f_{nons}(p',X)f'_{nons}(p,X) - \vec{g}_{nons}(p',X)\cdot\vec{g}_{nons}(p,X) \tag{25}$$

and

$$\left[\frac{\partial}{\partial T} + \frac{\vec{p}}{m}\cdot\vec{\nabla}_R + \frac{\partial U}{\partial X}\vec{\nabla}_p\right]\vec{g}_{nons}(p,X) = -2\pi n_i i\int\frac{d^3p'}{(2\pi)^3}(T_{pp'})^2[i\vec{g}_{nons}(p',X)\times\vec{g}_{nons}(p,X)$$
$$+f'_{nons}(p',X)\vec{g}_{nons}(p,X) + f_{nons}(p,X)\vec{g}'_{nons}(p',X) - f_{nons}(p',X)\vec{g}'_{nons}(p,X)$$
$$-f'_{nons}(p,X)\vec{g}_{nons}(p',X) - i\vec{g}_{nons}(p',X)\times\vec{g}'_{nons}(p,X)] \tag{26}$$

The differences between Eqs. (25), (26) corresponding to the spinor Boltzmann equation with non-Hermite distribution function and Eqs. (22), (23) corresponding to the spinor Boltzmann equation with Hermite distribution function are obvious. We should study the spin-dependent transport by the spinor

Boltzmann equation with Hermite distribution function, and obtain the observables by Eqs. (22) and (23).

3. Equations of Continuity

In the formalism of the spinor Boltzmann equation, the charge density $e\rho(R,T)$ and current density $e\vec{j}(R,T)$ are usually defined as[8,11]:

$$e\rho(R,T) = e\int dp\left(-\frac{\partial f^0}{\partial \varepsilon}\right) f_s(p,R,T), \tag{27}$$

and

$$e\vec{j}(R,T) = e\int dp\left(-\frac{\partial f^0}{\partial \varepsilon}\right)\frac{\vec{p}}{m} f_s(p,R,T), \tag{28}$$

They satisfy the equation of continuity which can be derived from the above spinor Boltzmann equation (22), where e is the electronic charge. Integrating over the momentum variable $\vec{p}$ on both sides of Eq.(22), we have

$$\begin{aligned}&\frac{\partial}{\partial T}\int dp\, f_s(p,X) + \vec{\nabla}_R\cdot\int dp\frac{\vec{p}}{m} f_s(p,X)\\ &= \pi n_i i\int dp\int\frac{d^3p'}{(2\pi)^3}(T_{pp'})^2[-\frac{2}{\pi}P\int_{-\infty}^{+\infty}\frac{f_s(p',X)}{p'-p}dp'\, f_s'(p',X)\\ &+2\vec{g}'(p',X)\vec{g}(p,X) - \frac{2}{\pi}f_s(p,X)P\int_{-\infty}^{+\infty}\frac{f_s'(p'',X)}{p''-p'}dp'' - \frac{2}{\pi}\vec{g}(p,X)\cdot P\int_{-\infty}^{+\infty}\frac{\vec{g}(p'',X)}{p''-p'}dp''\\ &-\frac{2}{\pi}f_s(p',X)P\int_{-\infty}^{+\infty}\frac{f_s'(p',X)}{p'-p}dp' - \frac{2}{\pi}P\int_{-\infty}^{+\infty}\frac{\vec{g}'(p',X)}{p'-p'}dp'\cdot\vec{g}(p',X)\\ &-\frac{2}{\pi}f_s'(p,X)P\int_{-\infty}^{+\infty}\frac{f_s'(p'',X)}{p''-p'}dp'' - \frac{2}{\pi}\vec{g}'(p,X)\cdot P\int_{-\infty}^{+\infty}\frac{\vec{g}(p'',X)}{p''-p'}dp'']\end{aligned} \tag{29}$$

It is just the equation of continuity for the charge density and charge current

$$\begin{aligned}&\frac{\partial}{\partial T}\rho(R,T) + \vec{\nabla}_R\cdot\vec{j}(R,T) = \pi n_i i\int dp\int\frac{d^3p'}{(2\pi)^3}(T_{pp'})^2[-\frac{2}{\pi}P\int_{-\infty}^{+\infty}\frac{f_s(p',X)}{p'-p}dp'\, f_s'(p',X)\\ &+2\vec{g}'(p',X)\vec{g}(p,X) - \frac{2}{\pi}f_s(p,X)P\int_{-\infty}^{+\infty}\frac{f_s'(p'',X)}{p''-p'}dp'' - \frac{2}{\pi}\vec{g}(p,X)\cdot P\int_{-\infty}^{+\infty}\frac{\vec{g}(p'',X)}{p''-p'}dp''\\ &-\frac{2}{\pi}f_s(p',X)P\int_{-\infty}^{+\infty}\frac{f_s'(p',X)}{p'-p}dp' - \frac{2}{\pi}P\int_{-\infty}^{+\infty}\frac{\vec{g}'(p',X)}{p'-p'}dp'\cdot\vec{g}(p',X)\\ &-\frac{2}{\pi}f_s'(p,X)P\int_{-\infty}^{+\infty}\frac{f_s'(p'',X)}{p''-p'}dp'' - \frac{2}{\pi}\vec{g}'(p,X)\cdot P\int_{-\infty}^{+\infty}\frac{\vec{g}(p'',X)}{p''-p'}dp'']\end{aligned} \tag{30}$$

Similarly, if we define spin accumulation by the vector distribution function as[8,11]:

$$\hbar\vec{m}(R,T) = \hbar\int dp\left(-\frac{\partial f^0}{\partial \varepsilon}\right)\vec{g}(p,R,T), \tag{31}$$

and define the spin current density as

$$\hbar\vec{j}_m(R,T)=\hbar\int dp\left(-\frac{\partial f^0}{\partial\varepsilon}\right)\frac{\vec{p}}{m}\vec{g}(p,R,T)\quad , \tag{32}$$

They also satisfy the corresponding equation of continuity

$$\begin{aligned}\frac{\partial}{\partial T}\vec{m}(R,T)+\vec{\nabla}_R\cdot\vec{j}_m(R,T)&=-\pi n_i i\int dp\int\frac{d^3p'}{(2\pi)^3}(T_{pp'})^2[-\frac{2}{\pi}P\int_{-\infty}^{+\infty}\frac{\vec{g}'(p",X)}{p"-p'}dp"\times\vec{g}(p,X)\\ &-\frac{2}{\pi}f_s'(p',X)P\int_{-\infty}^{+\infty}\frac{\vec{g}(p',X)}{p'-p}dp'+\frac{2}{\pi^2}P\int_{-\infty}^{+\infty}\frac{f_s(p',X)}{p'-p}dp'P\int_{-\infty}^{+\infty}\frac{\vec{g}'(p",X)}{p"-p'}dp"\\ &-\frac{2}{\pi}P\int_{-\infty}^{+\infty}\frac{f_s'(p",X)}{p"-p'}dp"\vec{g}(p,X)-\frac{2}{\pi}f_s(p,X)P\int_{-\infty}^{+\infty}\frac{\vec{g}'(p",X)}{p"-p'}dp"\\ &-\frac{2}{\pi^2}iP\int_{-\infty}^{+\infty}\frac{\vec{g}(p',X)}{p'-p}dp'\times P\int_{-\infty}^{+\infty}\frac{\vec{g}'(p",X)}{p"-p'}dp"+2\vec{g}(p',X)\times\vec{g}'(p,X)\\ &-\frac{2}{\pi}f_s'(p,X)\cdot P\int_{-\infty}^{+\infty}\frac{\vec{g}'(p",X)}{p"-p'}dp"]-\frac{2}{\pi}P\int_{-\infty}^{+\infty}\frac{f_s'(p",X)}{p"-p'}dp"\vec{g}(p',X)\\ &-\frac{2}{\pi}P\int_{-\infty}^{+\infty}\frac{f_s(p',X)}{p'-p}dp'\vec{g}'(p,X)-\frac{2}{\pi}f_s'(p,X)P\int_{-\infty}^{+\infty}\frac{\vec{g}(p',X)}{p'-p}dp'\\ &+\frac{2}{\pi^2}iP\int_{-\infty}^{+\infty}\frac{\vec{g}(p',X)}{p'-p}dp'\times P\int_{-\infty}^{+\infty}\frac{\vec{g}'(p",X)}{p"-p'}dp"]\end{aligned} \tag{33}$$

which can be derived by integrating over the momentum variable $\vec{p}$ on both sides of Eq.(23)

$$\begin{aligned}&\frac{\partial}{\partial T}\int dp\left(-\frac{\partial f^0}{\partial\varepsilon}\right)\vec{g}(p,R,T)+\vec{\nabla}_R\cdot\int dp\left(-\frac{\partial f^0}{\partial\varepsilon}\right)\frac{\vec{p}}{m}\vec{g}(p,R,T)\\ &=-\pi n_i i\int dp\int\frac{d^3p'}{(2\pi)^3}(T_{pp'})^2[-\frac{2}{\pi}P\int_{-\infty}^{+\infty}\frac{\vec{g}'(p",X)}{p"-p'}dp"\times\vec{g}(p,X)\\ &-\frac{2}{\pi}f_s'(p',X)P\int_{-\infty}^{+\infty}\frac{\vec{g}(p',X)}{p'-p}dp'+\frac{2}{\pi^2}P\int_{-\infty}^{+\infty}\frac{f_s(p',X)}{p'-p}dp'P\int_{-\infty}^{+\infty}\frac{\vec{g}'(p",X)}{p"-p'}dp"\\ &-\frac{2}{\pi}P\int_{-\infty}^{+\infty}\frac{f_s'(p",X)}{p"-p'}dp"\vec{g}(p,X)-\frac{2}{\pi}f_s(p,X)P\int_{-\infty}^{+\infty}\frac{\vec{g}'(p",X)}{p"-p'}dp"\\ &-\frac{2}{\pi^2}iP\int_{-\infty}^{+\infty}\frac{\vec{g}(p',X)}{p'-p}dp'\times P\int_{-\infty}^{+\infty}\frac{\vec{g}'(p",X)}{p"-p'}dp"+2\vec{g}(p',X)\times\vec{g}'(p,X)\\ &-\frac{2}{\pi}f_s'(p,X)\cdot P\int_{-\infty}^{+\infty}\frac{\vec{g}'(p",X)}{p"-p'}dp"]-\frac{2}{\pi}P\int_{-\infty}^{+\infty}\frac{f_s'(p",X)}{p"-p'}dp"\vec{g}(p',X)\\ &-\frac{2}{\pi}P\int_{-\infty}^{+\infty}\frac{f_s(p',X)}{p'-p}dp'\vec{g}'(p,X)-\frac{2}{\pi}f_s'(p,X)P\int_{-\infty}^{+\infty}\frac{\vec{g}(p',X)}{p'-p}dp'\\ &+\frac{2}{\pi^2}iP\int_{-\infty}^{+\infty}\frac{\vec{g}(p',X)}{p'-p}dp'\times P\int_{-\infty}^{+\infty}\frac{\vec{g}'(p",X)}{p"-p'}dp"]\end{aligned} \tag{34}$$

The above equation is exactly the equation of continuity for spin accumulation and spin current.

4. Example

In the next, we will illustrate the above physical quantities by a simple example. We study the quantum transport scattered by impurities with low concentration in a mesoscopic system of spintronics with a length of 10 nm. The self-energy in Eq.(8) can be simplified by Born approximation. Driven by an external electric field of 1mV/nm, the system will arrive at a steady state after a relaxation time, so we only investigate the transport in the steady state, where $\frac{\partial}{\partial T}f_s(p,X)=0$ and $\frac{\partial}{\partial T}\vec{g}(p,X)=0$.

We should make a momentum-independent relaxation time approximation to simplify the scattering terms in Eq.(22) and (23) [19], i.e., $T_{pp'}$ are treated as a constant, and the relaxation time τ is adopted as $\frac{1}{\tau}\approx(T_{pp'})^2$. Then we have

$$\left[\frac{\partial}{\partial T}+\frac{\vec{p}}{m}\cdot\vec{\nabla}_R+\frac{\partial U}{\partial X}\vec{\nabla}_p\right]f_s(p,X)=\pi n_i i[\frac{-\frac{2}{\pi}P\int_{-\infty}^{+\infty}\frac{f_s(p',X)}{p'-p}dp' < f_s'(p',X) >}{\tau}$$
$$+\frac{2 < \vec{g}'(p',X) > \vec{g}(p,X)-\frac{2}{\pi}f_s(p,X) < P\int_{-\infty}^{+\infty}\frac{f_s{'}(p'',X)}{p''-p'}dp'' >}{\tau}$$
$$+\frac{-\frac{2}{\pi}\vec{g}(p,X)\cdot < P\int_{-\infty}^{+\infty}\frac{\vec{g}(p'',X)}{p''-p'}dp'' > +\frac{2}{\pi} < f_s(p',X) > P\int_{-\infty}^{+\infty}\frac{f_s{'}(p',X)}{p'-p}dp'}{\tau}$$
$$\frac{-\frac{2}{\pi}P\int_{-\infty}^{+\infty}\frac{\vec{g}{'}(p',X)}{p'-p'}dp'\cdot < \vec{g}(p',X) > -\frac{2}{\pi}f_s'^{(p,X)} < P\int_{-\infty}^{+\infty}\frac{f_s{'}(p'',X)}{p''-p'}dp'' >}{\tau}$$
$$+\frac{-\frac{2}{\pi}\vec{g}'(p,X)\cdot < P\int_{-\infty}^{+\infty}\frac{\vec{g}(p'',X)}{p''-p'}dp'' >}{\tau}] \tag{35}$$

and

$$\left[\frac{\partial}{\partial T}+\frac{\vec{p}}{m}\cdot\vec{\nabla}_R+\frac{\partial U}{\partial X}\vec{\nabla}_p\right]\vec{g}(p,X)$$
$$=\pi n_i i\frac{-\frac{2}{\pi} < P\int_{-\infty}^{+\infty}\frac{\vec{g}{'}(p'',X)}{p''-p'}dp'' > \times\,\vec{g}(p,X)-\frac{2}{\pi} < f_s'(p',X) > P\int_{-\infty}^{+\infty}\frac{\vec{g}(p',X)}{p'-p}dp'}{\tau}+$$
$$\frac{\frac{2}{\pi^2} < P\int_{-\infty}^{+\infty}\frac{f_s(p',X)}{p'-p}dp' > P\int_{-\infty}^{+\infty}\frac{\vec{g}{'}(p'',X)}{p''-p'}dp''-\frac{2}{\pi} < P\int_{-\infty}^{+\infty}\frac{f_s{'}(p'',X)}{p''-p'}dp'' > \vec{g}(p,X)}{\tau}$$
$$+\frac{-\frac{2}{\pi} < f_s(p,X) > P\int_{-\infty}^{+\infty}\frac{\vec{g}{'}(p'',X)}{p''-p'}dp''-\frac{2}{\pi^2}i < P\int_{-\infty}^{+\infty}\frac{\vec{g}(p',X)}{p'-p}dp' > \times P\int_{-\infty}^{+\infty}\frac{\vec{g}{'}(p'',X)}{p''-p'}dp''}{\tau}$$
$$+\frac{2 < \vec{g}(p',X) > \times\,\vec{g}'(p,X)-\frac{2}{\pi} < f_s'(p,X > \cdot P\int_{-\infty}^{+\infty}\frac{\vec{g}{'}(p'',X)}{p''-p'}dp''}{\tau}$$
$$+\frac{-\frac{2}{\pi} < P\int_{-\infty}^{+\infty}\frac{f_s{'}(p'',X)}{p''-p'}dp'' > \vec{g}(p',X)-\frac{2}{\pi} < P\int_{-\infty}^{+\infty}\frac{f_s(p',X)}{p'-p}dp' > \vec{g}'(p,X)}{\tau}$$
$$+\frac{-\frac{2}{\pi} < f_s'(p,X) > P\int_{-\infty}^{+\infty}\frac{\vec{g}(p',X)}{p'-p}dp'+\frac{2}{\pi^2}i < P\int_{-\infty}^{+\infty}\frac{\vec{g}(p',X)}{p'-p}dp' > \times P\int_{-\infty}^{+\infty}\frac{\vec{g}{'}(p'',X)}{p''-p'}dp''}{\tau} \tag{36}$$

where $< \cdots >$ denotes the momentum integral over scalar and vector function.

The numerical results for charge density as a function of position are shown in Fig.1, where the momentum relaxation time is chosen as 0.01ps. The dashed line is the charge density with the non-Hermite distribution function, and it decreases oscillatory. The solid line is the charge density with Hermite distribution and decreases to a small value. Since the non-Hermite distribution function in Eq.(3) can be decomposed as the Hermite and anti-Hermite parts, the solid line corresponds to the real part of the charge density calculated by the non-Hermite distribution, and it is the reasonable physical quantity. An imaginary part exists for the charge density if we calculate it by the non-Hermite distribution function. It corresponds to the dashed line for the anti-Hermite distribution function and is unreasonable.

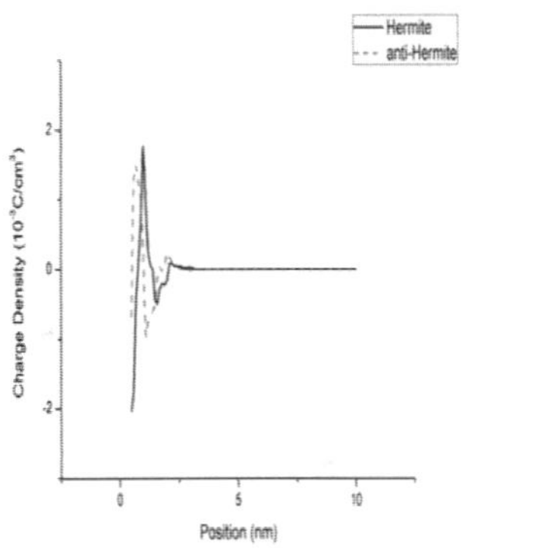

Figure 1. The charge density as a function of position. The momentum relaxation time is $\tau = 0.01ps$, and the electric field $E = 1mV / nm$.

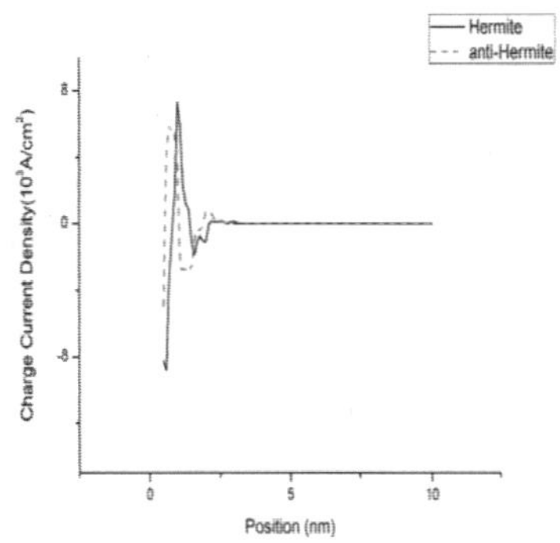

Figure 2. The charge current versus position. The momentum relaxation time is $\tau = 0.01ps$, and the electric field $E = 1mV / nm$.

In order to get perfect physical quantity, we should calculate them by the Hermite distribution function (4) instead of the non-Hermite distribution (3) or anti-Hermite distribution (5). A similar case can also be found in charge currents shown in Fig.2 or other physical quantities in Fig.3 –Fig.6. In Fig.2, the charge current corresponding to the anti-Hermite distribution function (dashed line) decreases oscillatory, and the charge current with Hermite distribution also decreases, it is reasonable for the transport with impurity scattering. It has a similar shape to the charge density due to the fact that only the electrons near the Fermi surface contribute to the charge transport. We illustrate the three components of spin accumulation corresponding to the Hermite distribution vs. position in Fig.3. For comparison, we also plot the spin accumulation with anti-Hermite distribution in Fig.4.

Since the spin currents and spin accumulation are all concerned with the momentum integral over the vector distribution function, the spin currents are mainly contributed by the electrons with Fermi velocity, so the spin currents have a similar shape to spin accumulations. Spin current is a tensor, and it has three components. When we only study the electron transport along the x-direction, we draw these three components for the Hermite distribution function in Fig.5, the spin current for the anti-Hermite distribution is shown in Fig.6, and the three components have phase delays between each other, whatever in Hermite or anti-Hermite case. The above figures show that the spinor Boltzmann equation with Hermite distribution function (4) can give reasonable results. We should utilize it to study the spin-polarized transport in mesoscopic transport instead of the spinor Boltzmann equation with non-

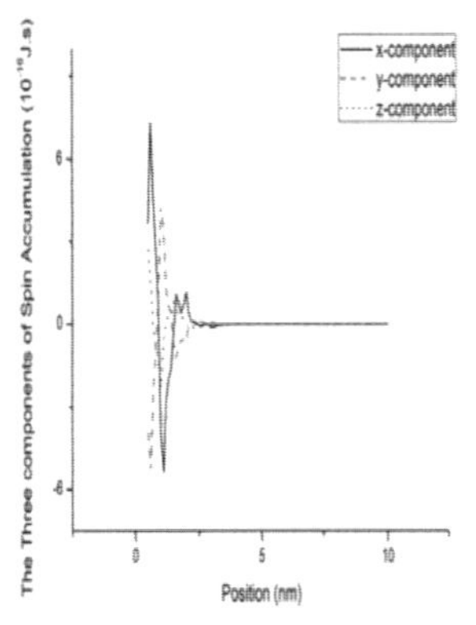

Figure 3. The three components of spin accumulation for the Hermite distribution function versus position. The momentum relaxation time is $\tau = 0.01ps$, and the electric field $E = 1mV / nm$.

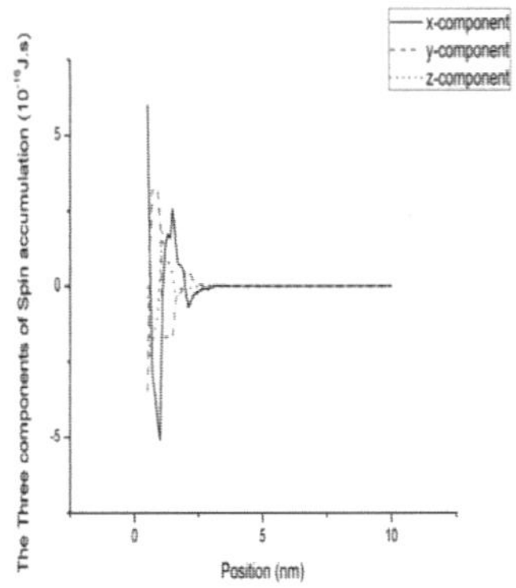

Figure 4. The three components of spin accumulation for the anti-Hermite distribution function versus position. The momentum relaxation time is $\tau = 0.01ps$, and the electric field $E = 1mV / nm$.

Hermite distribution (3) or anti-Hermite distribution (5).

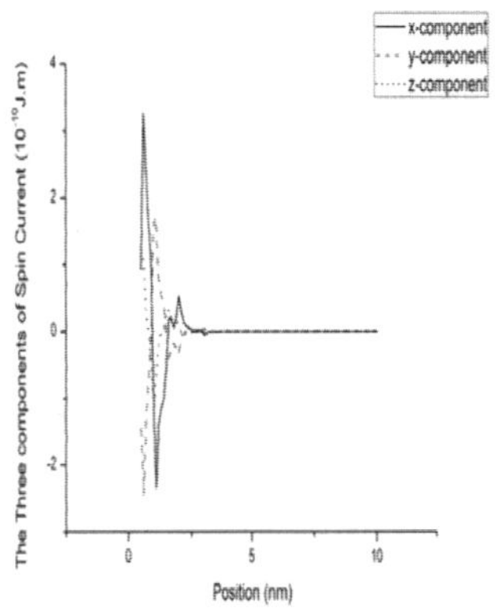

Figure 5. The three components of spin current for the Hermite distribution function versus position. The momentum relaxation time is $\tau = 0.01ps$, and the electric field $E = 1mV / nm$.

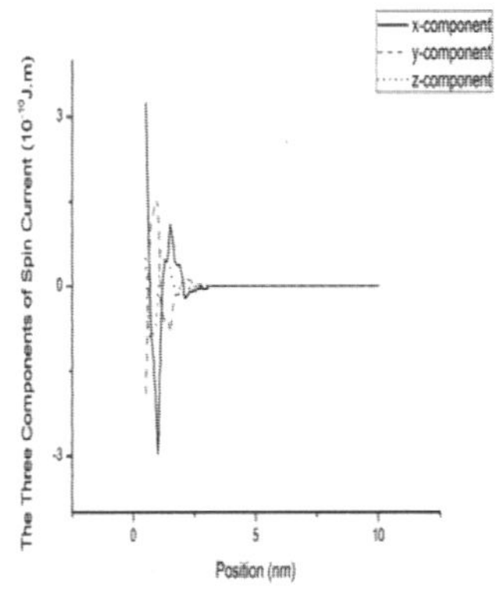

Figure 6. The three components of spin current for the anti-Hermite distribution function versus position. The momentum relaxation time is $\tau = 0.01ps$, and the electric field $E = 1mV / nm$.

5. Summary and Discussion

In this paper, we try to construct a Hermite distribution function that has more physical meaning than the usual non-Hermite spinor distribution function. Then we obtain the spinor Boltzmann equation satisfied by it. The equations of continuity for the charge density, current, spin accumulation, and spin current are derived. We show the numerical results for these physical quantities in Fig.1—Fig.6. For comparison, we also plot the results calculated by the anti-Hermite distribution function, which is unphysical quantities. In this example, only a single relaxation time τ is adopted in our calculation. In fact, we should introduce two different relaxation time, the momentum relaxation time τ and the spin-flip relaxation time τ_{sf} in the spin-dependent system. We only give a simplification in this manuscript.

If we don't consider the spin freedom in our discussion, the expression (4) and (5) will reduce to the usual Wigner distribution, and the matrices f and $\tilde{f}$ will become a c number f_c

$$f_c = ReG^< = ReG_t - ReG^R = ReG_t + \frac{1}{2}A(k,R) \quad (37)$$

and

$$\tilde{f}_c = ImG^< = ImG_t - ImG^R = ImG_t - \Gamma(k,\omega + i\eta), \quad (38)$$

where G_t is the time-ordered Green function, G^R is the retarded Green function, $A(k,R)$ is the spectral weight function, and $\Gamma(k,\omega + i\eta)$ is the damping function. It indicates that f has a relation with the spectral function, while $\tilde{f}$ corresponds to the damping function, so Eq.(37) and (38) are helpful for us to understand the physical meaning of the Hermite spinor distribution function and anti-Hermite distribution function. When we include the spin freedom in f and $\tilde{f}$, the diagonal elements $f_{\uparrow\uparrow}$ and $\tilde{f}_{\uparrow\uparrow}$ correspond to spectral function and damping function with up spin, $f_{\downarrow\downarrow}$ and $\tilde{f}_{\downarrow\downarrow}$ correspond to the spectral function and damping function with down spin, while the off-diagonal elements of and f and $\tilde{f}$ are concerned with the quantum effect of the system, i.e., the quantum interference oscillation in the transport procedure, which is an interesting open question that is worthwhile for future exploration.

Acknowledgments
This study is supported by the National Key R&D Program of China (Grant No. 2018FYA0305804), the Key Research Program of the Chinese Academy of Sciences (Grant No. XDPB08-3), and the Fundamental Research Funds for the Central Universities.

References
[1] R. E. Camley and J. Barnas, Phys. Rev. Lett. **63** (1989) 664-667.
[2] J. Barnas, A. Fuss, R. E. Camley, P. Grŭnberg, and W. Zinn, Phys. Rev. **B42** (1990) 8110-8120.
[3] T. Valet and A. Fert, Phys.Rev. **B48** (1993) 7099-7113.
[4] S. Zhang and P. M. Levy, Phys. Rev. **B57** (1998) 5336-5339.
[5] A. Shapiro and P. M. Levy, Phys. Rev. **B63** (2000) 014419-1— 014419-6.
[6] E. Wigner, Phys. Rev. **40**, (1932)749-759; L. P. Kadanoff and G. Baym, Quantum Statistical Mechanics (Benjamin, New York, 1962).
[7] G. D. Mahan, Phys. Rep. **145**, (1987) 251-318.
[8] J. Zhang, P. M. Levy, S. Zhang, and V. Antropov, Phys. Rev. Lett. **93**, (2004) 256602-1--256602-4.
[9] L. Sheng, D. Y. Xing, Z. D. Wang, and J. Dong, Phys. Rev. **B55**, (1997) 5908-5913.
[10] L. Sheng, H. Y. Teng, and D. Y. Xing, Phys. Rev. **B58**, (1998) 6428-6433.
[11] J. W. Zhang and P. M Levy, Physical Review **B71**, (2005) 184426-1—184426-11.
[12] Z. C. Wang, Euro. Phys. Jour. **B85**, (2012) 303-1--303-6.
[13] M. Hillery, R. F. O'Connel, M. O. Scully and E. P. Wigner, Physics Report, **106**, (1984) 121-171.
[14] K. Husimi, Proc. Phys. Math. Soc. Japan, **22**, (1940) 264-279.
[15] L. Cohen, J. Math. Phys. **7**, (1960) 781-793.
[16] H. Margenau and R. N. Hill, Prog. Theoret. Phys. (Kyoto), **22**, (1961) 722-735.
[17] Z. C.Wang and P. M. Levy, Science in China Series G**51**, (2008) 1178-1187.
[18] H. Smith and H. H. Jenson, "Transport Phenomena," (Clarendon Press, Oxford, 1989).
[19] S. Zhang, Phys. Rev. Lett. 85, (2000) 393-396.

IGBT Gate Waveform Acquisition Based on Compressed Sensing Technology

Yifan Wang[a], Mingquan Zeng[a], Jinlong Gong[a], Ming Sun[a], Yifan Wang[a], Li Luo[a], Jiansheng Hu*[b], Junjie Ma[b]

[a]State Grid Zhejiang Electric Power Co., Ltd. Electric Power Research Institute, Hangzhou, Zhejiang, China;

[b]XJ Electric Co., Ltd., Xuchang, Henan, China

* Corresponding author:2438767538@qq.com

Abstract: As the core device of power electronic equipment, IGBT (Insulated Gate Bipolar Transistor) is related to the reliability of the system, so its online health monitoring is particularly important. Due to the large amount of data and high sampling rate for IGBT online health monitoring, data transmission has become a major problem, the method of effectively reconstructing the original signal has become one of the current research hotspots. Based on the theory of compressed sensing, the acquisition of IGBT gate waveforms is studied in this paper. First, the theoretical principle of compressed sensing is expounded and analyzed, and the evaluation criteria for signal reconstruction performance are given. Then, from the theory and simulation, the sparseness of IGBT gate drive signals under different sparse bases and different measurement matrices are discussed and analyzed. Next, aiming at the shortcomings of the stagewise weak orthogonal matching pursuit algorithm, a SWOMP algorithm based on the backtracking strategy of different Sigmoid functions is proposed. Simulation and experimental results show that the improved SWOMP4 algorithm has the best reconstruction effect. Finally, the SWOMP4 algorithm is applied to the process of IGBT gate drive signal reconstruction.
This paper mainly collects and reconstructs the IGBT gate drive signal through the improved SWOMP4 algorithm. The designed algorithm has high reconstruction accuracy and can be used in the IGBT module online health status system.

1. INTRODUCTION

With the rapid development of power electronics, more and more power devices are widely used in the industrial fields, epecially in heavy-duty applications in urban distribution networks[1]. Among the power devices, due to the characteristics of low driver power, high current capability, IGBT is one of the most widely used high-power devices[2]. For the converter, IGBT plays a very important role in its interior, but due to environmental factors, system transient problems, overload and other situations, IGBT is easily damaged[3]. In order to make the overall system operate normally, we can monitor the operation of the IGBT module. But the amount of data for the online health monitoring is huge, the sampling rate is high, so data transmission has become the main problem. It is important look for a method that can transmit the signal with low sampling rate, less transmission data and then reconstruct the original signal at the receiving end.

Wang et al. proposed a filtering-based compressed signal acquisition method. It solves the problem that the random measurement matrix will make hardware implementation very difficult, which is beneficial to the storage and processing of data. At the same time, the Toeplitz measurement matrix itself satisfies the RIP (Restricted Isometry Property) and can better restore the original at the receiving end[4]. Yu et al. proposed a method for acoustic signal acquisition using compressed sensing technology. The original signal is distributed in the sparse domain, the sparse basis is selected as the DFT basis, the measurement matrix is selected as a Gaussian random matrix, and the recovery algorithm adopts the orthogonal matching pursuit algorithm. After that, the reconstruction error between the original signal and the reconstructed signal is analyzed in detail, as well as the maximum packet loss rate of the measurement node. It is concluded that the compressed sensing technology can meet its requirements in terms of completely reconstructing the audio signal[5]. Jiang et al. completed the hardware design of the ECG signal based on the compressed sensing theory. For the problem of a small amount of energy coefficient of the signal, they proposed to use an uncertain demodulation method to achieve the sparse characteristic of the original signal. The parameter settings of some components in the actual circuit is imperfect, which results in a certain difference between the simulation and the actual experiment[6]. Zhang et al. studied the measurement of weak signals by the compressed sensing theory. For the problem that the signal cannot be accurately determined during reconstruction due to one measurement, the method of introducing two measurements in the selection of the measurement matrix is proposed. The reconstruction probability of the signal is improved, and the reconstruction effect is better[7].

The traditional IGBT state detection method achieves the purpose of monitoring the IGBT by comparing the original data with the existing data[8]. Once a continuous parameter abnormality occurs, emergency measures will be initiated to ensure that the entire system is not damaged[9]. The operating status of the IGBT can be obtained by monitoring the gate driving signal of the IGBT, but there is a problem with a large amount of data in direct transmission. To reduce the amount of data transmission, this paper applies the compressed sensing technology to the collection and reconstruction of the IGBT gate signal, hoping to reduce the amount of data transmission volume.

2. THEORY

The signal processing of compressed sensing includes three steps. First, a sparse transformation is performed on the original compressible signal to extract data that can reflect the characteristic parameters of the entire signal. Second, the original signal is dimensionally reduced through the measurement matrix. After that, the transmission volume of the entire data is greatly reduced. Finally, the original signal is restored at the receiving end.

2.1 Compression sensing

The original signal f is projected to an N-dimensional basis matrix $\Psi = [\Psi_1, \Psi_2, \ldots, \Psi_N]$, as shown in Formula (1). The dimension is reduce off under the measurement matrix $\Phi \in R^{MN}$, and obtain the observed signal $y \in R^M$, its comprehensive expression is Formula (1),

$$f = \sum_{i=1}^{N} x_i \Psi_i \quad or \quad f = \Psi x \tag{1}$$

When it is known that N>>M, the solution result of Equation (1) is indeterminate, that is, the solution result has infinite solutions. However, for the original signal f, it has the K-sparse characteristic, which is mathematically expressed as K<<M; the original signal f can be solved by the measured value y, and the mathematical formula selected is the way to solve the optimal l_0 norm, such as Formulae (2) and (3).

$$(P_0) \begin{matrix} min||f||_0 \\ s.t. y = \Phi f \end{matrix} \tag{2}$$

$$y = \Phi f = \Phi \Psi x \tag{3}$$

Let $A^{CS} = \Phi\Psi$, then $y = A^{CS}x$, where A^{CS} is the sensing matrix.

For compressive sensing theory, the sparseness of a signal is the premise of compressed sensing. Therefore, finding a suitable standard sparse basis Ψ can ensure that the original signal can be restored at the receiving end. In the process of signal reconstruction, RIP[10] is an important condition for reconstruction in the compressed sensing process. This condition is the key to the reconstruction of CS theory. The definition of the RIP condition is as Formula (4):

$$(1-\delta_k)\|x\|_2^2 \le \|Ax\|_2^2 \le (1+\delta_k)\|x\|_2^2 \tag{4}$$

Where $\delta_k \in (0,1)$, it can be known from Formula (3) that the composition of A consists of two parts, namely the measurement matrix and the sparse basis. Incoherence is understood as follows: the column vector $\{\Psi_i\}_{i=1}^{N}$ which makes the signal show a small amount of energy coefficient and the row vector $\{\Phi_i\}_{i=1}^{M}$, which reduces the transmission amount of the signal during the transmission process M, the two of them cannot be sparsely represented by each other, and the definition of coherence is as Formula (5):

$$\mu(\Phi,\Psi) = \sqrt{N}\cdot max|\langle\Phi_i,\Psi_j\rangle|, 1 \le i \le M, 1 \le j \le N \tag{5}$$

Where $\mu(\Phi,\Psi) \in [1,\sqrt{N}]$, if the numerical result of μ(Φ,Ψ) is smaller, it means that the correlation between Ψ and Φ is smaller; when μ(Φ,Ψ), if the numerical calculation result is close to 1, it means that in the CS process, the original signal can be well restored at the receiving end with only a small amount of measurement values during signal transmission. The most used uncertain measurement matrices are Gauss random measurement matrixes, Bernoulli random measurement matrix and so on. The commonly used measurement matrices obtained by deterministic methods mainly include Hadamard measurement matrix, rotation moment measurement matrix[11], etc.

2.2 Stagewise weak orthogonal matching pursuit

At present, CS algorithms are mainly divided into two categories: one is known sparsity, and the other is unknown sparsity. MP and OMP are matching pursuit algorithms with known sparsity, but they can only ensure that the atoms found each time meet the characteristics of local optimality, cannot guarantee that all the selected atoms also meet the requirements, so neither is the best option[12]. StOMP and SAMP are matching pursuit algorithms under blind sparsity, in which StOMP does not need to know the number of a small number of energy coefficients exhibited by the original signal in advance but has requirements on the selection of the threshold. In this paper, MSE (Mean square error), SNR (Signal to Noise Ratio), AE (Absolute error), RE (Relative error) are used to evaluate the pros and cons of IGBT electrical signal reconstruction performance.

2.3 Improved SWOMP4 algorithm

At present, in order to solve the problem that the use of a fixed threshold in the SWOMP algorithm reduces the reconstruction performance of the algorithm, this paper improves the fixed threshold, introduces the idea of backtracking, and adds different sigmoid functions. The selection strategy of atoms is shown in Formula (6):

$$|\hat{x}_{t2}| \ge D_t \cdot max|\hat{x}_{t1}| \tag{6}$$

where D_t is the soft threshold calculated by different sigmoid functions in the t-th iteration, $|\hat{x}_{t1}|$ is the label number of the best atom found after a selection, and the selected label corresponding to $|\hat{x}_{t2}|$.

The principle of the improved SWOMP algorithm is as follows: first, according to the threshold screening formula, items larger than the threshold are selected, and the selected column numbers that meet the requirements is put into J_0, and Λ_t is updated, as shown in Formula (7):

$$\Lambda_t = \Lambda_{t-1} \cup J_0 \tag{7}$$

Here, Λ_t is the index set updated by the (t-1)-th iteration, and then the least squares method is used to solve the estimated signal, as shown in Formula (8):

$$\hat{x}_{t1} = argmin\,\theta \, \|y - A_t\theta_t\|_{2=(A_t^T A_t^{-1})^{A_t^T y}} \tag{8}$$

Next, the idea of backtracking is introduced, and different sigmoid functions are added to form a set J_1 of the newly selected column numbers, as shown in Figure 1. Among them, D_t selects the above six sigmoid functions for screening, and the index set Λ_{t2} is updated.

$$\Lambda_{t2} = \Lambda_{t-1} \cup J_1 \tag{9}$$

The least squares method is used to solve the estimated signal, namely:

$$\hat{x}_{t2} = argmin\theta\|y - A_{t2}\theta_t\|_{2=(A_{t2}^T A_{t2}^{-1})} A_{t2}^T y \tag{10}$$

The iteration is repeated until the number of iterations is greater than 10 or the updated residual is 0.

As Figure 1 shows, the range of the abscissa x is the real number domain, and the value of the ordinate f(x) is [-1, 1]. When the number of iterations increases, the function value gradually approaches 1.

The SWOMP algorithm based on the backtracking strategies of different sigmoid functions makes up for the defects of low reconstruction accuracy and poor stability of the SWOMP algorithm. By adding a soft threshold, the atoms are screened twice, and the adaptive selection of atoms to update the support set is realized. The number of iterations is reduced, and the original signal is better recovered.

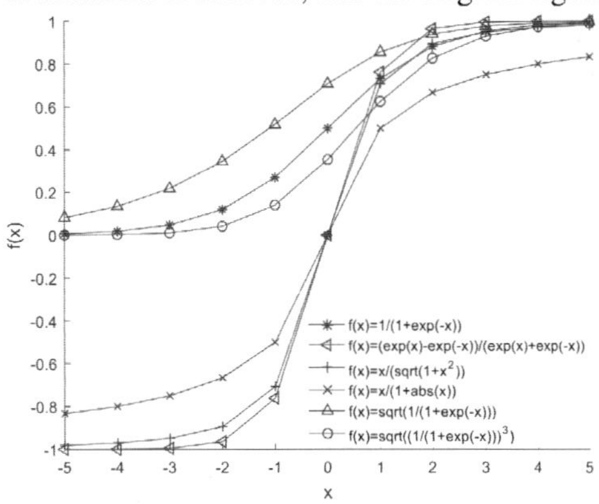

Figure 1. Graphs of various sigmoid functions

3. METHODOLOGY

3.1 Sparse base selection

A good sparse basis and measurement matrix are more conducive to signal compression, resulting in greatly improving performance and efficiency of signal reconstruction[13]. This paper mainly studies the compression method of IGBT electrical signal based on compressive sensing. IGBT selects Infineon 3300 V/1500 A IGBT, FZ1500R33HE3. V_{CE}=1200 V, I_C=200 A. In the experiment, the original signal adopts the waveform of the IGBT gate drive voltage V_{GE}. This signal has a length of 5000 and a frequency of 6.25 MHz. The sparse basis adopts the DCT transform basis, the DST transform basis, and the DFT transform basis respectively. To make the results observed in the whole experiment stable, each group of experiments in this paper is repeated 100 times and then the average value of the experimental results is calculated. As shown in Figure 2, The signal energy of the DCT transform and DST transform is concentrated in the low-frequency region, which conforms to the actual operation law.

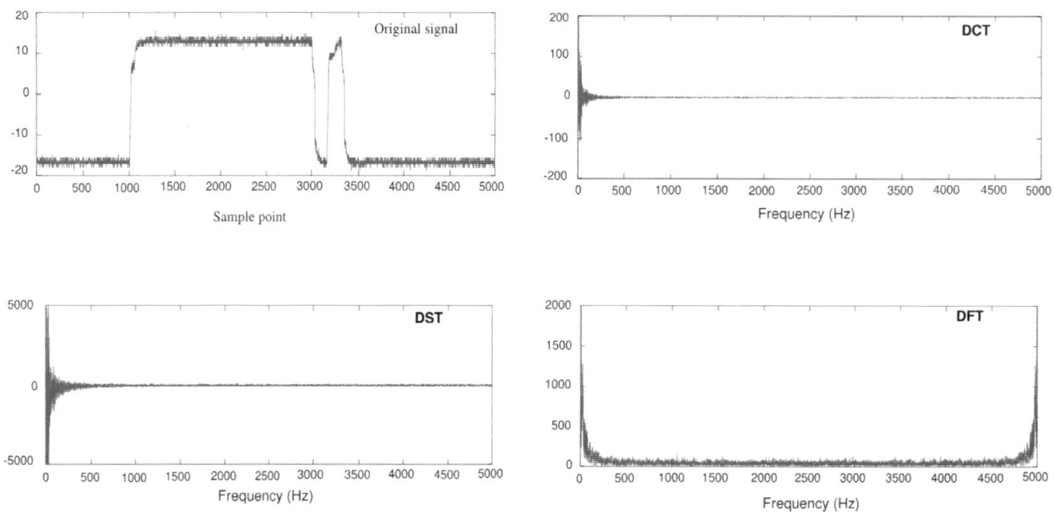

Figure 2. Energy signal distribution of DCT, DST, DFT

To further determine the number of sparseness, the threshold value of the sparse coefficient is set: q=10, and the sparse coefficients of the IGBT gate drive voltage signal under different sparse bases can be obtained, as shown in Table 1. In theory, the smaller the sparsity, the better the sparsity, so for the IGBT gate drive voltage, the sparsity under the DCT base is the best.

Table 1. The sparse coefficient of the signal under different sparse bases

	DCT	DST	DFT
Sparse coefficient	75	4052	4798

To continue to study the influence of different sparse bases on signal sparseness, the original signal of this experiment adopts the same signal as in Experiment 1, the signal reconstruction algorithm adopts the OMP algorithm[14]. The sparse coefficient is 75, and the sparse coefficient threshold of DCT base is 10. The DST base is 650, and the DFT base is 750. The experiment is repeated for 100 times to average the results, the indicators of signal reconstruction are compared with different sparse bases, and different sampling rates M/N are taken to verify the performance. The statistical data of the mean square error, signal-to-noise ratio, absolute error, and relative error of each sparse base pair signal are shown in Table 2.

Table 2. MSE, SNR, AE, and RE of different sparse bases with the sampling rate

M/N	DCT+OMP				DST+OMP				DFT+OMP			
	MSE	SNR	AE	RE	MSE	SNR	AE	RE	MSE	SNR	AE	RE
0.1	3.340	42.119	128.983	0.1221	6.218	35.904	175.980	0.1669	2.377	45.479	108.924	0.1033
0.2	1.550	49.725	88.023	0.0835	3.020	43.062	122.840	0.1167	1.814	48.149	95.231	0.0903
0.3	1.428	50.542	84.488	0.0801	2.671	44.278	115.564	0.1098	1.737	48.584	93.184	0.0884
0.4	1.388	50.825	83.303	0.0790	2.598	44.557	113.961	0.1083	1.704	48.775	92.293	0.0876
0.5	1.367	50.977	82.668	0.0784	2.548	44.751	112.861	0.1073	1.686	48.877	91.819	0.0871
0.6	1.354	51.073	82.272	0.0780	2.513	44.887	112.097	0.1065	1.675	48.946	91.507	0.0869
0.7	1.346	51.131	82.036	0.0778	2.501	44.935	111.827	0.1063	1.666	49.000	91.256	0.0866

When the sampling rate is lower than 0.2, the reconstruction performance of the DFT base signal is the best. When the sampling rate is [0.2-0.7], the reconstruction performance of the DCT base signal is the best. With the change in sampling rate, DCT-based and DFT-based have significant advantages in IGBT electrical signal reconstruction accuracy and stability.

3.2 Measurement matrix selection

To reduce the dimension of the original signal, the quality of the introduced measurement matrix is directly related to the reconstruction of the signal at the receiving end. But at the same time, it must be ensured that the transmission value after the measurement matrix can retain the important information of the original signal, which guarantees that the original signal can be accurately reconstructed later. A measurement matrix suitable for the electrical signal of the IGBT will be searched. The comparative statistics of MSE, SNR, absolute error, and relative error of each measurement matrix to the signal are shown in Table 3. When the sampling rate is lower than 0.2, the effect of the sparse random measurement matrix in restoring the original signal is the best. When the ratio is [0.2-0.7], the performance of the three reconstructed original signals is similar.

Table 3. MSE, SNR, AE, and RE of different measurement matrices as a function of the sampling rate

M/N	Gaussian				Bernoulli				Sparse			
	MSE	SNR	AE	RE	MSE	SNR	AE	RE	MSE	SNR	AE	RE
0.1	3.340	42.119	128.983	0.1221	3.234	42.491	126.760	0.1200	3.193	42.587	126.053	0.1193
0.2	1.550	49.725	88.023	0.0835	1.553	49.706	88.101	0.0835	1.578	49.547	88.818	0.0842
0.3	1.428	50.542	84.488	0.0801	1.427	50.551	84.452	0.0801	1.423	50.573	84.360	0.0800
0.4	1.388	50.825	83.303	0.0790	1.391	50.801	83.403	0.0790	1.386	50.837	83.248	0.0789
0.5	1.367	50.977	82.668	0.0784	1.368	50.967	82.711	0.0784	1.363	51.003	82.561	0.0783
0.6	1.354	51.073	82.272	0.0780	1.350	51.098	82.171	0.0779	1.353	51.080	82.244	0.0780
0.7	1.343	51.142	81.979	0.0778	1.345	51.141	81.995	0.0778	1.340	51.147	81.967	0.0777

3.3 Weak threshold selection

The SWOMP algorithm based on backtracking strategies of different sigmoid functions can effectively improve the reconstruction performance of the algorithm. Since the weak threshold in the SWOMP algorithm is fixed and the range is (0,1], the selection of atoms is selected by the weak threshold each time. When the threshold is too large, fewer atoms are selected in each round, so it takes many iterations of the algorithm to support an index set close to sparseness. When the threshold is too small, a lot of irrelevant atoms will be selected each time, resulting in a waste of resources. At the same time, it will cause the wrong atoms to be received when the signal is reconstructed. As a result, errors occur, and the reconstruction performance of the algorithm is reduced. Therefore, choosing an appropriate weak threshold is the key to the study of the SWOMP algorithm. To solve the problem that the use of a fixed threshold in the SWOMP algorithm reduces the reconstruction performance of the algorithm, this paper improves the fixed threshold. At the same time, the idea of backtracking is introduced and different sigmoid functions are added, as shown in Figure 1.

3.4 SWOMP4 algorithm steps

The SWOMP algorithm based on the backtracking strategies of different sigmoid functions proposed in this paper is a method of adding soft thresholds for atomic secondary screening.

Input:(1) M×N-dimensional sensing matrix A=ΦΨ; (2) The measured value y of N×1 dimension; (3) The number of iterations S, the threshold parameter α.

Output:(1) Signal sparse representation sparse estimation $\hat{\theta}$; (2) N×1-dimensional residual $r_s = y - A_s\hat{\theta}_s$.

Step 1: Let $r_0 = y$, $\Lambda_0 = \emptyset$, put the selected ones that meet the requirements into $\Lambda_0 = \emptyset$, and the initial number of cycles is counted as t=1.

Step 2:Calculate $u = \left|\langle r_{t-1}, a_j\rangle\right|(1 \le j \le N)$, select from u larger than the threshold $Th = \alpha max\{abs(u)\}, 0 < \alpha \le 1$ value, the column serial number j of A corresponding to these values constitutes J_0.

Step 3:Update the index set $\Lambda_t = \Lambda_{t-1} \cup J_0, A_t = A_{t-1} \cup a_j, (\forall j \in J_0)$; if $\Lambda_t = \Lambda_{t-1}$ (that is, no new column is selected), then stop the iteration and go to step 7.

Step 4:$\hat{x}_{t1} = argmin\theta\|y - A_t\theta_t\|_{2=(A_t^T A_t^{-1})}A_t^T y$ by the least square method.

Step 5: Calculate the soft threshold D_t, find out$|\hat{x}_{t2}| \geq D_t.max|\hat{x}_{t1}|$, the column number j of A corresponding to $|\hat{x}_{t2}|$ constitutes J_1.

Step 6: Update the index set $\Lambda_{t2} = \Lambda_{t-1} \cup J_1, A_{t2} = A_{t-1} \cup a_j, (\forall j \in J_1)$.

Step 7: Obtain $\hat{x}_{t2} = argmin\theta\|y - A_{t2}\theta_t\|_{2=(A_{t2}^T A_{t2}^{-1})}A_{t2}^T y$ by the least square method.

Step 8: Update the residual $r_t = y - A_{t2}\hat{x}_{t2}$.

Step 9: Judge whether t>S or the residual is 0, if so, stop the iteration; otherwise, t=t+1, return to step 2. Step 10: Refactor to get $\hat{x}_{t2}$.

After the experiment, the following data are obtained. When the hard threshold parameter is 0.5 ,the sampling rate is 0.1 in the process of analyzing MSE, SNR, absolute error, and relative error, it is found that the reconstruction effect of SWOMP1 is the best, and the corresponding values are:4.291, 39.882, 145.201, 0.138; when the sampling rate is 0.2, the reconstruction effect of SWOMP3 is the best, and the corresponding values are:1.796, 48.370, 94.457, 0.089; when the sampling rate is 0.3, the reconstruction effect of SWOMP4 is the best found by comparison, and the corresponding values are:1.091, 53.251, 73.815, 0.070; when the sampling rate is 0.4, the reconstruction effect of SWOMP4 is still the best, and the corresponding values are:0.815, 56.160, 63.808, 0.060; when the sampling rate is 0.5, the reconstruction effect of SWOMP1 is the best, and the corresponding values are:0.662, 58.233, 57.521, 0.054; when the sampling rate is 0.6, the reconstruction effect of SWOMP6 is the best, and the corresponding values are:0.568, 59.754, 53.313, 0.050; when the sampling rate is 0.7, the reconstruction effect of SWOMP4 is the best, and the corresponding values are:0.498, 61.078, 49.897, and 0.046, respectively. In short, under the same sampling rate, the reconstruction effect of the improved SWOMP algorithm is better than that of the SWOMP algorithm, and the recovery effect of the SWOMP4 and SWOMP3 algorithms is the best, which verifies that the improved SWOMP algorithm is conducive to signal reconstruction.

4. RESULTS

The whole experiment includes data preparation, data compression and data reconstruction. Among them, data preparation and compression are completed by single chip microcomputer, data reconstruction is completed by MATLAB. In the reconstruction selection conditions, the setting range of the sampling rate is 0.1 to 0.7; the sparse basis include the DCT basis, DST basis, and DFT basis; the measurement matrix is: the Gauss random measurement matrix, Bernoulli random measurement matrix, Sparse random measurement matrix; the reconstruction algorithms include OMP, StOMP, SWOMP, SWOMP1, SWOMP2, SWOMP3, SWOMP4, SWOMP5, SWOMP6.

Table 4. MSE, SNR, AE, and RE of different measurement matrices as a function of the sampling rate

M/N	SWOMP				SWOMP4			
	MSE	SNR	AE	RE	MSE	SNR	AE	RE
0.1	23.477	22.5521	216.692	0.338046	21.7546	23.3144	208.588	0.318405
0.2	4.90402	38.212	99.0356	0.148306	4.56961	38.9183	95.5992	0.143635
0.3	2.73521	44.0505	73.9623	0.109438	2.18416	46.3003	66.0932	0.0995628
0.4	1.44904	50.4036	53.8338	0.0801742	1.35153	51.1002	51.991	0.0774769
0.5	1.10173	53.1438	46.941	0.0700778	1.0009	54.1036	44.7416	0.066826
0.6	0.823413	56.0556	40.5811	0.0607169	0.806178	56.2671	40.1542	0.0600436
0.7	0.648469	58.444	36.013	0.0538456	0.575166	59.6436	33.9165	0.0507413

The improved SWOMP4 algorithm is compared with the traditional SWOMP algorithm. The experimental results are shown in Table 4. The larger the MSE, AE, and RE, the worse the recovery effect of the signal; the SNR, on the contrary, the larger the SNR, the better the recovery effect of the signal. The experimental results show that under the same sampling rate, the values of MSE, AE and RE of the SWOMP4 algorithm are all smaller than those of the SWOMP algorithm, but the SNR values are all larger than those of the SWOMP algorithm. The reconstruction effect of the SWOMP4 algorithm is better than that of the SWOMP algorithm. From the reconstructed signal graph, for the same original signal and the same sampling rate, the reconstructed signal of the SWOMP algorithm has more glitches than the reconstructed signal of the SWOMP4 algorithm, and the reconstruction effect of the SWOMP algorithm is worse than that of the SWOMP4 algorithm. With the increase of the sampling rate, the reconstruction effect of the signal is getting better and better. Through experiments, it is verified that the SWOMP4 algorithm has a higher reconstruction effect than the SWOMP algorithm, and the correctness of the improved algorithm are also proved.

5. CONCLUSION

With the rapid and stable development of electric power and computer science and technology, IGBT is of great significance to the entire system. Online health monitoring is particularly important, and its safe operation is related to the reliability of the entire equipment. Due to the large amount of data and high sampling rate of IGBT online health monitoring, data transmission has become a major problem. Based on compressive sensing theory, this paper proposes a SWOMP algorithm based on backtracking strategies of different sigmoid functions, and studies the compression and reconstruction of IGBT electrical signals based on compressive sensing. The sparseness and reconstruction effect of electrical signals are discussed and analyzed, which provides preconditions for the subsequent research on IGBT electrical signal reconstruction. The whole experiment includes data preparation, data compression and data reconstruction, which verifies the correctness of the improved algorithm.

ACKNOWLEDGEMENTS

This paper is supported by State Grid Zhejiang Electric Power Co., Ltd. Technology Project (5211DS21N00A).

REFERENCES

[1] Jiangjian, Loujian, "Research on the improvement of power supply capacity of urban distribution network based on load reorganization," Zhejiang Electric Power, Papers 2 (1), 9-15 (2020).

[2] Tang Guangfu, Panghui, H., "Development and application of advanced AC / DC transmission technology in China," Chinese Journal of electrical engineering, Papers 36 (7), 1760-1771 (2016).

[3] Guzman H, F Barrero, Duran M J., "IGBT-Gating for Failure Effect on a Fault-Tolerant Predictive Current-Controlled Five-Phase Induction Motor Drive," Industrial Electronics IEEE Transactions, Paper 62(1), 15-20 (2015).

[4] Wang Tianjing, Zheng Baoyu, Yang Zhen., "Compressed sensing signal acquisition scheme based on filtering," Journal of instrumentation, Paper 34 (003), 573-581 (2013).

[5] Yu Kai, Li Yuanshi, W., "Sound signal acquisition method based on compressed sensing," Journal of instrumentation, Paper 33 (1), 105-112 (2012).

[6] Jiang Hao, Qian Hui, X., "Implementation of ECG signal acquisition circuit based on compressed sensing," Internet of things technology, Paper 7 (12), 20-23 (2017).

[7] Wang Xiping, Ding Xiangkuan, Yao Fang, Tang Shengxue, L., "Overview of IGBT module failure mechanism and condition monitoring," China power, Paper 52 (09), 61-72 (2019).

[8] Wang Z,Tian B,Qiao W., "Online aging monitoring for IGBT modules using case temperature,"IEEE Transactions on Industrial Electronics, Paper 63(2), 1168-1178 (2016).

[9] Bing G,Fan Y,Chen M., "A temperature gradient based condition estimation method for IGBT module," IEEE Transactions on Power Electronics, Paper 32(3), 1-10 (2016).

[10] Barranca Victor J., "Neural network learning of improved compressive sensing sampling and receptive field structure," Neurocomputing, Paper 34 (3), 455-458 (2021).

[11] Huqin., "Research on physical layer security scheme based on compressed sensing," Xi'an University of Electronic Science and technology, 10-32 (2020).
[12] Vehkapera M,Kabashima Y,Chatterjee S., "Analysis of Regularized LS Reconstruction and Random Matrix Ensembles in Compressed Sensing," IEEE Transactions on Information Theory, Paper 62(4), 2100-2124 (2016).
[13] Lan Pu,Zhang Jiangtao,Xia Kewen,Zhou Qiao,He Ziping., "Research on Improvement of Stagewise Weak Orthogonal Matching Pursuit Algorithm," Pattern Recognition and Image Analysis, Paper 29(4), 795-802 (2019).
[14] Li S, Wei S, Jin L., "Research on measurement matrix based on compressed sensing theory," IEEE International Conference on Control Science & Systems Engineering, pp. 716-719 (2017).

Effect of H/F Adsorption on the Electronic Structure of TH-$Ge_{0.5}Si_{0.5}C_2$-zz

GuiHong Luo, Ying Yang*, RuKun Li, MengYang Zhang, WenJing Zhao

School of Automation and Information Engineering, Xi'an University of Technology, Xi'an, Shaanxi Province, 710048, P. R. China

*Corresponding author: yangy@xaut.edu.cn

Abstract: A new material TH-$Ge_{0.5}Si_{0.5}C_2$-zz has been discovered, which has a direct bandgap and has a broad application prospect in photoelectric devices. The effect of H/F atom adsorption on the structural parameters and electrical properties of TH-$Ge_{0.5}Si_{0.5}C_2$-zz was studied in this paper. In a nutshell,bandgap of TH-$Ge_{0.5}Si_{0.5}C_2$-zz can be notable increased by H atom adsorption, and the work function and electron affinity energy are decreased. While the bandgap of TH-$Ge_{0.5}Si_{0.5}C_2$-zz cannot be opened by the full adsorption of the F atom.

1. INTRODUCTION

The latest proposed TH-SiC_2 is a half metallic material with a bandgap of -0.01eV, which may limit its applications. TH-GeC_2 has a direct bandgap of 0.87eV [1]. The alloy TH-$Ge_{0.5}Si_{0.5}C_2$-zz formed by TH-GeC_2 and TH-SiC_2 can not only improve the bandgap problem of TH-SiC_2 but also can be used in middle and far infrared detection because it has a bandgap of 0.59eV. However, the bandgap width of TH-$Ge_{0.5}Si_{0.5}C_2$-zz is very narrow. In order to broaden the practical application range of TH-$Ge_{0.5}Si_{0.5}C_2$-zz and be suitable for visible light detection, the H/F atom adsorption method was adopted to change its band structure in this paper.

The method of H/F atom adsorption is adopted to change the properties of materials because two-dimensional materials have a huge specific surface area. So it can change the properties of two-dimensional materials more effectively. In fact, surface functionalization by adsorption of H/F atoms has been applied to change the band gaps of graphene, g-SiC, and g-GeC [2-6]. Among them, fluorinated graphene (FG) has a wider energy gap and other excellent properties and has been prepared[7]. Therefore, to make TH-$Ge_{0.5}Si_{0.5}C_2$-zz show a good application prospect in the field of electronic devices, this paper hopes to adjust the electronic properties of TH-$Ge_{0.5}Si_{0.5}C_2$-zz by adsorbing H / F atoms.

The effect of H/F atom adsorption on the structural parameters and electrical properties of TH-$Ge_{0.5}Si_{0.5}C_2$-zz is theoretically investigeted, in this article. The results are a theoretical guidance on the functionalization of TH-$Ge_{0.5}Si_{0.5}C_2$-zz, and further expand the application field of this material in electronic devices.

2. COMPUTATIONAL METHODS

In this paper, the PAW projected plane-wave pseudopotential of the VASP package [8][9] is used for geometric optimization calculation by using the exchange-correlation energy functional (PBE) in the generalized gradient approximation (GGA). All calculations use the same truncation energy 500eV, the energy convergence criterion is 1×10^{-5}eV, and the Monkhorst-Pack grid is used to relax the atomic structure of the interface at 8×6×1 K-points centered on Γ. The geometric relaxation takes force as the convergence criterion, and the geometric relaxation terminates when the force on each ion decreases to less than 1×10^{-4}eV/Å. For the purpose of avoiding coupling between crystal layers, the thickness of vacuum layer must be sufficient. In this paper, 20Å vacuum layers are set in the vertical direction.

3. RESULTS AND DISCUSSION

3.1 Adsorption H atoms

The structure of TH-$Ge_{0.5}Si_{0.5}C_2$-zz and H/F adsorbed TH-$Ge_{0.5}Si_{0.5}C_2$-zz is shown in Fig. 1. It can be seen from Fig. 1 (a) that there are 12 atoms in one cell of TH-$Ge_{0.5}Si_{0.5}C_2$-zz, of which 8 C atoms are sp^2 hybrids, 2 Ge atoms are sp^3 hybrids and two Si atoms are sp^3 hybrids. 1/4 adsorption coverage represent that C1 and C3 are adsorbed by H atoms. 1/2 adsorption coverage means that 4 sp^2 C atoms on the same side are adsorbed by H atoms. 1 adsorption coverage indicates that 8 sp^2 C atoms are adsorbed by H atoms(Fig. 1 (b-d)).

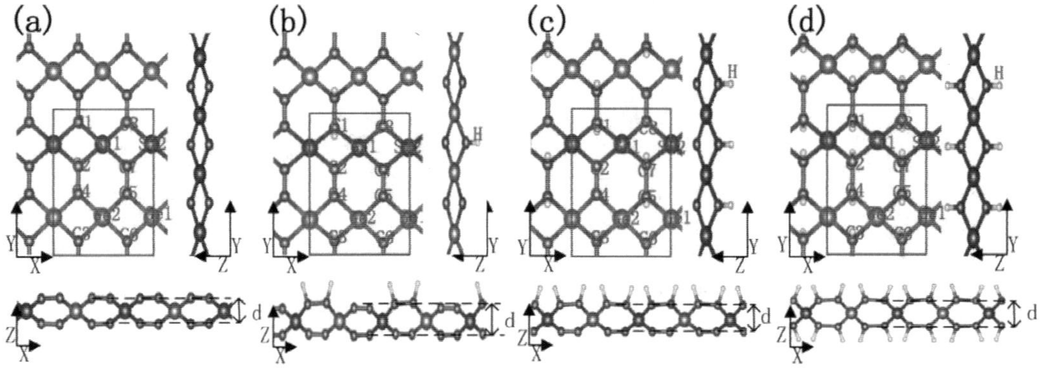

Fig. 1 Structure of TH-$Ge_{0.5}Si_{0.5}C_2$-zz, (a) Pristin of TH-$Ge_{0.5}Si_{0.5}C_2$-zz, (b) Coverage of 1/4, (c) Coverage of 1/2, (d) Coverage of 1

The lattice constants of TH-$Ge_{0.5}Si_{0.5}C_2$-zz adsorbing H atoms are shown in Table 1. With the increase of the coverage of H, the lattice parameters *a* and *b* decrease, and the thickness *d* increases.

Table 1. TH-$Ge_{0.5}Si_{0.5}C_2$-zz lattice constants of adsorbed H atoms

Surface coverage	*a*(Å)	*b*(Å)	*d*(Å)
Pristine	5.75	7.24	1.42
1/4	5.67	7.21	1.77
1/2	5.61	7.19	1.75
1	5.46	7.20	2.06

The bandgap of th-$Ge_{0.5}Si_{0.5}C_2$-zz is 0.59eV(HSE06), showing the direct band gap semiconductor property(Fig. 2 (a)). The bandgap of th-$Ge_{0.5}Si_{0.5}C_2$-zz increases with the increase of adsorption coverage(Fig .2(b)). When the adsorption coverage is 1/4, 1/2, and 1, th-$Ge_{0.5}Si_{0.5}C_2$-zz is a semiconductor.

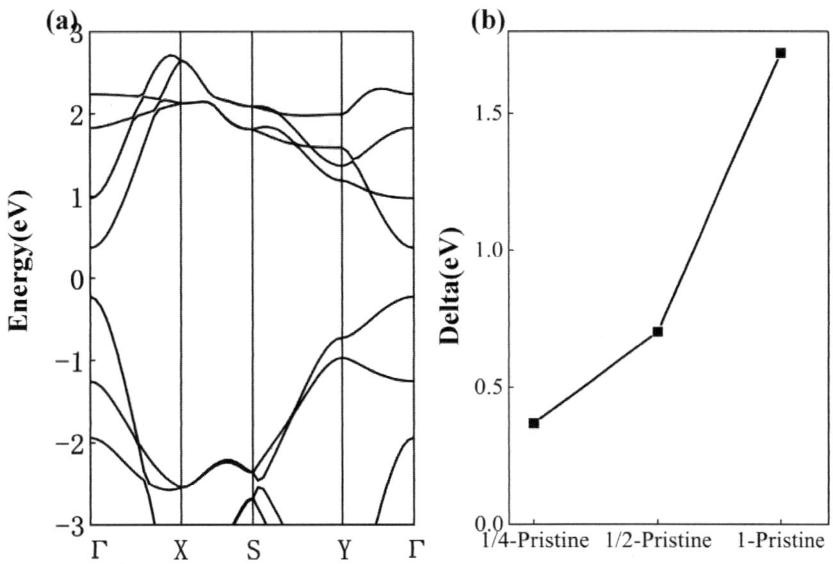

Fig. 2 (a) The band structure of TH-$Ge_{0.5}Si_{0.5}C_2$-zz predicted by HSE, (b) H adsorption coverage are 1/4, 1/2, and 1 band gap increments, respectively.

Hydrogenation can also change electron affinity energy and the work function of TH-$Ge_{0.5}Si_{0.5}C_2$-zz, in addition to adjusting the bandgap. Calculate the work function of semiconductor material [10] using Formula (1). The E_0 is vacuum energy, E_F is the Fermi level. The electron affinity energy can be calculated using Formula (2), where χ is the electron affinity energy, E_0 is the vacuum energy level, and E_C is the energy level at the bottom of conduction band.

$$W = E_0 - E_F \tag{1}$$

and

$$\chi = E_0 - E_C \tag{2}$$

For H adsorption, the work function of TH-$Ge_{0.5}Si_{0.5}C_2$-zz decreases as the increase of adsorption coverage, as shown in Fig. 3. Since C is more electronegative than H, electrons are transferred from H to C, resulting in an increase in electron concentration as the adsorption coverage of H increases. resulting in an increase in the Fermi energy level. So this eventually makes work function decreases with the increase of the adsorption coverage. The minimum work function is 2.63eV.

In terms of H adsorption, compared with pristine structure, the electron affinity energy decreases with the increase of adsorption coverage. While the adsorption coverage is 1, its electron affinity is only 1.35eV.

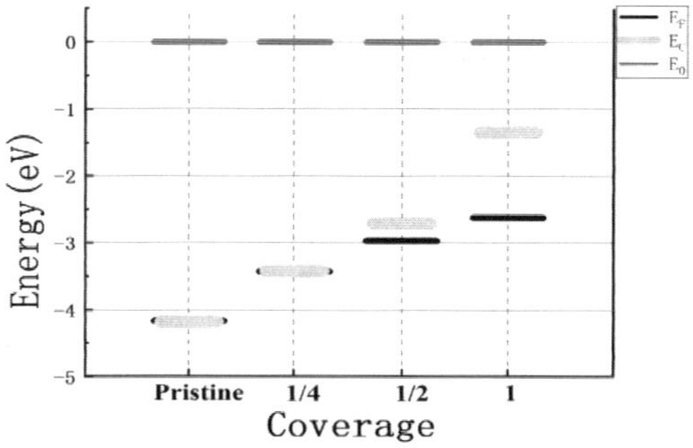

Fig. 3 The work function and electron affinity energy of hydrogenate-TH-$Ge_{0.5}Si_{0.5}C_2$-zz

The vacuum energy level is specified to be 0 making Fermi energy levels of the semiconductor material are aligned at the VBM.

3.2 Full adsorption of H/F atoms

Besides, studying the effect of H adsorption on TH-$Ge_{0.5}Si_{0.5}C_2$-zz with different adsorption coverage, this paper also investigates the effect of F adsorption on TH-$Ge_{0.5}Si_{0.5}C_2$-zz. The results suggest that compared with the full adsorption of the H atom (Fig. 4 (a)), the gap width of TH-$Ge_{0.5}Si_{0.5}C_2$-zz decreases to 0.2eV when the H atom is adsorbed on one edge and the F atom is adsorbed on the other edge (Fig. 4 (c)). However, the bandgap width of TH-$Ge_{0.5}Si_{0.5}C_2$-zz of full adsorbed F atom is the smallest (Fig. 4 (b)).

It was found that the bandgap of TH-$Ge_{0.5}Si_{0.5}C_2$-zz could not be opened by the full adsorption of F atoms (Fig.4 (b)). As can be seen from the density of state(DOS) diagram, when H absorbs TH-$Ge_{0.5}Si_{0.5}C_2$-zz completely, the H atom makes no contribution to the valence band maximum(VBM) and little contribution to the conduction band minimum(CBM). However, the contribution of the F atom to VBM and CBM is second only to that of the C atom when the TH-$Ge_{0.5}Si_{0.5}C_2$-zz of F is completely adsorbed. It can also be clearly seen from the electron cloud density distribution that the VBM and CBM of the full adsorption of F on TH-$Ge_{0.5}Si_{0.5}C_2$-zz are mainly contributed by C atoms and F atoms (Fig.4 (b)). This is because the bond energy of the C-F bond is more intense than that for the C-H, which results in the decrease of the band gap of the full adsorption F atom.

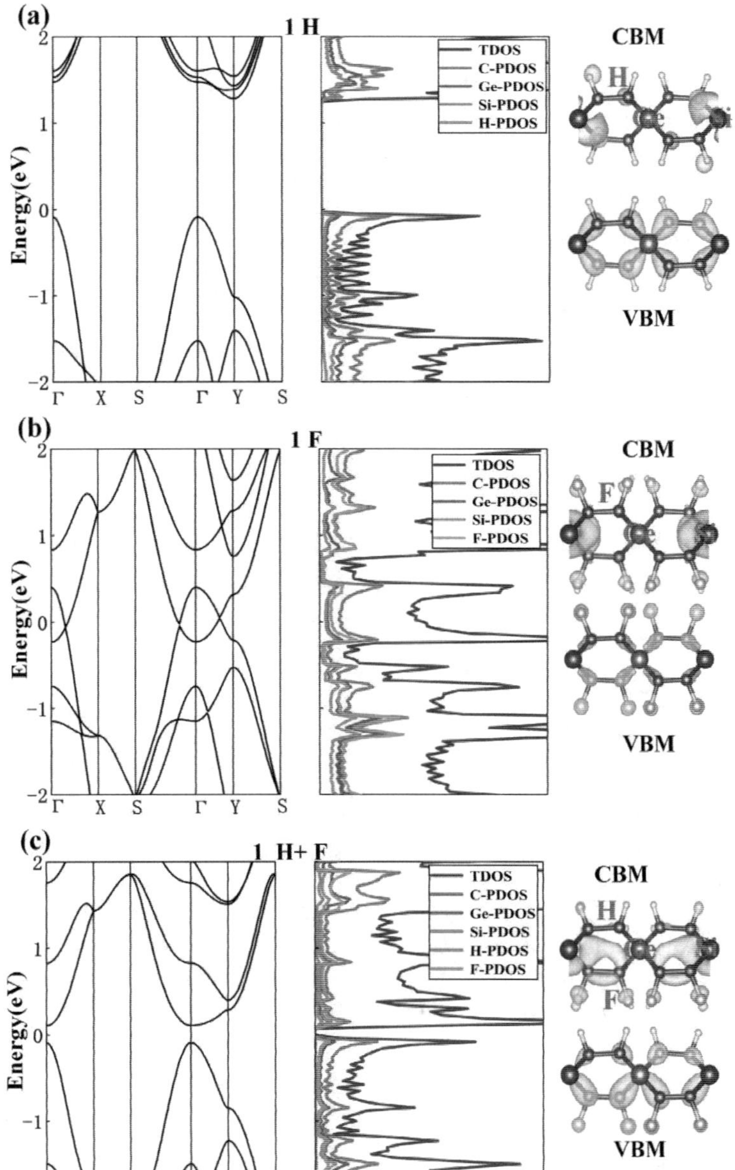

Fig. 4 Band structure of full adsorption of H/F on TH-$Ge_{0.5}Si_{0.5}C_2$-zz, fermi energy level is set as 0 eV, full adsorption of H/F on TH-$Ge_{0.5}Si_{0.5}C_2$-zz, and the electron density contour plots. (Black line in the diagram represents the total density of state (TDOS) of adsorbed atoms. The blue line represents the partial density of state (PDOS) of the Ge, the red line represents the C, the green line represents the Si, the purple line represents the H, and the yellow line represents the F) (a) full adsorption of H atoms, (b) full adsorption of F atoms, (c) H atom is adsorbed on one edge of TH-$Ge_{0.5}Si_{0.5}C_2$-zz and F atom is adsorbed on the other edge of TH-$Ge_{0.5}Si_{0.5}C_2$-zz.

4. CONCLUSIONS

The results show that the gap width of TH-$Ge_{0.5}Si_{0.5}C_2$-zz enlarges with the increase of H adsorption. Adsorption of the H atom affects not only the gap width, but also modulates the electron affinity energy and work function. As the amount of H adsorption increases, TH-$Ge_{0.5}Si_{0.5}C_2$-zz work function and electron affinity energy are decreasing. The lower the electron affinity energy, the more likely it is that electrons will be released. Therefore, TH-$Ge_{0.5}Si_{0.5}C_2$-zz adsorbed by H atoms can be used to manufacture field emission devices. However, compared with the full adsorption of H atoms, full adsorption of F cannot widen the band gap of TH-$Ge_{0.5}Si_{0.5}C_2$-zz, this is because C-F bond energy is stronger than C-H bond energy, resulting in a smaller band gap.

ACKNOWLEDGMENT

This work acknowledged funding from the National Natural Science Foundation of China (Grant No.51207128 and 62174134) and the Shaanxi Innovation Capability Support Project (Grant No. 2021TD-25).

REFERENCES

[1] Kilic, M.E. and K. Lee, Novel two-dimensional Group-IV carbides containing C2 dimers: Sizable direct band gap, high carrier mobility, and anisotropic properties for nanoelectronics. Carbon, 2021. 181: p. 421-432.

[2] Drissi, L.B., F.Z. Ramadan and N.B.J. Kanga, Fluorination-control of electronic and magnetic properties in GeC-hybrid. Chemical Physics Letters, 2016. 659: p. 148-153.

[3] Houssa, M., et al., Electronic properties of hydrogenated silicene and germanene. Applied Physics Letters, 2011. 98(22): p. 223107.

[4] Ma, Y., et al., Magnetic properties of the semifluorinated and semihydrogenated 2D sheets of group-IV and III-V binary compounds. Applied Surface Science, 2011. 257(17): p. 7845-7850.

[5] Hoat, D.M., et al., Transition from indirect to direct band gap in SiC monolayer by chemical functionalization: A first principles study. Superlattices and Microstructures, 2020. 137: p. 106320.

[6] Vu, T.V., et al., Surface functionalization of GeC monolayer with F and Cl: Electronic and optical properties. Superlattices and Microstructures, 2020. 137: p. 106359.

[7] WANG Xu, C. Xinyu and F. Kun, Preparation and application of fluorinated graphene based on direct fluorination technology. Journal of Solid Rocket Technology, 2021.

[8] Kresse, G. and J. Furthmüller, Efficiency of ab-initio total energy calculations for metals and semiconductors using a plane-wave basis set. Computational materials science, 1996. 6(1): p. 15-50.

[9] Zunger, J.P.P.A., Self-interaction correction to density-functional approximations for many-electron systems. PHYSICAL REVIEW B, 1980. 23: p. 10.

[10] Arefi, H.H. and G. Fagas, Chemical Trends in the Work Function of Modified Si(111) Surfaces: A DFT Study. The Journal of Physical Chemistry C, 2014. 118(26): p. 14346-14354.

Controllable Potassium Iodide-Assisted Preparation of Two-dimensional SnS_2 for Hydrogen Evolution Reaction

Qi. Chao. Yang[a], Jun. Qi. Wang[a], Ke. Yan. Li[a], Jun. Jie. Qi*[a]

[a]School of Materials Science and Engineering, University of Science and Technology Beijing, Beijing 100083, P.R. China

* Corresponding author: junjieqi@ustb.edu.cn

Abstract: The controllable preparation of Two-dimensional transition metal chalcogenides (TMDCs) is very important. However, the uncontrollable growth of SnS_2 by chemical vapor deposition (CVD) and the unsatisfactory performance of hydrogen evolution reaction (HER) based on SnS_2 have hindered its applications. Herein, we reported the direct synthesis of high-quality 1T-SnS_2 with various morphologies via potassium iodide-assisted CVD. On the basis of CVD technology, We have developed process technology to grow different shapes of SnS_2 from dot line to fishbone controlled by growth conditions. The growth mechanism of SnS_2 was studied, and their HER performance with different morphologies of SnS_2 was investigated. The growth trend of 2D SnS_2 provides a reference for the controllable preparation of TMDCs materials. These results demonstrate that TMDCs can adjust the performance of HER through shape control.

1. INTRODUCTION

Two-dimensional materials have become the most promising candidate materials to replace silicon materials. Among them, transition metal chalcogenide (TMDC) is the most studied 2D semiconductor. The chemical formula is MX_2, which is composed of two layers of chalcogenide atom plane (X = S, Se, Te) and one layer of transition metal atom plane (M = Mo, W, Re, Sn, etc.). The transition metal atom plane is in the middle and forms a sandwich structure[1].

Transition metal disulfide compounds (TMDs) have many excellent properties: high carrier mobility[2][3], edge active sites[4], good flexibility[5] and thermal stability[6]. It has been widely studied and applied in photoelectric detectors[7][8], FET[9][10], flexible devices[5][11], super capacitors[12][13][14], lithium-ion batteries[15][16][17][18] and hydrogen evolution reaction[19][20][21][22][23][24]. Among them, MoS_2 is the most widely studied one, which has been studied and applied in the above fields. Although MoS_2 has shown its significance in various applications and basic attribute research, other TMDC[25][26][27][28][29] may also furnish important ways for furnish mechanisms, new devices or new applications.

Tin disulfide (SnS_2) have a future in 2D TMDCs materials. With a T-phase structure, Sn atoms are coordinated by six S atoms. The single layer and block are both indirect band gaps (2.0–2.6 eV)[30]. SnS_2 has great potential in gas sensors[31][32], CO_2 reduction reaction[33][34][35], field effect transistors[36], and photoelectric catalysis[37][38][39][40]. Energy range of conduction band and valence band of SnS_2 exceeds the redox potential of water, which makes it expected to be used for water decomposition[41]. Mechanical stripping[42], liquid stripping[43], and CVD[36] are usually used to prepare two-dimensional SnS_2 for laboratory research. CVD is a economic method to manufacture TMDs by controlling growth conditions. Adding halogen salts can increase the vapor pressure of the intermediate, reduce the sublimation temperature of precursor[44]. Potassium halide is more suitable to obtain 1T phase SnS_2 than sodium halide.

Although some researchers have successfully synthesized two-dimensional SnS_2 nanosheets by a CVD method, the precise and controllable preparation of layers and shapes is difficult, the size is small, and the nucleation growth law and mechanism are still unclear. Therefore, the controllable growth and nucleation growth of 2D SnS_2 need to be further explored.

Herein this paper report the 2D SnS_2 controllable growth on silicon dioxide/silicon substrate by adding potassium iodide via CVD. Through regulate various growth conditions for instance temperature, holding time, airflow rate and precursor distance, 2D SnS_2 with various morphologies and different layers were prepared. The relationship between the layers and Raman peaks was established, and then the layers of the prepared materials could be qualitatively determined by Raman peaks. In the preparation process, the growth trend of two-dimensional SnS_2 was analyzed, and the dot linear and linear two-dimensional SnS_2 as the intermediate form was found. Finally, the fishbone SnS_2 with more branches showed improved hydrogen evolution property, Tafel slope is 138 mV dec^{-1} and overpotential is 435 mV, indicating its potential applicability in HER. In addition, the good performance and controllable methods of 2D SnS_2 will enable us to release its potential in other fields, such as gas sensors, oxygen evolution reactions, carbon dioxide reduction, optoelectronic devices, etc.

2. RESULTS AND DISCUSSION

In order to obtain highly uniform, large size and various morphologies of single molecular layer SnS_2, SnS_2 was synthesized on silicon dioxide/silicon substrate by CVD with tin dioxide sulfur precursor. CVD device shown in Figure 1a, 2D SnS_2 can be deposited on a silica/silicon substrate by grinding potassium iodide into powder and mixing it fully with tin dioxide. As Figure 1b, two-dimensional SnS_2 is sandwich structure. The two-layer S atomic layer is on both sides, and the middle interlayer is a layer of Sn atoms. Optical images show regular windmill shapes and uniform colors, indicating that the grown SnS_2 has good uniformity and continuity.

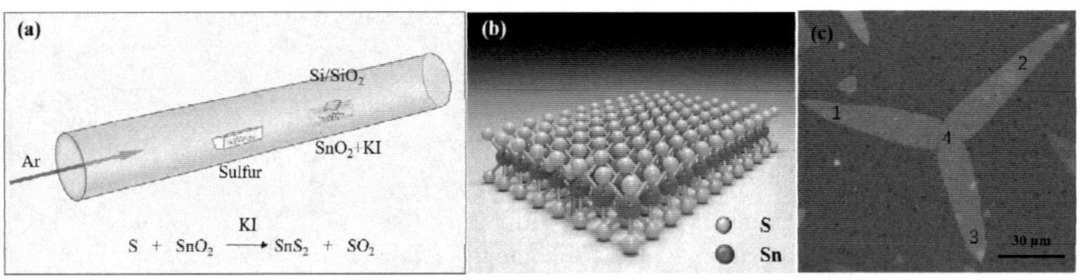

Figure 1. Preparation of two-dimensional SnS_2 (a) CVD schematic of two-dimensional SnS_2. (b) Structure scale of two-dimensional SnS_2: 30μm. (c) Typical optical image of two-dimensional SnS_2

By accurately adjusting the holding time, the intermediate state in the growth process of the material can be obtained, and the growth trend diagram of the windmill-like two-dimensional SnS_2 is obtained. As shown in Figure 2a, when the holding time is very short, some nucleation points are regularly arranged in a line to form a windmill point line. As the holding time increases, the nucleation points are connected to form a windmill line, and finally, grow into a regular windmill SnS_2. In addition, in addition to the regular windmill shape, two-dimensional SnS_2 with different morphologies (Figure 2b) were successfully prepared by accurately adjusting various growth parameters (triangle, burr windmill shape, fishbone shape and dendritic shape). Thus, two-dimensional SnS_2 materials with various morphologies can be fully prepared.

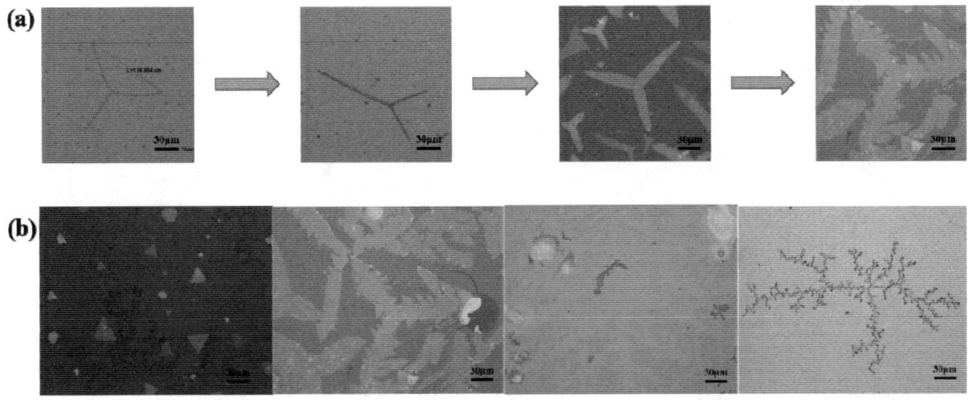

Figure 2. 2D SnS_2 with various morphologies (a) Growth trend diagram of 2D regular windmill-like SnS_2. (b) Optical microscope image scale of 2D SnS_2 with different morphologies.

Further qualitative analysis of the prepared two-dimensional windmill-like SnS_2 was carried out by Raman spectroscopy. Four points of the two-dimensional windmill-like SnS_2 were evenly selected by 532 nm laser for Raman and PL tests. The selection of points is shown in Figure 3a, respectively, at the three corners and the middle of the windmill. From the Raman spectra of Figure 3b, it can be seen that two-dimensional SnS_2 has a characteristic peak at about 311.5 cm^{-1}, conform to the A_{1g} peak.

SnS_2 light absorption coefficient of more than 10^4 cm^{-1} and indirect band gap of 2.08-2.44 eV. Theoretical calculations show that SnS_2 in 1-4 layers is an indirect band gap. The PL characteristic peak of the prepared 2D SnS_2 sample shown in Figure 3f is 2.13 eV. Atomic force microscopy images show that SnS_2 is uniform in color and the thickness is about 1.6 nm, which proves that the prepared material is two layers. At the same time, the Raman spectra and PL spectra of the four points have little difference, indicating that the prepared two-dimensional SnS_2 samples have good uniformity and high quality.

A multi-layer two-dimensional SnS_2stacking position is selected, using Raman spectroscopy to characterize different layers of SnS_2, as shown in Figure 3g, respectively 1,2,3,4 layer and multilayer. Like Figure 3i, the Raman peaks of two-dimensional SnS_2 with different layers have changed to some extent. As the number of layers increases, SnS_2 characteristic peaks have a slight blue shift and the peak intensity increases. By characterizing the Raman spectra of different layers, the relationship between the layers and Raman peaks can be roughly determined, and then the layers of SnS_2 can be approximately determined by Raman spectroscopy.

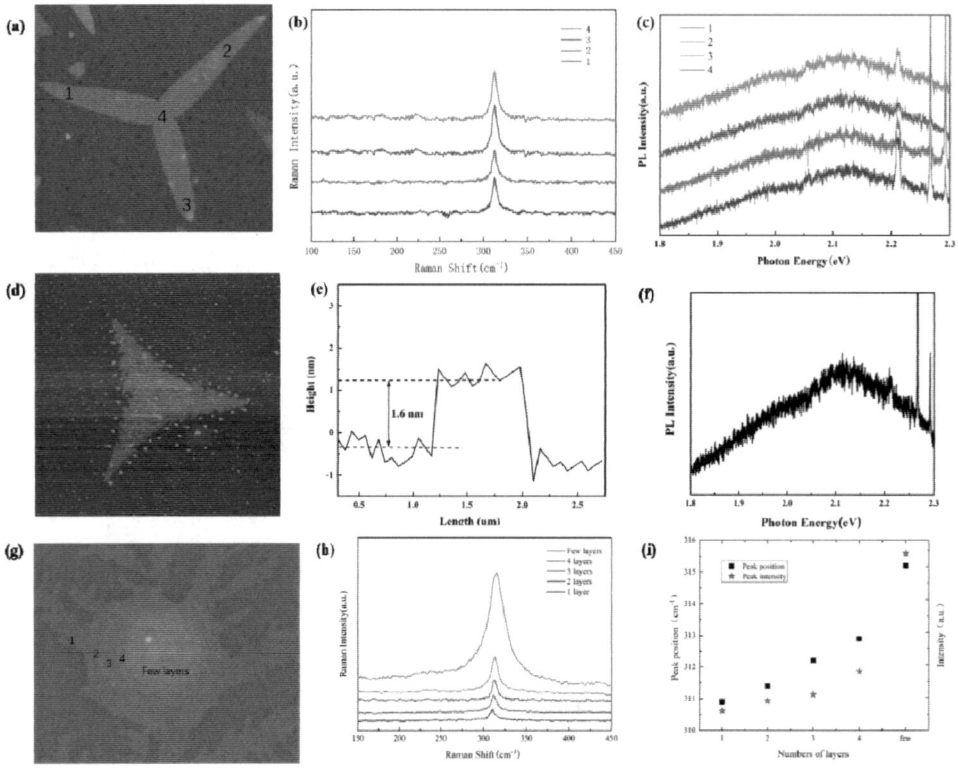

Figure 3. (a) Scale of windmill-like SnS_2 optical microscope. (b) Raman map of four points of SnS_2. (c) PL map of four points of SnS_2. (d) Atomic force microscope. Scale bar of two-dimensional SnS_2. (e) Thickness schematic map of two-dimensional SnS_2. (f) Typical PL map of two-dimensional SnS_2. (g) Optical microscope map of two-dimensional SnS_2 with different layers. (h) Raman map of two-dimensional SnS_2 with different layers. (i) Peak position (left longitudinal axis) and Peak strength (right longitudinal axis) of A1g Raman peak of two-dimensional SnS_2 and the relationship between layers.

Use XPS analysis to further determine the constitute and chemical states of the 2D SnS_2 grown. Typical XPS spectra of SnS_2 are shown in Figure 4a. Due to spin-orbit coupling, the Sn 3d peak splits into two characteristic peaks ($3d_{3/2}$ and $3d_{5/2}$) with 495.8 eV and 487.2 eV, respectively, indicating that it is in an Sn^{4+} state. For THE XPS image of S, as Figure 4c shown, two characteristic peaks corresponding to $2p_{3/2}$ and $2p_{1/2}$, respectively 161.8eV and 162.9eV, confirming that S is in S^{2-} state.

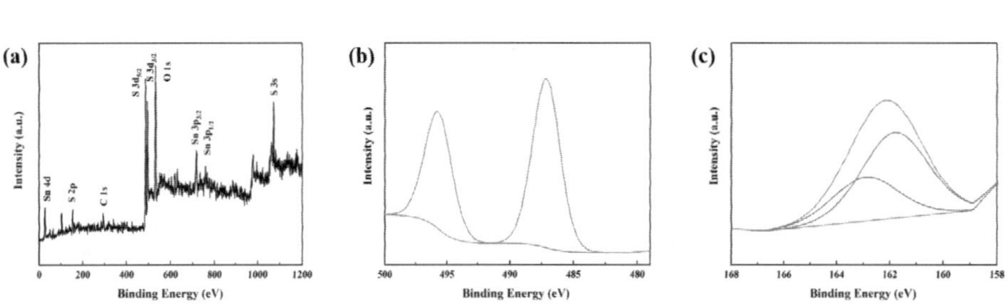

Figure 4. (a) XPS full spectrum. (b) XPS characteristic peak of Sn 3d. (c) XPS characteristic peak of S 2p.

Figure 5(a) (b) is the LSV curve and Tafel slope of two-dimensional SnS_2 with different shapes. It can be seen from the figure that the overpotentials of the triangle, windmill and fishbone SnS_2 at η=10 mA cm^{-2} are 572 mV, 494 mV and 435 mV, respectively. The Tafel slopes were 163 mV dec^{-1}, 150 mV dec^{-1} and 138 mV dec^{-1}, respectively. The LSV curve and Tafel slope showed that SnS_2 with fishbone shape had the best catalytic activity, followed by windmill shape and triangle shape. The reason is that electrocatalytic performance of two-dimensional SnS_2 comes from the edge active site, and the base surface has no catalytic activity. In the unit area, the fishbone shape has a longer edge, followed by the windmill shape, and finally the triangle. Therefore, the shape has a great influence on its catalytic activity, and the shape engineering research of two-dimensional materials is of great significance. Shape controllable preparation can indirectly affect the performance of materials and then can control the performance by adjusting the shape. Figure 5c shows the impedance (EIS) diagrams of two-dimensional SnS_2 with different shapes. The interfacial charge transfer ability of the catalyst can be obtained by analyzing the EIS diagram. The solution impedance (R_s) of the three shapes of 2D SnS_2 is almost equal, charge transfer impedance (R_{ct}) is quite diverse. The arc diameter of the fishbone SnS_2 impedance diagram is the smallest, indicating that the charge transfer ability of fishbone SnS_2 is the strongest, which is also one of the reasons why fishbone SnS_2 has the best electrocatalytic activity. The arc diameter of the impedance diagram from large to small is a triangle, windmill and fishbone, so the charge transfer ability triangle < windmill < fishbone. The trend of electrocatalytic activity of EIS is the same as that of η=10mA cm^{-2} and Tafel slope reflected by LSV.

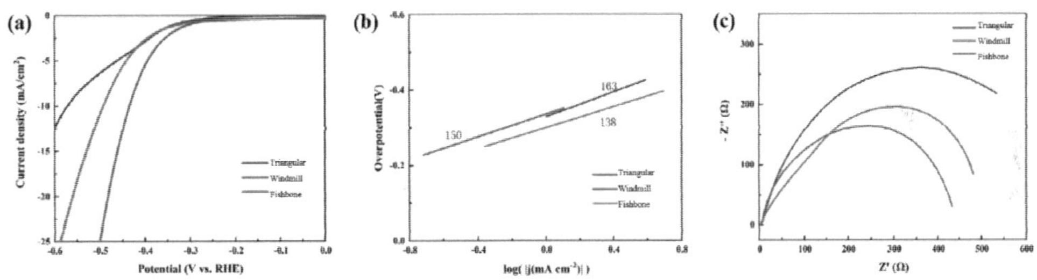

Figure 5. (a) LSV curves for hydrogen evolution of triangular, windmill-shaped and fishbone-shaped SnS_2. (b) Tafel slope (over potential vs. log | current density |). (c) corresponding EIS diagrams for HER AC impedance.

3. CONCLUSION

In summary, two-dimensional SnS_2 with various shapes was prepared by the potassium iodide-assisted CVD method. The morphology of SnS_2 was controlled by adjusting the parameters of the preparation process. Compared with regular triangular and windmill-shaped SnS_2, fishbone-shaped SnS_2 has more branches. The branches of SnS_2 improved the performance of HER. The hydrogen evolution overpotential was reduced by 24% and the Tafel slope was reduced by 15.4%. We investigate the growth trend of SnS_2, and we found the dot linear and linear SnS_2 as the intermediate form. The research on shape regulation of SnS_2 provides a way for controllable preparation of TMDs and their application in electrochemistry and electronics.

4. METHODOLOGY

2D SnS_2 with various shapes was prepared by the CVD method. Two-dimensional SnS_2 with different shapes was grown by two-temperature atmospheric CVD. The silica /silicon substrate was washed with concentrated H_2SO_4/H_2O_2 (3:1), acetone, anhydrous ethanol and deionized water, respectively. The quartz crucible containing sulfur powder is placed upstream of the CVD tube furnace. The tin dioxide (5 mg, 0.3 mg, ≥99.5 %) and the ground potassium iodide powder (2 mg) are fully mixed and put into the glass boat. SiO_2/Si substrates are placed above and then placed downstream.

The CVD tube furnace was initial ventilated with 200 sccm argon for 30 min to discharge the impurity gas. Heating the second temperature zone to 720-800°C, hold for 5-15 minutes, Natural cooling.

The working electrode was prepared by wet transfer of two-dimensional SnS_2. Firstly, SiO_2/Si PMMA with 2D SnS_2 samples was rotated at 2000 r/min for 1 min on a leveling machine and then cured at 180°C for 1 min. For the second time, it was rotated at 1500 r/min for 1 min and heated at 180 °C for 1 min. After rotary coating, use a small knife to cut the PMMA film around, conducive to etching. The sample was etched in HF solution (HF: H_2O = 1:5). After the sample was separated from the substrate silicon, the conductive glassy carbon sheet cleaned in the previous step was used to remove the sample and then heated at 200 °C for 15 min. Finally, PMMA glue was removed by acetone and isoacetone and dried to prepare a two-dimensional SnS_2-supported glassy carbon electrode.

Characterization. Optical microscope (Olympus BX51M) used to characterize the surface topography of samples. The surface morphology and sample thickness were characterized by AFM. Raman and PL spectral images were obtained using a Raman spectrometer (JY-HR800). The elemental valence states of two-dimensional SnS_2 were analyzed by XPS.

Testing of HER performance. The 2D SnS_2 of different shapes grown by CVD were transferred to glassy carbon plates as working electrodes. LSV, Use three-electrode system with an electrochemical workstation and H_2SO_4 solution (0.5 M) performed Tafel and EIS tests. Use Ag/AgCl and Pt electrodes as reference electrodes and counter electrodes, respectively.

ACKNOWLEDGEMENTS

This research was supported by Beijing Natural Science Foundation (2202030), the National Defense Pre-Research Foundation of China (No. 41422050303), and the Program of Introducing Talents of Discipline to Universities (B14003).

REFERENCES

[1] Ovchinnikov, D., Manzeli, S., Pasquier, D. et al. "2D transition metal dichalcogenides". *Nat. Rev. Mater.*, 2(8), (2017).

[2] Xu, X. Gao, Q., Zhang, Z., et al. "Scalable high-performance radio frequency electronics based on large domain bilayer MoS2". Nat Commun., 9(1), 4778 (2018).

[3] Shen, T., Xu, L. Li, F., et al. "High mobility monolayer MoS_2 transistors and its charge transport behaviour under E-beam irradiation". *J. Mater. Sci.*, 55(29), 14315-25 (2020).

[4] Salamone, M. M. Li S, Wang S, et al. "Edge-Enriched 2D MoS_2 Thin Films Grown by Chemical Vapor Deposition for Enhanced Catalytic Performance". *ACS Catal.*, 7(1), 877-86(2016).

[5] Li, F., Xu, L. Shen, T., et al. "Strain Improving the Performance of a Flexible Monolayer MoS_2 Photodetector". *Adv. Electron. Mater.*, 5(12), (2019).

[6] Fan, W., Wang, X., Fan, Z. et al. "Substrate modified thermal stability of mono- and few-layer MoS_2". *Nanoscale*, 10(7), 3540-6 (2018).

[7] Qi, J., Bai, F., Li, F. et al. "A High-Performance Self-Powered Photodetector Based on Monolayer MoS_2/Perovskite Heterostructures". *Adv. Mater. Interfaces*, 5(6), (2018).

[8] Li, F., Shen, T., Zhang, Z. et al. "High-Performance Broadband Photodetector Based on Monolayer MoS_2 Hybridized with Environment-Friendly $CuInSe_2$ Quantum Dots". *ACS Appl. Mater. Interfaces*, 12(49), 54927-35 (2022).

[9] Shu, W., Li, Z., Li, Q. et al. "Nondegenerate P-Type In-Doped SnS_2 Monolayer Transistor". *Adv. Electron. Mater.*, 7(7), (2021).

[10] Palacios, P., Reato, E., Uzlu, B. et al. "Zero-Bias Power-Detector Circuits based on MoS_2 Field-Effect Transistors on Wafer-Scale Flexible Substrates". *Adv. Mater.*, 2108469 (2022).

[11] Wang, S., Tang, X., Liang, Y. et al. "High-performance, self-powered flexible MoS_2 photodetectors with asymmetric van der Waals gaps". *Phys. Chem. Chem. Phys.*, 24(12), 7323-30 (2022).

[12] R. J., Ramalingam, Konikkara, N., Al-lohedan, H. et al. "Synthesis of MoS_2 nanoparticle deposited graphene/mesoporous MnOx nanocomposite for high-performance supercapacitor application". *Int. J. Hydrogen Energ.*, 43(36), 17121-31 (2018).

[13] Lee, S., Hwang, J., Kim, D. et al. "Oxygen incorporated in 1T/2H hybrid MoS_2 nanoflowers prepared from molybdenum blue solution for asymmetric supercapacitor applications". *Chem. Eng. J.*, 419 (2021).

[14] Yan, X., Wang, D., Zhou, C. et al. "A free-standing electrode based on 2D SnS_2 nanoplates@ 3D carbon foam for high-performance supercapacitors". *Int. J. of Energy. Res.*, 44(11), 542-54 (2022).

[15] Wang, H., Deng, C., Wang, S. et al "Clarifying the lithium storage behavior of MoS2 with in situ electrochemical impedance spectroscopy". *J. Mater. Chem. A*, 9(28), 15734-43 (2021).

[16] Chen, H., Ke, G., He, J. et al. "Ultrathin MoS_2 anchored on 3D carbon skeleton containing SnS quantum dots as a high-performance anode for advanced lithium-ion batteries". *Chem. Eng. J.*, 403 (2021).

[17] Wang, K., Xu, Y., Yao, Z. et al. "In Situ, Atomic-Resolution Observation of Lithiation and Sodiation of WS_2 Nanoflakes: Implications for Lithium-Ion and Sodium-Ion Batteries". *Small*, 17(24), 2100637 (2021).

[18] S. A., Cui, Z., He, Zhu, J. et al. "Tailoring the Void Space Using Nanoreactors on Carbon Fibers to Confine SnS_2 Nanosheets for Ultrastable Lithium/Sodium-Ion Batteries". *Small Methods*, 2101484 (2021).

[19] Ling, N., Lee, Y., Kim, D. et al. "Heterophase Boundary for Active Hydrogen Evolution in $MoTe_2$". *Adv. Funct. Mater.*, 32(10), (2021).

[20] Gu, Q., Li, Y., Johannessen, B. et al. "Synergistic Pt doping and phase conversion engineering in two-dimensional MoS_2 for efficient hydrogen evolution". *Nano Energy*, 84 (2021).

[21] Gao, M., Sun, L., Jing, Z. et al. "1 T-Phase Enriched P doped WS_2 nanosphere for highly efficient electrochemical hydrogen evolution reaction". *Chem. Eng. J.*, 429 (2022).

[22] Zhao, R., Wu, J., Xiang, H. et al. "Exposing highly active (100) facet on an SnS_2/SnO_2 electrocatalyst to boost efficient hydrogen evolution". *Appl. Catal. B-Environ.*, 292 (2021).

[23] Zhai, P., Chang, Y., Hou, J. et al. "Excellent HER and OER Catalyzing Performance of Se-Vacancies in Defects-Engineered $PtSe_2$: From Simulation to Experiment". *Adv. Energy Mater.*, 12(1), (2021).

[24] Chen, Chang, Y. S., C. Y., Ho, C. J. et al. "Surface electron accumulation and enhanced hydrogen evolution reaction in $MoSe_2$ basal planes". *Nano Energy*, 84 (2021).

[25] Roy, Kovalska, E., P. K., Antonatos, N. et al. "Photocatalytic activity of twist-angle stacked 2D TaS_2". *npj 2D Mater. Appl.*, 5(1), (2021).

[26] Yin, K., Xie, D., Yang, Z. J. et al. "Polarization-perceptual anisotropic two-dimensional ReS_2 neuro-transistor with reconfigurable neuromorphic vision". *Mater. Horiz.*,9(5), 1448-1459 (2022).

[27] Tao, L., Zhou, Y., Chen, Z. et al. "Defect Etching of Phase-Transition-Assisted CVD-Grown 2H-$MoTe_2$". *Small*, 17(32), 2102146 (2021).

[28] Chou, S. A., Yang, P. K., Hou, C. H. et al. "Tin disulfide piezoelectric nanogenerators for biomechanical energy harvesting and intelligent human-robot interface applications". *Nano Energy*, 75 (2022).

[29] S. J., Kim, Park, S., Yun, Y. I. et al. "Tailoring Domain Morphology in Monolayer $NbSe_2$ and WxNb1-xSe_2 Heterostructure". *ACS Nano*, 14(7), 8784-92 (2020).

[30] Gong, Y., Ye, G., Lei, S. et al. "Synthesis of large-scale atomic-layer SnS2 through chemical vapor deposition". *Nano Res.*, 10(7), 2386-94 (2017).

[31] Liu, Y., Zhang, B., Liang, T. et al. "Activating the Basal Plane of Defective SnS_2 Nanosheets for NH_3 Gas Sensing". *ACS Appl. Nano Mater.*, 3(5), 4642-53 (2020).

[32] Behera, S., Das, B., Satpati, B. et al. "Layered SnS_2/porous nickel foil based Schottky junction: An excellent ammonia sensor at room temperature". *J Hazard Mater.*, 428, 128252 (2022).
[33] Lai, S., Xu, J., Hu, M. et al. "Semimetal 1H-SnS_2 Enables High-Efficiency Electroreduction of CO_2 to CO". *Small Methods*, 4(10), (2020).
[34] Kumar, D. P., Rangappa, A. P., Wang, J. et al. "In situ growth of Ag_2S quantum dots on SnS_2 nanosheets with enhanced charge separation efficiency and CO_2 reduction performance". *J. Mater. Chem. A*, 10(13), 7291-7299 (2022).
[35] Zhao X., Yin, S., Jiang, E. et al. "Boosting water decomposition by sulfur vacancies for efficient CO_2 photoreduction". *Energ. Environ. Sci.*, 15(4), 1556-1562 (2022).
[36] Zhang, P., Xu, L., Jiang, H. et al. Boosting water decomposition Large-Scale Growth and Field-Effect Transistors Electrical Engineering of Atomic-Layer SnS_2". *Small*, 15(46), 1904116 (2019).
[37] Wang, Z. L., Lv, Y. R., Yang, Y. X. et al. "Tin bisulfide nanoplates anchored onto flower-like bismuth tungstate nanosheets for enhancement in the photocatalytic degradation of organic pollutant". *J Hazard Mater.*, 432, 128665 (2022).
[38] Wang, X., Chen, Y., Lao, M. et al. "Electrocatalytically inactive SnS_2 promotes water adsorption/dissociation on molybdenum dichalcogenides for accelerated alkaline hydrogen evolution". *Nano Energy*, 64 (2019).
[39] Xue, X. X., Shao, G., Zhou. X. et al. "Shape-Engineered Synthesis of Atomically Thin 1T-SnS_2 Catalyzed by Potassium Halides". *ACS Nano*, 13(7), 8265-74 (2019).
[40] Chen, M., Wang, B., Lv, J. et al. "Improved hydrogen evolution with SnS_2 quantum dot-incorporated black Si photocathode". *Dalton T.*, 50(38), 13329-36 (2021).
[41] Masroor, M., Giri, B., Yan, T. et al. "Balancing Light Absorption and Charge Transport in Vertical SnS_2 Nanoflake Photoanodes with Stepped Layers and Large Intrinsic Mobility". *Adv. Energy. Mater.*, 9(31), (2019).
[42] Xia, C., Liu, J., Li, H. et al. "High on/off ratio photosensitive field-effect transistors based on few-layer SnS_2". *Nanotechnology*, 27(34), 34LT01 (2016).
[43] Wang, L., Gao, P., Zhang, Y. Y. et al. "High-Resolution Tracking Asymmetric Lithium Insertion and Extraction and Local Structure Ordering in SnS_2". *Nano Lett.*, 16(9), 5582-8 (2016).
[44] Lin, Y. C., Li, S., Zhao, W. et al. "Vapour-liquid-solid growth of monolayer MoS_2 nanoribbons". *Nat. Mater.*, 17(6), 535-42 (2018).

Degradation Behavior of High Power Semiconductor Modules by Low-Temperature Sintering Technology

Shancan Fu[1,*], Guoliang Wang[1], and Feng Xiao[1]

[1] College of Engineering and Technology, Tianjin Agriculture University, Tianjin, PR China

[*] Corresponding author. *E-mail address*: fushancan@163.com

Abstract. This paper presents a degradation behavior of high-power semiconductor modules, i.e., IGBT, with sintered silver by power cycle test and thermal cycle test. The electrical characteristics, such as on-state voltage drop, thermal resistance and leakage current, were studied. For power cycle test, the IGBT module still has an acceptable electrical performance after the 30 K power cycles. No noticeable changes can be found in the die-attach layer, while the DBC-attach layer begins to fail after 30 K power cycles. For the thermal cycling test, the electrical characteristics of the IGBT module were only tested within the first 300 cycles because the Al_2O_3 DBC substrate failed first before the failure of the bonding layers. Finally, finite element methods (FEM) were used to analyze thermal and mechanical stress distribution and failure mechanism of the bonding layers.

1. Introduction

Insulated gate bipolar transistor module has been widely used to various fields of high power applications because of their attractive properties, such as high-current handling capability and high breakdown resistance [1-3]. However, IGBT module still encounter many challenges with demand of high degree integration of electronic products. In addition, the high voltage and high current of IGBT modules also cause large power consumption. Therefore, junction temperature of high power semiconductor device is rising faster and faster. For example, wide bandgap semiconductor devices, e.g., SiC and GaN, have the intrinsic high-temperature capability and can theoretically work as high as 500°C [4].

Although the traditional lead-free die-attach materials are widely used in electronic packaging, these materials cannot meet demand of high temperature and high density semiconductor modules. Therefore, people have to apply lead-based solders to these modules, but these devices that contain lead can seriously pollute the environment. Nowadays, nanosilver paste [5] has received more and more attention because of its excellent performance, such as lower sintering temperature, higher melting point and thermal conductivity. It is also used as a replacement for lead-free or leaded chip connection materials in high-temperature applications.

In the last decades, researchers have done a lot of work about the degradation of IGBT modules using traditional lead-free or leaded chip connection materials, such as Sn-Ag and Sn-Pb solders. The degradation of semiconductor devices mainly related to thermo-mechanical stresses which are result of thermal expansion deformation [6]. During the operation, the junction temperature mainly depends on the power losses of IGBT modules. Different output power leads to different power loss so this module endures temperature cycles. As we know, the IGBT module has a multilayer structure, which is fabricated by different kinds of materials, such as silicon chips, chip connection material, and copper based DBC substrates. With fluctuation of temperature, these materials are easily subjected to thermal and mechanical stress concentration caused by different thermal expansion coefficients of their connection layers [7]. As a result, the cyclic stress can break connection layers in the semiconductor module, leading to the fatigue of entire modules potentially.

However, the degradation of the semiconductor module with sintered silver as chip connection material still lacks detailed data analysis. This paper presents a detailed analysis of this novel IGBT module by power cycle test and thermal cycle test. The electrical characteristics, such as leakage current, thermal resistance and on-state voltage drop, were measured. Thermal resistance of semiconductor module was detailed discussed at different thermal cycles. Finally, FEM was adopted to analyze thermal and mechanical stress distribution and failure mechanism of the bonding layers, i.e., the die-attach layer and DBC-attach layer.

2. Materials and methods

The 1200 V/150 A IGBTchip and 1200 V/150 A diode were used to high power IGBT device. The DBC substrates were coated with silver film. Firstly, one IGBT chip and one diode chip were attached to a DBC substrate by sintered nanosilver. Secondly, the emitter path between the substrate and the chips was connected by Al based wire bonding. Thirdly, we welded two DBC substrates to one baseplate by soldered SAC305. Figure 1 shows the cross-section of a 1200 V/150 A IGBT semiconductor module with sintered silver.

2.1. Finite element analysis model

The purpose of using FEM is to study the effect of thermal cycle test and power cycle test on thermal and mechanical stress distribution in bonding layers. Chip connection layer was made of sintered silver, while the DBC-attach layer was made of soldered SAC305. The geometry of the module is a multilayered structure, which consists of the silicon chip, die-attach layer, Al_2O_3 DBC substrate, DBC-attach layer, and copper baseplate. In order to calculate conveniently, a quarter 3-D model for thermal and mechanical coupling analysis was created by ANSYS software, which is shown in Figure 2. Sintered nanosilver and soldered SAC305 are thus calculated by a viscoplastic model based on the Anand model, while remaining materials are set as isotropic elastic material. The parameters of the bonding layers were listed in Table 1.

Table 1. Anand module parameters of sintered silver(I) and soldered SAC305 (II).

Set	S_0 (MPa)	Q/R (K)	A (s^{-1})	ξ	m	h_0 (MPa)	$\hat{S}$ (MPa)	n	a
I	2.93	5706.3	9.81	12	0.6572	14600	101.7	0.003	1.0
II	45.9	7460	5.87	2	0.0942	9350	58.3	0.015	1.5

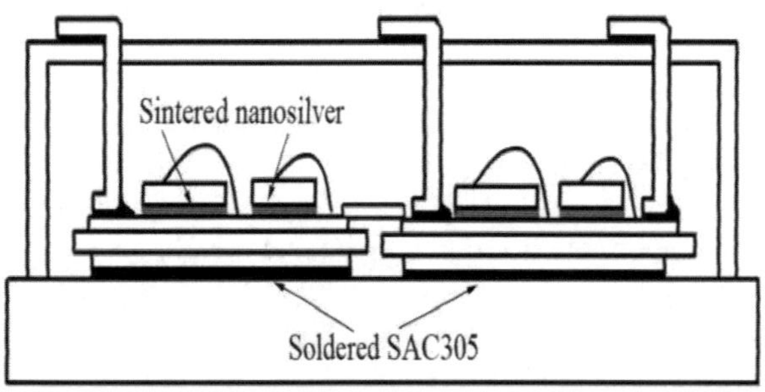

Figure 1. Cross section structure of the IGBT module.

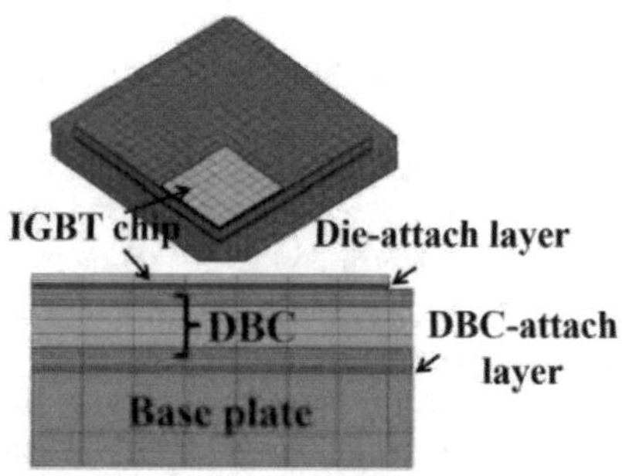

Figure 2. A quarter 3-D model of the IGBT module.

2.2. Power cycle test and thermal cycle test

For the power cycle test, power loss is generally output by power device itself. Each power cycle lasts a longer time, such as 5 minutes, so that there is ample time to transfer lost heat from the power chip to baseplate. Figure 3 shows the heat transfer process of the IGBT module. The heat begins to diffuse from power chip to baseplate through bonding layers. The collector current injected in the module was 150 A, while the collector-emitter voltage (V_{ce}) was kept between 1.8 V and 2.0 V during the test. We chose a power chip of 236 W with the purpose of setting junction temperature change (ΔT_{junc}) to 100°C. Case temperature (T_{case}) was obtained by means of thermocouple measurements. The maximum junction temperature (T_{junc}, max) was calculated as follows:

$$T_{junc} = \frac{\Delta V_{ce}}{k} + T_{case} = \frac{V_{ce\_junc} - V_{ce\_case}}{k} + T_{case} \quad (1)$$

where V_{ce_junc} is the voltage between collector and emitter at maximum junction temperature, and V_{ce_case} is the voltage between the collector and emitter at minimum temperature of the cooling plate. k is a fixed value. The junction temperature varies from 25°C to 125°C, and the case temperature ranges from 25°C to 105°C. For the thermal cycling test, the IGBT modules were thermally cycled in a thermal cycling chamber. Figure 4 shows the temperature profile for the thermal cycling test. The minimum and maximum temperature conditions are fixed at -55°C and 150°C, respectively. Each cycle lasts for about 10 min, and the hot and cold temperature duration time is 30 min.

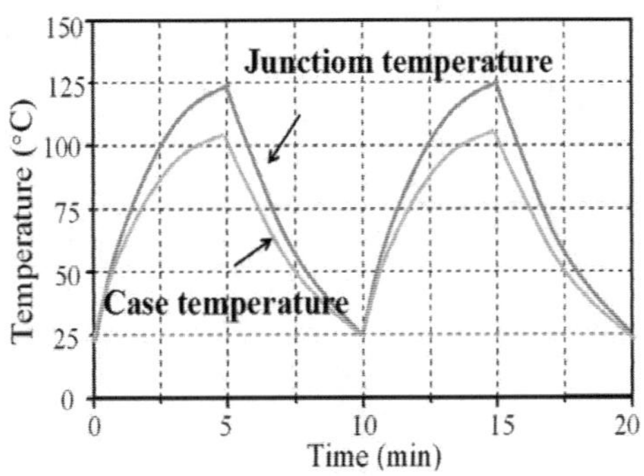

Figure 3. Schematic for the power cycling test.

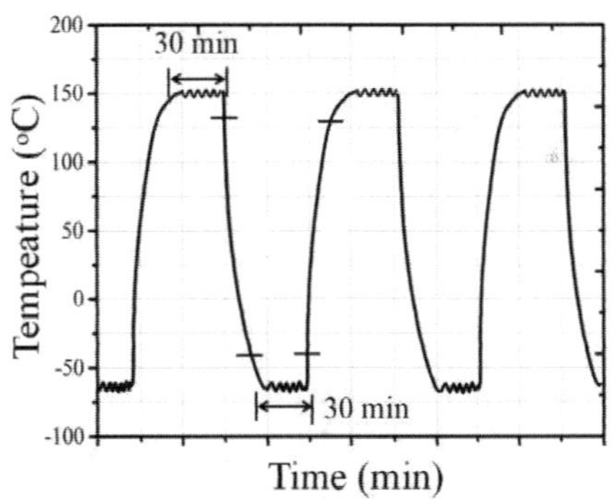

Figure 4. Temperature profile for the thermal cycling test.

3. Results and discussion

3.1. Power cycle test

To study the effect of power cycle test on high power IGBT module with sintered silver chip connection material, basic static electrical characteristics, such as leakage current and on-state voltage drop, were detailed studied. Figure 5 shows the on-state voltage drop of high power IGBT module versus power cycles. The test results show that on-state voltage drop of high power IGBT also slightly increases with number of power cycles. The reason for increase of on-state voltage drop may be due to increase of thermal resistance of high power module [8]. Fortunately, the on-state voltage drop still has not exceeded the threshold limit of 5%, indicating that the IGBT chip can still work in the long-term running. In addition, the power cycling test has led to a small increase in leakage current because of decrease of the collector-emitter resistance. As the power cycles, the IGBT chips were heated by power losses generated in power semiconductor module themselves. As result, the collector-emitter resistance decreases with increasing the junction temperature because the number of electrons through the gap depends on the temperature and the specific intrinsic material [9].

Figure 6 shows the equivalent plastic strain accumulated over the number of power cycles at the corner of bonding layers. When the power module is switched on and off, cyclic strain will appear in connection layer, which is the main factor that determines lifetime of power semiconductor modules [10]. After the fifth power cycling, the equivalent plastic strain of the DBC-attach layer increases to 6×10^{-4}, which is 14% higher than that of the die-attach layer. It seems that the sintered nanosilver can be considered as an excellent chip connection material for power semiconductor modules with long-term reliability.

No noticeable changes can be found in the die-attach layer, while the DBC-attach layer begins to fail after 30 K power cycles. It is indicated that the sintered nanosilver as a die-attach layer shows excellent long-term mechanical integrity in the long-term run. The failure of IGBT module should likely be mainly related to the DBC-attach layer. The fatigue of bonding layers also promotes an increase in thermal resistance. After 30K power cycles, the thermal resistance increases to 0.243°C/W, which is 5% higher than that of the initial stage. For the DBC-attach layer, the voids have a negative effect influence on the large-area solder joints due to the formation of intermetallic compounds. For the die-attach layer, however, void growth may be possibly stopped by the pores among sintered silver particles.

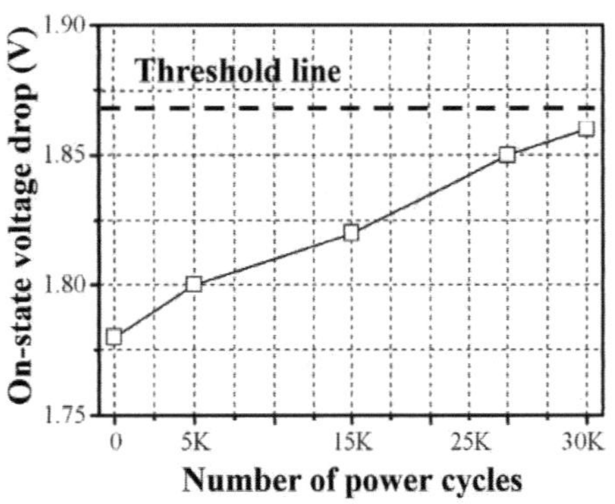

Figure 5. The on-state voltage drop of the IGBT module by power cycles.

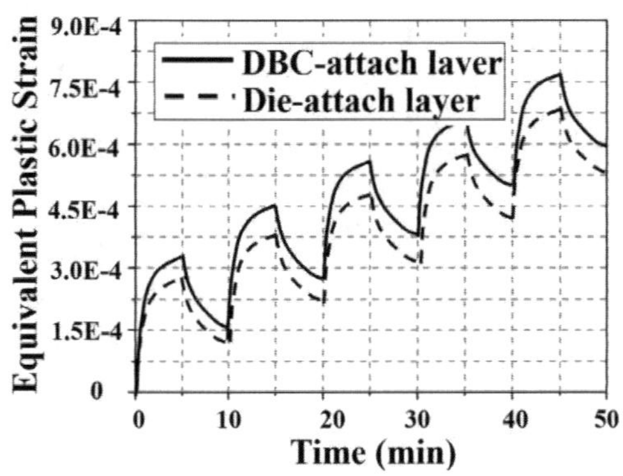

Figure 6. Equivalent plastic strain accumulated by power cycles.

3.2. Thermal cycle test

In thermal cycle test, the electrical characteristics of the IGBT module were only tested within the first 300 cycles because the Al_2O_3 DBC substrate failed first before the failure of the bonding layers. The failure of the Al_2O_3 DBC substrate may be due to the oxidation reaction at the interface between alumina and copper laminates [11]. In our further work, we will try to use an improved DBC substrate that is covered by a CTE-matched encapsulant layer with high thermal reliability [12]. Figure 7 shows the leakage current and on-state voltage drop of high power semiconductor module versus number of thermal cycles. After 300 thermal cycles, the leakage current increases to 0.75 μA, while the on-state voltage drop increases to 1.84 V. As expected, the increase of on-state voltage drop does not exceed threshold limit of 5%.

Figure 8 shows the thermal resistance versus the number of thermal cycles. The thermal resistance is 0.251°C/W at the 300th thermal cycle, which is 8% higher than that of the initial state. The high thermal resistance causes higher temperature fluctuations within power semiconductor devices [13]. Extra heat can cause rapid void growth and the formation of cracks in the bonding layer due to the thermal and mechanical stress. With a constant power loss generated by the power device, the bonding layers easily fail and deteriorate the heat dissipation, resulting in the degradation of the entire module.

The cyclic thermo-mechanical stress and strain of the bonding layers were also analyzed by FEM during the thermal cycling test. Figure 9 shows the stress distribution between the die-attach layer and the DBC-attach layer. After the 100th power cycle, the maximum von Mises stress of DBC attachment increases by 1.02 times, while one of the die-attach layers increases only by 0.056 times. The reason may be that cyclic thermo-mechanical stresses play important role in the formation of voids. Large thermo-mechanical stresses may break the connection layer, especially the interface between soldered SAC305 and substrate. Voids have also been considered the defects in electronic assembly, which may degrade the mechanical robustness of bonding layers and consequently affect the lifetime of power semiconductor module.

Figure 10 shows that the von Mises strain in connection layer is greatly influenced by number of thermal cycles. Maximum von Mises strain increases in thermal cycle test. After 100 cycles of thermal cycling, maximum von Mises strain of DBC-attach layer is 1.9 times higher than the one of the die-attach layer. Maximum von Mises strain of DBC attachment is 2.2% at the 100th cycle. It can be indicated that the cracks may first occur at the corners of the DBC-attach layer. Microcracks at the interface between connection layer and substrate may lead to rapid crack propagation in the connection layer [14].

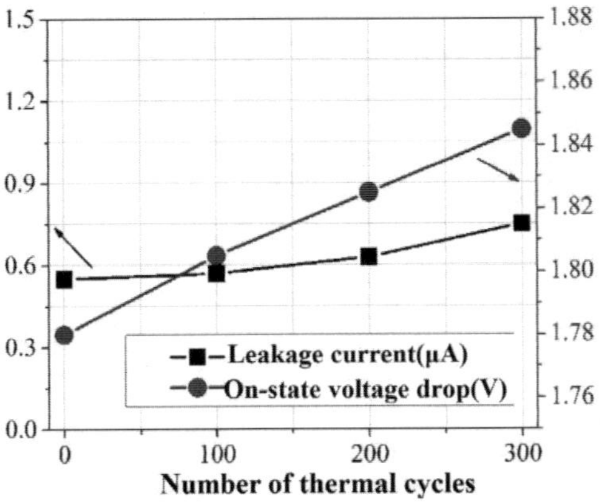

Figure 7. Leakage current and on-state voltage drop versus thermal cycles.

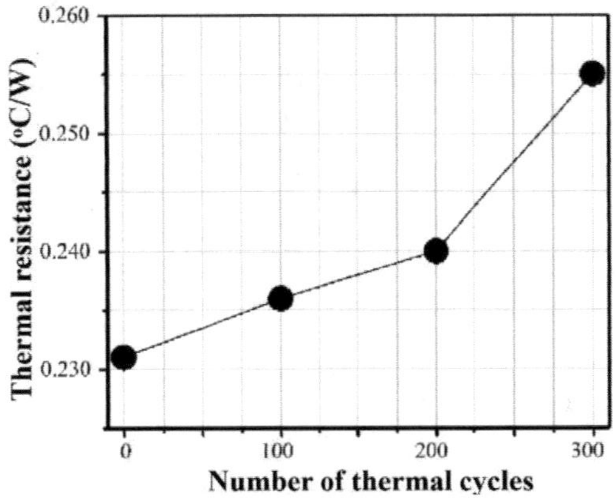

Figure 8. Thermal resistance at different thermal cycles.

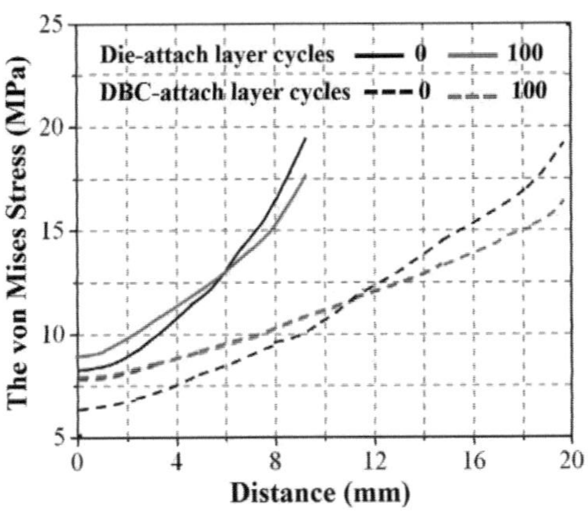

Figure 9. Von-Mises stress of the boning layers by thermal cycles.

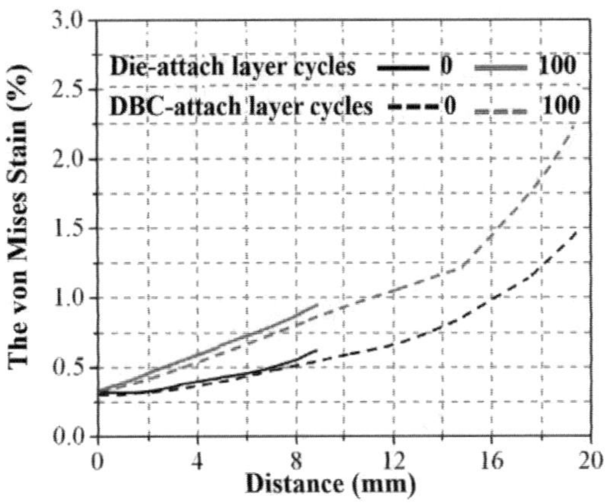

Figure 10. Total von Mises strain of the bonding layers by thermal cycles.

4. Conclusion

We investigated the degradation behavior of high power IGBT module with sintered silver by power cycle test and thermal cycle test. The electrical characteristics of high power IGBT module, such as leakage current, thermal resistance and on-state voltage drop, were studied. For power cycle test, the module still has an acceptable electrical performance after the 30 K power cycles. There is no noticeable change that can be found in chip connection layer, while DBC-attach layer begins to fail after 30 K power cycles. FEM was used to evaluate thermo-mechanical stress and strain of connection layers. Large cyclic stress of the DBC-attach layer may be the reason for the failure of bonding layers. For the thermal cycling test, the Al_2O_3 DBC substrate failed first before the failure of the bonding layers after 300 thermal cycles. The equivalent plastic strain of the DBC-attach layer is 4.1×10^{-3}, which is 17% higher than that of the die-attach layer. This study provides theoretical guidance for the application of an IGBT module with nanosilver paste in high-temperature packaging.

Acknowledgments
The authors are grateful for the financial support of Scientific Research Project of the Tianjin Education Commission (No. 2018KJ187).

References

[1] Richard M, Chen L, Pickemt V, Naayagi T and Ji B 2019 *IEEE T. Power Electr.* **34** 9158
[2] Yang Y Y, Zhang Q H and Zhang P J 2021 *IEEE T. Ind. Appl.* **57** 685
[3] Moussodji J, Kociniewski T and Khatir Z 2013 *Microelectron. Reliab.* **53** 1728
[4] Mohsenzade S, Zarghani M and Kaboli S 2021 *IEEE T. Ind. Electron.* **68** 7900
[5] Wissam S, Stéphane A, Cyril B, Régis M and Eric W 2013 *Microelectron. Reliab.* **53** 1620
[6] Khatir Z and Lefebvre S 2004 *Microelectron. Reliab.* **44** 935
[7] Smet V, Forest F and Lefebvre S 2011 *IEEE T. Ind. Electron.* **58** 4935
[8] Jiang J, Dang M and Xu P 2006 *Appl. Phys. Lett.* **89** 182902
[9] Abdi B, Ranjbar H, Gharehpetian B and Milimonfared J 2009 *IEEE T. Ind. Electron.* **56** 2135
[10].Gholini M, Ranjbar A A, Javidan M and Hosseinpour AA 2021 *Energy Rep.* **7** 6850
[11] Xu L, Wang C, Zhou Y, Qian F and Liu S 2016 *Microelectron. Reliab.* **56** 108
[12] Ning P Q, Lai R X and Huff D 2010 *IEEE T. Power Electron.* **25** 20
[13] Zheng G, Ngo T and Lu G Q 2015 *Microelectron. Reliab.* **55** 2581
[14] Ma M Y, Guo W S, Zhan M Y, Yang S Y and Li F 2020 *Energy Rep.* **9** 558

Transient characteristics emulated in modified graphene-oxide solid electrolyte gated synaptic transistors

Hong-Fang Qi*, Juan Wen

School of Intelligent Manufacturing, Wuhan Huaxia Institute of Technology, Wuhan, 430223, China

*765065675@qq.com

Abstract. To develop neuromorphic computational systems and achieve artificial intelligence, it is crucial to realize neuromorphic devices that mimic basic neural behavior. The electrolyte-gated transistors (EGTs) were fabricated using a 3-triethoxysilylpropylamine modified graphene-oxide (KH550-GO) electrolyte. The transistor exhibited good performance owing to the EDL effect of the KH550-GO electrolyte. A resistor-load inverter was also achieved with a high voltage gain of 12.4, and have low operating voltage. Furthermore, the crucial short-term synaptic plasticity(STP) functions, including excitatory post-synaptic current (EPSC) and high-pass filtering, were realized. More importantly, adaptation was emulated in our devices.

1. Introduction

A human brain includes hundreds of billions of neurons connected by synapses, giving it unique computational capabilities, such as high-speed computations, low power consumption, and fault-tolerant characteristics[1-2]. In the signal processing process, two neurons are connected by synapses, and synaptic weights is critical to synaptic activity. Ionic fluxes in neurons and synapses are responsible for these neural activities[3-4]. Bioinspired devices simulating synaptic plasticity have garnered momentous research attention over the past few years. In pursuit of this direction, several important synaptic functions were mimicked using material systems based on two-terminal memristors[5] and three-terminal transistors[6-7]. Furhermore, EGTs provide promising modulation methods for channel carrier transport, which are analogous to ionic flux in the synapses. Thus, EGTs are useful devices for artificial synapses[8-9]. Recently, oxide transistors gated by methyl cellulose were prepared for the applications of the artificial synapse. Even more recently, EGTs based on KH550-GO electrolytes have been proposed as artificial synapse transistors for exploiting excellent performances of synapse transistors[11].

In this paper, EGTs gated by the KH550-GO electrolyte have been proposed, which have good proton conductivity due to the hydrolysis reaction of $-SiOC_2H_5$ bonds to Si-OH bonds. The synaptic transistors exhibited good transistor performances owing to the extremely high EDL capacitance. In addition, a resistance loaded inverter with high voltage gain was demonstrated. A detailed discussion of neuromorphic functions was conducted, consisting of EPSC and high-pass filtering. In addition, adaptation was discovered in our devices. The synaptic transistor based on the KH550-GO realizes the transient characteristics, which paves the way for the development of artificial intelligence chip.

2. Experimental Section

First, GO powders were dispersed in dimethylformamide (2 mg/mL)GO solution. The KH550 solution was mixed in GO solution(1:20), which was mixed with dilute ethyl alcohol at room temperature.

After 2 h, the KH550-GO solution was successfully prepared. The KH550-GO solid electrolyte film was spin-coated on ITO class. Then, patterned ITO electrodes were sputtered onto the KH550-GO film. It was possible to self-assemble a thin ITO channel between two patterned electrodes. KH550-GO gated synaptic transistors were then obtained using this technique, as observed in Figure1(a). The ITO electrodes and self-assembled ITO channel layer can be considered a pre-synapse; the EPSC in the self-assembled ITO channel layer can be acted as synaptic weight; the ITO bottle electrodes act as post-synapse. Thus, an artificial synapse was demonstrated, as expressed in Figure1(b). The capacitance and impedence of the KH550-GO devices were measured by a Solartron 1260 impedance analyzer. Using Keithley 4200SCS semiconductor parameter analyzer, the transient characteristics of the ITO synaptic transistors were texted. During measurement, the voltage and spike stimuli bring to bear on the ITO bottle electrodes, and the EPSC can be detected in the drain electrodes with the grounded source electrodes. All experimental data were proposed in the following figures by origin.

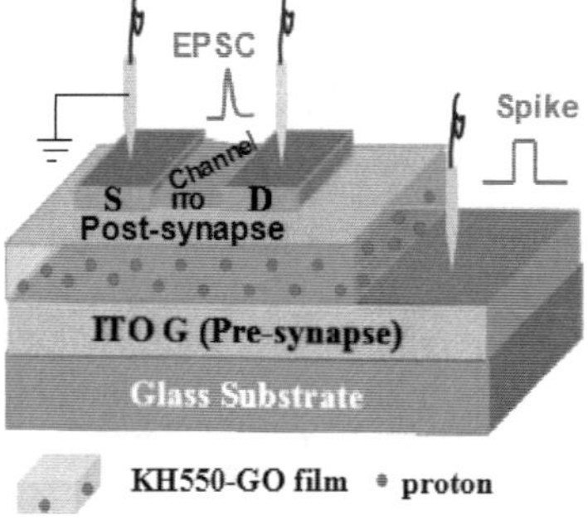

(a)The schematic diagram of ITO synaptic transistor gated by KH550-GO electrolyte.

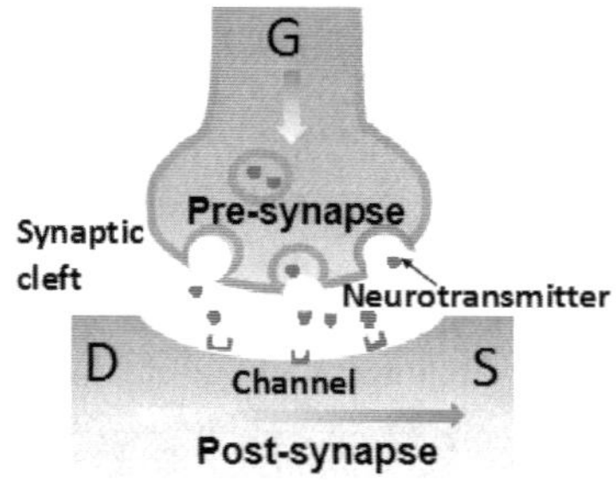

(b)The schematic diagram of a biological synapse

Figure 1. The schematic diagram of the experiment

3. Results and Discussions

Figure 2(a) expresses that the capacitance of the sandwich structure varies with the frequency. The illustration in Figure 2(a) indicates the structure diagram during measurement. The voltage was applied to the ITO bottle electrodes, and the channel current can be detected in the ITO top electrodes (drain electrodes). A high capacitance (C) observed at 1.0 Hz is ~3.1 μF/cm^2. It can be expressed by the formation of EDL at the KH550-GO/ITO junction. Figure 2(b) indicates the impedance spectroscopy of the KH550-GO film. When the y-coordinate is 0, the actual impedance value (R) is estimated to be about 463 Ω. In addition, proton conductivity(σ) can be obtained according to the formula:

$$\sigma = \frac{D}{A(R - R_0)} \tag{1}$$

where D is the thickness of the function lay (5 μm), A is the ITO electrode surface area (1.5 × 10^{-3} cm^2), and R_0 is the ITO electrodes' resistance (30 Ω). Thus, σ is calculated at 7.7 × 10^{-4} S/cm. A high proton conductivity plays an important role in EDL formation. Figure 2(c) expresses the I_{ds}-V_{gs} transfer curve with a V_{read} of 1.5 V. During sweeping, the KH550-GO solid electrolyte displays a 0.2 V hysteresis window on account of its presence of mobile protons. The device reveals excellent electrical properties consisting of a large $I_{on/off}$ of 1.0 × 10^6 and a low subthreshold swing(SS) of 89.8 mV/dec.According to the x-axis intercept of the square root of the I_{ds}-V_{gs} curve, the threshold voltage (V_{th}) of 0.09 V can be obtained. Therefore, the migration rate (μ) in the saturation region is obtained by the formula:

$$I_{ds} = \mu \frac{CW}{2L} (V_{gs} - V_{th})^2 \tag{2}$$

where L (80 μm) is the channel width, W (1000 μm) is the channel length, and C is the specific capacitance. The μ is calculated to be ~29.9 $cm^2/(V\bullet s)$. Figure 2(d) indicates the output characteristics with V_{gs} range from -0.2 to 1 V. When V_{ds} is low, the curves are close to linear, demonstating that the interface of the device is Ohmic contact. However, when V_{ds} is high, the I_{ds} of the ITO transistor is becoming too saturated. On the whole, the transistor exhibits good electrical characteristics.

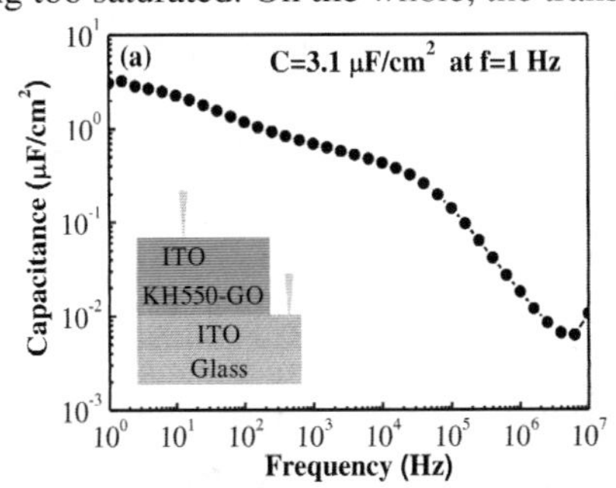

(a) Frequency-dependent specific capacitance of the KH550-GO electrolyte film. Inset: The schematic diagram of the testing structure.

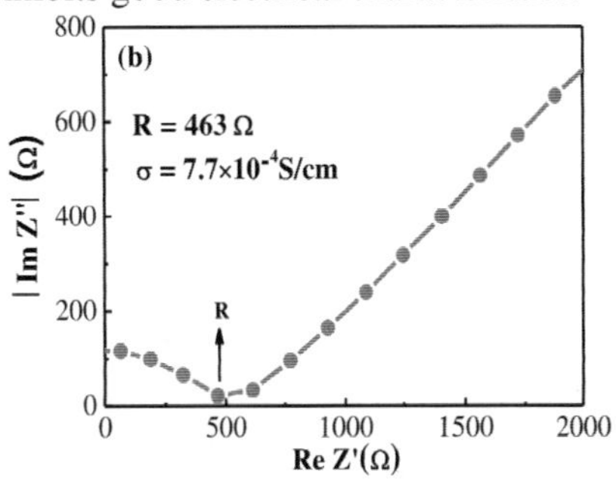

(b) Cole-Cole plot of KH550-GO electrolyte film.

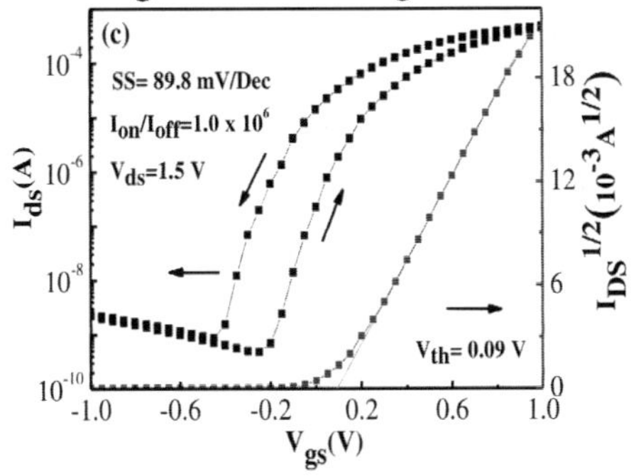

(c) Transfer curves (I_{ds} vs V_{gs}) of synaptic gated by KH550-GO electrolyte

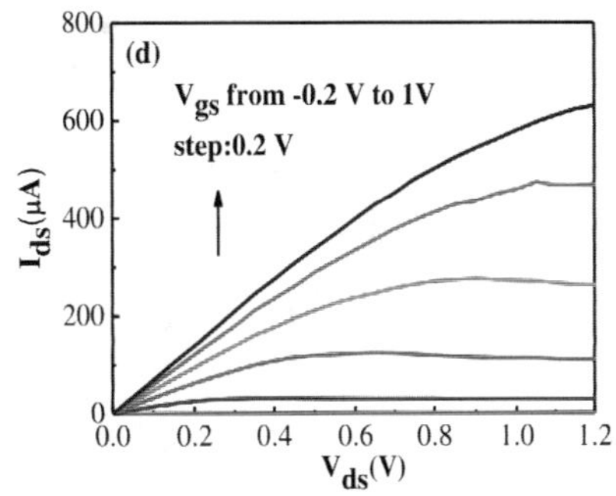

(d) Output curves(I_{ds} vs V_{ds}) of synaptic transistor gated by KH550-GO electrolyte

Figure 2. The electrical properties of KH550-GO solid electrolyte film

In series with a 4.0-MΩ resistor, this transistor can act as a resistor-loaded inverter, as observed in the illustration in Figure 3(a). Figure 3(a) indicates the transfer performances of this inverter. The V_{DD} ranges from 0.5 to 2 V with the V_{in} which was applied on the ITO bottle electrodes scanning from −0.2 to 0.6 V in steps of 0.02 V. It is noteworthy that obvious inverter responses are obtained. When V_{in} does not have the V_{th} of the ITO transistor, the current flows through the ITO transistor are minimal. The working state of the ITO transistor is off state, and the ITO transistor bears most of the voltage. The output voltage (V_{out}) which is detected in the drain electrodes, changes to a "high" level. In contrast, when V_{in} exceeds V_{th}, the current through the ITO transistor increases suddenly, V_{out} switch to a "low" level. However, the V_{out} switch near V_{in} = 0V, and the transition width that the inverter needs to be completely switched increases with increasing V_{DD}, as exhibited in the illustration in Figure 3(a). The voltage gain (-dV_{out}/dV_{in}) increases from 6.2 to 24 V when V_{DD} is varied from 0.5 to 2 V. Figure 3(b) reveals the dynamic action of the resistor-loaded inverter, when V_{DD} is 1.0 V. It can be observed that the device has good inverter performance and pulse respond performances applying square pulse (-1 V/1 V, 5 Hz). A relaxation time is observed in V_{out} when V_{in} switches between −1 and 1 V. Dynamic actions of the inverter at a higher frequency are obtained (results not shown here). The dynamic actions become weaker with the increased frequency. The inherent limitations of proton migration limit the operating speed of our EGT. Therefore, low frequency is suitable for this inverter operation and is inappropriate for traditional logic circuit applications. It is interesting to notice that the frequency of synaptic responses in biological nervous networks ranges from several hertz to tens of hertz. Thus, the proton migration in our EGT is favorable for synaptic device applications. With the proton migrations, STPs are excepted to be mimicked.

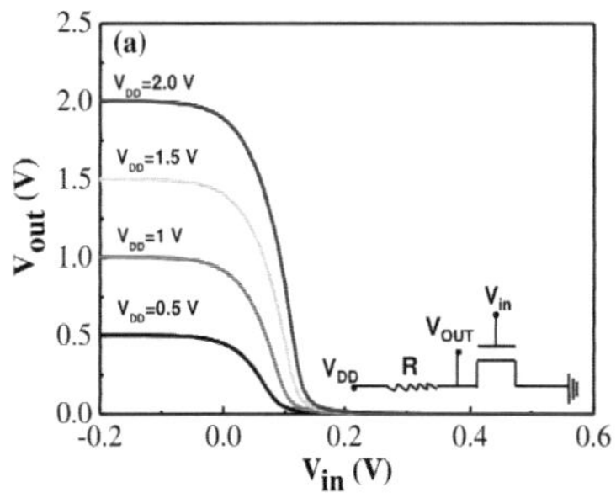

(a) The static DC voltage transfer characteristic (VTC) curves of the resistor-loaded inverter

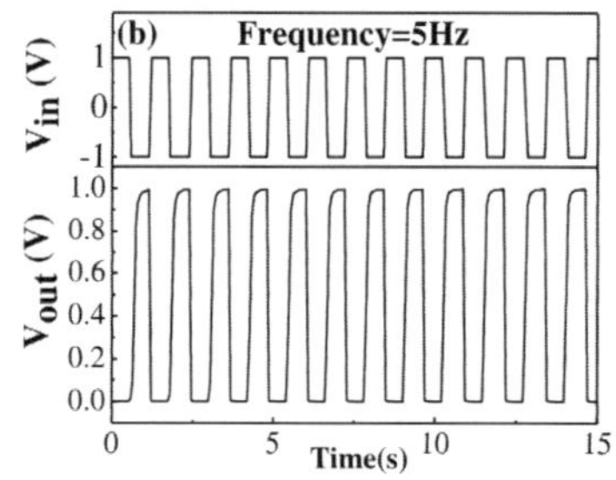

(b) The dynamic action of the resistor-loaded inverter

Figure 3. The performance of the resistor-loaded inverter

In a biological synapse, a neuron transports its information via an action potential. An external spike will trigger a change in the membrane potential, which will gradually decay back to the resting state after the spike. The variation in the external spike can also affect the peak value of the membrane potential. Therefore, the biological synapse can accomplish signal transmission. Similarly, The ITO synaptic transistors are proposed to simulate this characteristic. An EPSC response is observed after appling a input pulse (0.5 V, 10 ms), as shown in Figure 4(a) inset. Furthermore, when the spike amplitude ranges from 0.5 to 2.5 V, when spike width is fixed to 10 ms, the absolute EPSC values(A) extracted from EPSC responses enlarge from 6.8 to 343.1 μA, as expressed in Figure 4(a). Figure 4(b) illuminates a series of EPSC responses to several pulse voltages with different spike widths with a spike amplitude of 0.5 V. The absolute EPSC(A) augments from ~7.57 to ~23.98 μA when the width varies from 10 to 200 ms. These results indicate that the EPSC responses depend on the spike amplitude and width. It can be explained that more protons will move toward and accumulate on the KH550-GO/ITO channel interface with a larger spike amplitude and width.

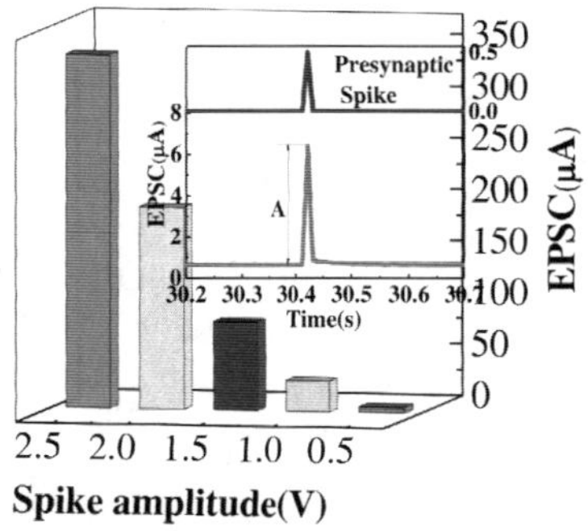

(a) Spike amplitude dependent absolute value of the EPSC

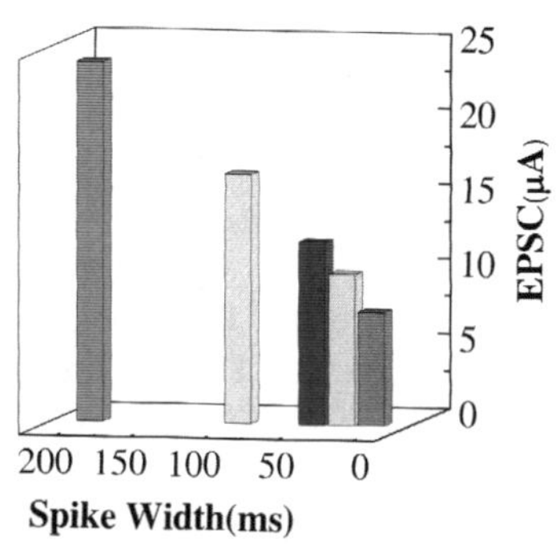

(b) Spike width-dependent absolute value of the EPSC

Figure 4. The synaptic plasticities

In neuroscience, the synaptic plasticities indicate the synaptic efficacy changes on a short timescale and endow biological synapses with information-processing capacities. According to different signal frequencies, synapses can count as dynamic filters for information preprocessing. Temporally coupled short-term synaptic depression is useful to low-pass filtering and short-term synaptic facilitation is good for high-pass filtering. Because PPF behaviors were observed in our previous synaptic transistors, high-pass temporal filtering is expected[11]. The inset in Figure 5 illustrates the EPSCs of the oxide synaptic transistor applying pulses with different frequencies on the pre-synapse. Each pulse includes five pulses (0.5 V, 20 ms). The EPSCs augment with the increased number of stimuli due to short-term facilitation. Furthermore, the gain (A5/A1), defined as the ratio of the first absolute EPSC amplitude (A1) and fifth absolute EPSC amplitude (A5), depends on the frequency, as observed in Figure 5. Distinctly, the gain augments from ~1.1 to ~1.7 when the frequency increases from 2 to 33 Hz. Such behavior is related to the accumulative effect of the protons when the mobile protons have insufficient time to move to the equilibrium position with the increasing stimulus frequency. Therefore, synaptic transistors can realize dynamic high pass filters.

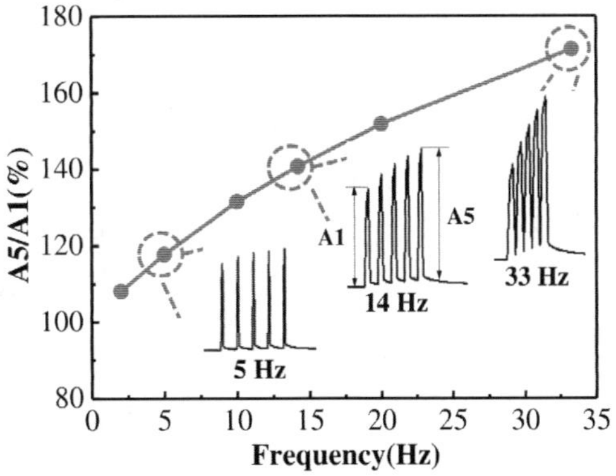

Figure 5. The EPSCs of the oxide synaptic transistor

Sensory adaptation is the response of the sensory system to continuous stimuli over time. When perceiving strong and constant stimuli, the sensory neuron immediately responds and expresses gradual adaption to persistent stimuli. It is common in visual, auditory, and tactile sense. The ITO synapses are expected to mimic sensory adaptation after realizing STP. In this case, the EPSC response is measured by applying 100 trains of stimulis (0.5 V, 10 ms, 5 Hz), as observed in Figure 6. We can observe that EPSC amplitude A0 initially responds to repeated input stimuli and gradually adapts after a few (10-12) pulses. It accounts for the limited protons in the gate dielectric. After removing the pulse train, the EPSCs recover to the initial state. Hence, the adaptation is STM.

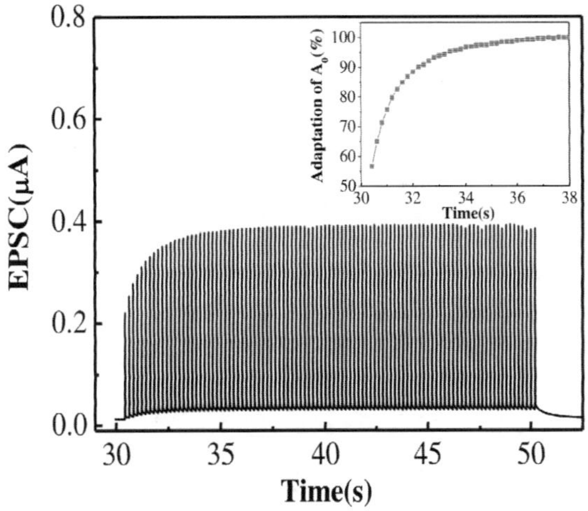

Figure 6. The EPSC response is measured by applying 100 trains of pulses

4. Conclusions

In summary, EGT based on KH550-GO solid electrolytes has been fabricated with a high $I_{on/off}$ ratio of ~1.0×10^6 and high σ of 29.9 $cm^2/(V{\cdot}s)$. A resistor-loaded inverter was realized by connecting this device and a *R* in series with a high voltage gain of 12.4. In addition, the basic STP functions, such as EPSC and high-pass filtering, were realized. Especially, adaptation was mimicked in our devices. The imitation of transient characteristics in ITO-based synaptic transistors is crucial to the hardware implementation of neuromorphic computation.

Acknowledgments
This work was supported by the Natural Science Foundation of Hubei Province (2019CFC911).

References

[1] D. B. Chklovskii, "Synaptic connectivity and neuronal morphology: two sides of the same coin," Neuron, Vol. 43, No. 5, pp. 609-617, Sep. 2004. DOI: 10.1016/j.neuron.2004.08.012
[2] D. V. Buonomano and W. Maass, "State-dependent computations: spatiotemporal processing in cortical networks," Nat. Rev. Neurosci., Vol. 10, No. 2, pp. 113-125, Jan. 2009. DOI: 10.1038/nrn2558
[3] P. Gkoupidenis, N. Schaefer, X. Strakosas, J. A. Fairfield and G. G. Malliara, "Synaptic plasticity functions in an organic electrochemical transistor," Appl. Phys. Lett., Vol. 107, No. 26, pp. 1156-1157, Dec. 2015. DOI:10.1063/1.4938553
[4] G. Voglis and N. Tavernarakis, "The role of synaptic ion channels in synaptic plasticity," EMBO Rep., Vol. 7, No. 11, pp. 1104-1110, Nov. 2006. DOI: 10.1038/sj.embor.7400830
[5] Z. G. Xiao and J. S. Huang, "Energy-efficient hybrid perovskite memristors and synaptic devices" Adv. Electron. Mater. Vol. 2, No. 7, pp. 1600100, 2016. DOI: 10.1002/aelm.201600100
[6] F. Alibart, D. Vuilaume, C. Novembre, D. Guerin, S. Pleutin, S. Lenfant, K. Lmimouni, and C. Gamrat, "An organic nanoparticle transistor behaving as a biological spiking synapse," Adv. Funct. Mater., Vol. 20, No. 2, pp. 330-337, Jan. 2010. DOI: 10.1002/adfm.200901335
[7] Y. Kaneko, Y. Nishitani and M. Ueda, "Ferroelectric artificial synapses for recognition of a multi-shaded image," IEEE Trans. Electron Devices, Vol. 61, No. 8, pp. 2827-2833, Aug. 2014. DOI: 10.1109/TED.2014.2331707
[8] K. Kim, C. L. Chen, Q. Truong, A. M. Shen, and Y. Chen, "A Carbon Nanotube Synapse with Dynamic Logic and Learning," Adv. Mater., Vol. 25, No. 12, pp. 1693-1698, Mar. 2013. DOI: 10.1002/adma.201203116
[9] P. Gkoupidenis, N. Schaefer, B. Garlan, and G. G. Malliaras, "Neuromorphic functions in PEDOT: PSS organic electrochemical transistors," Adv. Mater., Vol. 27, No. 44, pp. 7176-7180, Nov. 2015.DOI: 10.1002/adma.201503674
[10] L. Q. Guo, J. Wen, J. N. Ding, C. J. Wan, and G. Q. Cheng, "Excitatory post-synaptic potential mimicked in indium-zinc-oxide synaptic transistors gated by methylcellulose solid electrolyte," Sci. Rep., Vol. 6, pp. 38578-1-38578-7, Dec. 2016. DOI: 10.1038/srep38578
[11] L. Q. Guo, J. Wen, G. Q. Cheng, N. Y. Yuan, and J. N. Ding, "Synaptic behaviors mimicked in indium-zinc-oxide transistors gated by high-proton-conducting graphene oxide-based composite solid electrolytes," J. Mater. Chem. C., Vol. 4, No. 41, pp. 9762-6970, Nov. 2016. DOI: 10.1039/C6TC02228F

Journal of Physics: Conference Series **2370** (2022) 012014 doi:10.1088/1742-6596/2370/1/012014

Effect of gamma irradiation on GaN lateral Schottky barrier diodes

Chao Duan*[a], Zhaoxi Wu [a], Meng Meng [a], Shenglei Zhao [a], Xu Wang [a], Qingkui Yu [a]

[a]China Aerospace Components Engineering Center, Beijing 100094, China

[*] Corresponding author: super_duan@126.com

Abstract: The effect of gamma irradiation on GaN lateral Schottky barrier diodes (SBDs) has been investigated. Six gamma irradiation conditions were carried out, namely 100/500 k rad, 1 M rad, 100/500 k rad and 100°C, and 1 M rad and 100 °C. All the irradiated devices have a similar parameter variation tendency. There is no change for turn-on voltage. The on-state resistance is reduced from 5.32~5.53 Ω to 5.10~5.26 Ω, while the reverse leakage at -100 V is increased from 3.53~4.99 μA to 4.13~5.79 μA. The interface information extracted is improved slightly, demonstrating the anode/GaN interface is improved, and the passivation/AlGaN interface may be degraded. The slight variation of irradiated performance demonstrates that the GaN lateral SBDs are hardly affected by gamma irradiation.

1. INTRODUCTION

GaN semiconductor is promising in high-frequency and high-voltage applications, benefitted from the large bandgap material. SBD is a kind of useful device in many applications. GaN lateral SBDs demonstrate excellent performance compared with conventional SBDs due to the high critical electric field and mobility. 1.93 kV breakdown voltage and low turn-on voltage were obtained in GaN SBDs with dual field plates [1]. Oxygen plasma treatment was used in GaN lateral SBD to reduce the leakage and current collapse. P-GaN termination was proposed in GaN multi-channel SBDs to modulate electric field, and a remarkably high BV of 3.3 kV has been achieved [2]. SBDs with two-four channels were demonstrated to obtain low on-state resistance. Effects of annealing and work function on device performance were investigated. Anode metal tungsten has been utilized in GaN SBDs, demonstrating excellent microwave power rectification performance in 2.45/5.8 GHz systems. GaN lateral SBDs possess many advantages, including high BV, low resistance, low turn-on voltage, and small size, in space and aeronautic power applications. Numerous studies have been reported in the irradiation effect of GaN high-electron-mobility transistors (HEMTs) and vertical SBDs [3], [4]. However, there is no research on the irradiation effect of GaN lateral SBDs investigated until now.

The effect of gamma irradiation has been investigated on GaN lateral SBDs in this paper. Six gamma irradiation conditions, including 100 k rad, 500 k read, 1 M rad, 100 k rad and 100 °C, 500 k rad and 100 °C, and 1 M rad and 100 °C, have been carried out. The turn-on voltages V_{ON}, on-state resistances R_{on}, reverse leakages $I_{reverse}$, barrier height ϕ_{B}, and ideality factor η are monitored before and after gamma irradiation. Experiments demonstrate the excellent radiation-tolerant capability of GaN lateral SBDs.

2. DEVICE STRUCTURE AND EXPERIMENTS

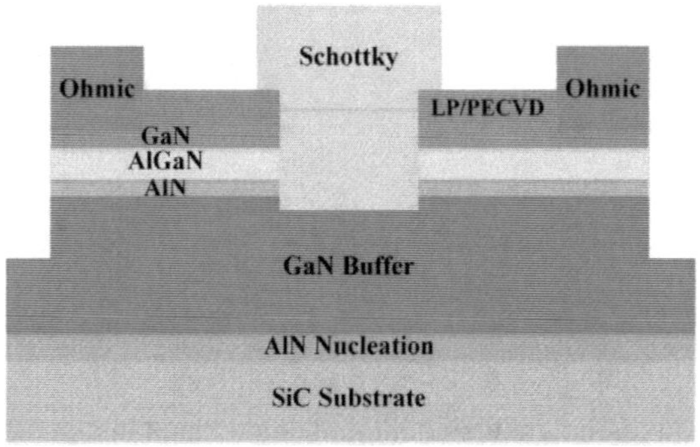

Figure 1 AlGaN/GaN lateral Schottky barrier diodes

AlGaN/GaN lateral Schottky barrier diodes structure is drawn in Fig. 1. The GaN materials were grown on SiC substrate through MOCVD system. The epilayers includes AlN nucleation layer, 1.5 μm GaN buffer layer, AlGaN barrier layer and GaN cap layer. The 2DEG density is 9.1×10^{12} cm^{-2} and mobility is 2071 $cm^2/V\cdot s$. The device isolation process was carried out firstly. The passivation layers include LPCVD Si_3N_4 and PECVD SiO_2. The SiN insulator can passivate the surface states, and the SiO insulator can undertake the high electric field. The GaN cap/AlGaN barrier/AlN insertion layers under anode were recessed, and tungsten metal was deposited as the anode metal to contact with the GaN materials. The cathode electrode is an ohmic contact. The anode-cathode spacing is 4 μm, and all the devices are packaged. Gamma irradiation was carried out with 1.25 MeV ^{60}Co source at room temperature and 100 °C.

3. RESULTS AND DISCUSSION

Six gamma irradiation conditions were carried out on GaN lateral SBDs, namely 100/500 k rad, 1 M rad, 100/500 k rad and 100°C, and 1 M rad and 100 °C. All the irradiated devices have the similar parameter variation tendency. Current-voltage characteristics of irradiated devices are drawn in Fig. 2 and Fig. 3, and the device parameters are extracted in Table 1 with the six irradiation conditions. The device characteristics before gamma irradiation are very close to each other for all irradiated devices. The turn-on voltages V_{ON} are 0.44~0.45 V, the on-state resistances R_{on} are 5.32~5.53 Ω, and the reverse leakages $I_{reverse}$ at -100 V are 3.53~4.99 μA.

Table 1. Device parameters extracted from *I-V* curves.

Devices	V_{ON} (V)	R_{on} (Ω)	$I_{reverse}$ (μA)	Barrier Height ϕ_B (eV)	Ideality factor η
Device 1, before 100 K	0.44	5.53	4.79	0.66	1.52
Device 1, after 100K	0.44	5.26	5.18	0.67	1.48
Device 2, before 500k	0.44	5.32	4.99	0.68	1.45
Device 2 after 500k	0.44	5.23	5.79	0.69	1.42
Device 3, before 1 M	0.44	5.35	4.40	0.66	1.50
Device 3, after 1 M	0.44	5.10	5.20	0.67	1.48
Device 4 before 100K and 100°C	0.44	5.50	4.76	0.66	1.50
Device 4 after 100K and 100°C	0.44	5.24	5.38	0.67	1.48
Device 5 before 500K and 100°C	0.44	5.32	3.53	0.68	1.47

Device 5 after 500K and 100°C	0.44	5.23	4.13	0.68	1.45
Device 6 before 100K and 100°C	0.45	5.44	3.88	0.68	1.48
Device 6 after 100K and 100°C	0.45	5.20	4.66	0.68	1.45

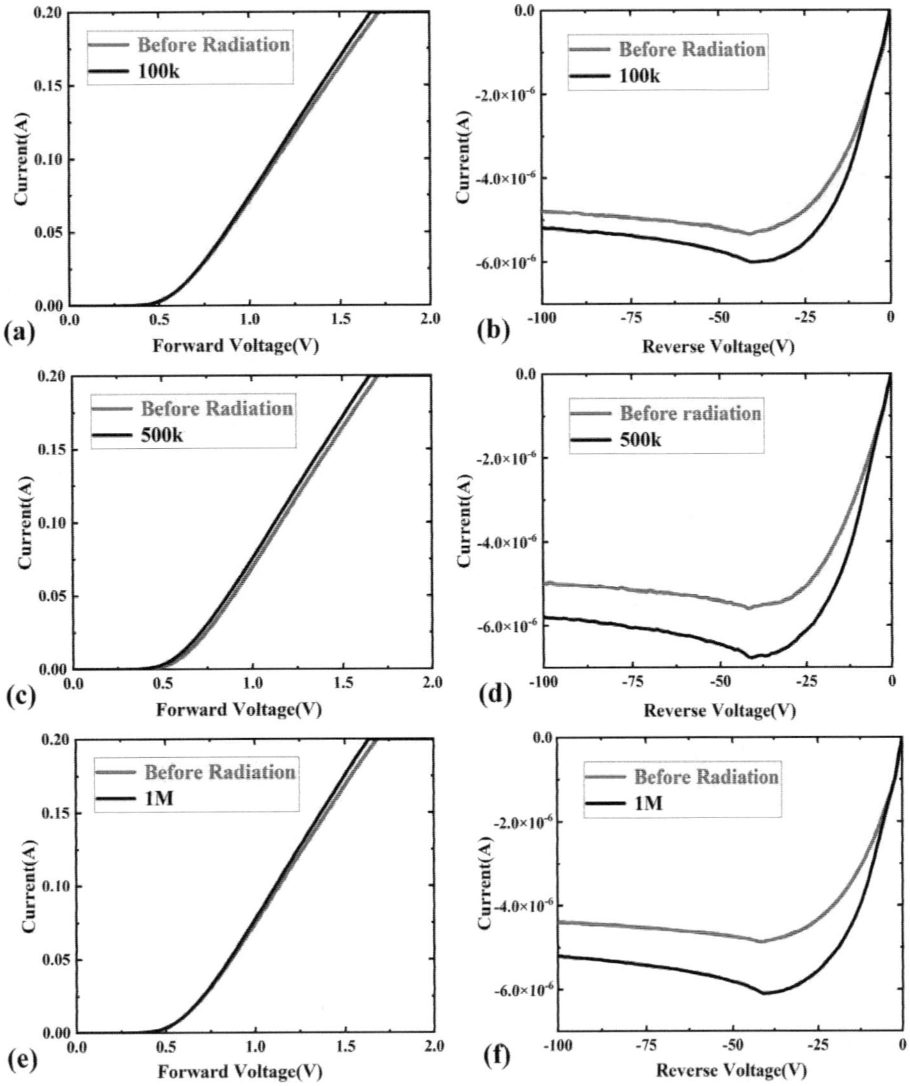

Figure 2 Current-voltage curves for SBDs with 100 k rad, 500k rad, 1 M rad.

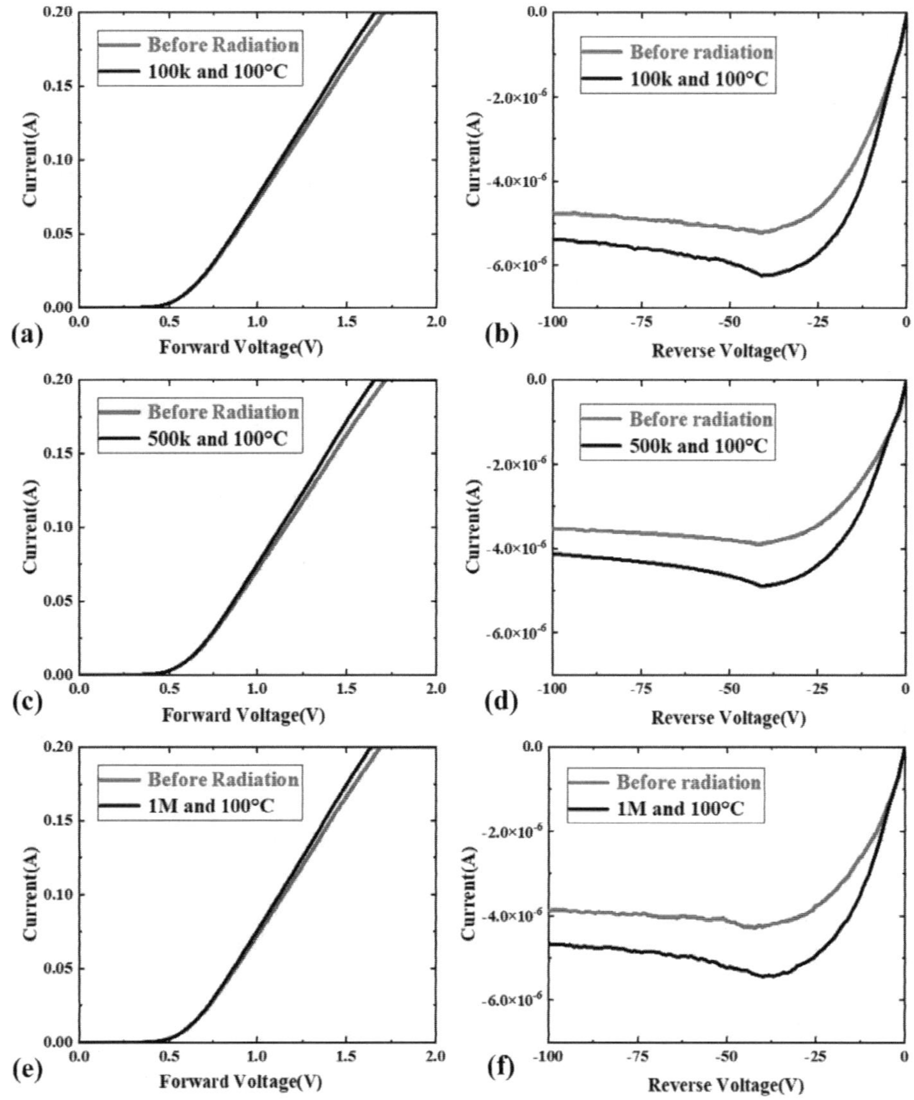

Figure 3 Current-voltage curves for SBDs with 100 k rad/100 °C, 500 k rad/100 °C, 1 M rad/100 °C.

After gamma irradiation, there is no change for V_{ON}, and V_{ON} for irradiated devices are 0.44~0.45 V. After 100 k rad, R_{on} is reduced from 5.53 Ω to 5.26 Ω. R_{on} for 500 k rad, 1 M rad, 100 k rad and 100 °C, 500 k rad and 100 °C, and 1 M rad and 100 °C irradiation are reduced from 5.32/5.35/5.50/5.32/5.44 Ω to 5.23/5.10/5.24/5.23/5.20 Ω, resulting in larger $I_{forward}$. $I_{reverse}$ increases from 4.79/4.99/4.40/4.76/3.53/3.88 μA to 5.18/5.79/5.20/5.38/4.13/4.66 μA for 100/500 k rad, 1 M rad, 100/500 k rad and 100 °C, and 1 M rad and 100 °C, respectively. Device performance dose not degrade further with the more serious irradiation condition. In order to investigate the variation mechanism, ϕ_B and η are obtained from forward current-voltage curves. ϕ_B increases slightly, and η is reduced 1.48~1.52 to 1.45~1.48, indicating that anode/GaN interface is improved after gamma irradiation. The reduced R_{on}, increased forward current, may result from the irradiated anode/GaN interface. The reduced R_{on} may result from the reduced anode/GaN interface resistance. Trap states may be introduced in the passivation/GaN interface, and increased $I_{reverse}$ may result from the more trap states in the irradiated passivation/GaN interface. The mechanism will be investigated in future work.

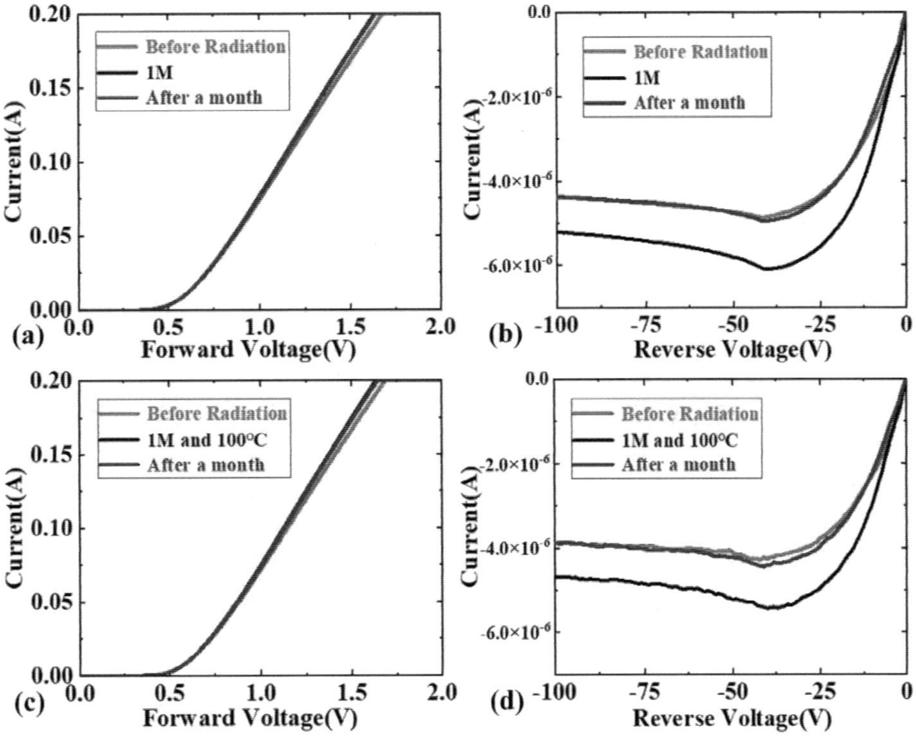

Figure 4 Current-voltage curves for 1M rad and 1 M rad/100 °C gamma irradiation.

A month later, we tested the characteristics of the device again. Current-voltage curves of irradiated devices are drawn in Fig. 4. There is slight variation for the forward curve, but the reverse curve is similar to that before irradiation. It may result from that anode/GaN interface is improved permanently after gamma irradiation, so there is no change for forward current-voltage curve after a month. And the trap state in the passivation/GaN interface reduced after a month, indicating that the passivation/GaN interface is restored to the state of before irradiation, so the reverse *I-V* characteristics is similar to that before irradiation. This indicates that the device has recovery ability after irradiation.

4. CONCLUSION

We have investigated the gamma irradiated GaN lateral SBDs. For 100 k~1 M rad and room/100℃ temperature, there is no change for turn-on voltage, and slight reduction of on-state resistance and slight increase of reverse leakage are observed. ϕ_B and η indicate that anode/GaN interface is improved after gamma irradiation. The reduced R_{on}, increased forward current, and increased $I_{reverse}$ may result from the irradiated anode/GaN passivation/GaN interface. In general, experimental results demonstrate the GaN lateral SBDs are promising gamma irradiation-tolerant devices for space and aeronautic power applications.

REFERENCES

[1] M. Zhu, B. Song, M. Qi, et al., "1.9-kV AlGaN/GaN lateral Schottky barrier diodes on silicon," IEEE Electron Device Lett., 36(4), 735-737 (2015).
[2] M. Xiao, Y. Ma, K. Cheng, et al., "3.3 kV multi-channel AlGaN/GaN Schottky barrier diodes with p-GaN termination," IEEE Electron Device Lett., 41(8), 1177-1180 (2020).

[3] Z. Bian, K. Su, J. Zhang, et al., "Gamma irradiation impact on GaN quasi-vertical Schottky barrier diodes," J. Phys. D: Appl. Phys., 53, 045103 (2020).

[4] X. Sun, O. I. Saadat, J. Chen, et al., "Total-ionizing-dose radiation effects in AlGaN/GaN HEMTs and MOS-HEMTs, " IEEE Trans. Nucl. Sci., 60(6), 4074-4079 (2013).

Synaptic devices based on gate-all-around InAs nanowire field effect transistor

Wei Luo, Chaofei Zha, Xia Zhang*, Xin Yan, and Xiaomin Ren

State Key Laboratory of Information Photonics and Optical Communications, Beijing University of Posts and Telecommunications, Beijing 100876, China

*xzhang@bupt.edu.cn

Abstract: In this paper, we proposed a gate-all-around InAs nanowire field effect transistor (GAA InAs NW FET) that can simulate synaptic behaviors such as short-term potentiation (STP) and long-term potentiation (LTP). The native oxide layer (In_2O_3) on the surface of InAs NW serves as a charge trapping layer for storing information. The transition from short-term potentiation (STP) to long-term potentiation (LTP) can be achieved by properly adjusting the gate voltage. Due to enhanced gate controllability, the GAA InAs NW FETs are expected to be widely used and promising in neuromorphic systems and networks.

1. INTRODUCTION

Nowadays, with the rise of artificial intelligence (AI), the computing power of current computers on the basis of the von Neumann architecture is facing a challenge. The processor and memory of computers on the basis of von Neuman architecture are separated, leading to low computing efficiency. In contrast with von Neumann computing, neuromorphic computing on the basis of the human brain offers the benefits of self-adaptive learning, high parallel computation, and low energy consumption [1, 2]. FETs based on III-V compound semiconductors have outstanding material properties and excellent repression of short-channel effects, making them one of the candidates for future high-speed applications [3]. Semiconductor nanowires offer many irreplaceable advantages over traditional thin-film materials and promise to improve the performance of FETs greatly. Previously, many silicon-based transistors were used to simulate synaptic functions [4]. InAs is a direct band gap semiconductor which has a band gap of 0.36 eV. It has a high intrinsic carrier concentration due to its relatively narrow band gap. The high surface-to-volume ratio of InAs nanowires increases the potential for electron trapping, exhibiting a stronger charge trapping effect. Current studies on InAs NW FET have been reported, but most of them achieve synaptic function based on top-gate FET [5], and there is a lack of studies on GAA FETs. GAA FETs show better performance due to due to their enhanced gate controllability [6].

In this paper, we propose that Synaptic devices based on GAA FETs. The electrical properties of the device are studied by Sentaurus TCAD, and the traps in the native oxide layer of InAs nanowires are used to capture and release charges. Biological synaptic properties are simulated, such as STP and LTP. By controlling the gate voltage, the transition from STP to LTP transition is achieved.

2. DEVICE STRUCTURE AND SIMULATION METHODOLOGY

Figure 1 (a) and (b) show the 3D schematic diagram and 2D cross-section diagram of the GAA InAs NW FET, respectively. Figure 1 (b) shows that the InAs nanowire is covered with an In_2O_3 oxide layer.

For the purpose of simulating the possible trapping of electrons in the native oxide layer, sufficient traps were introduced as trap centers to trap electrons travelling through the surface barrier. It is reported that the trap levels of the In_2O_3 layer were uniformly set to 0.1 eV, and the concentration was set to 10^{19} cm^{-3}[7]. The GAA structure is adopted for better gate control, whose main parameters of the device are shown in Table 1.

The device is simulated by 3D Sentaurus TCAD, and the electrical properties of the device are obtained by self-consistent methods of solving the carrier continuity equation and the Poisson equation. The Auger and SRH models were employed to consider the production and recombination mechanisms, which is the main factor of electron motion in GAA FETs. The band gap narrowing models, mobility degradation caused by impurity scattering, and thermodynamic and high field models should also be considered. The location of the oxide layer forming trap on the InAs nanowire surface is reported to store and release surface charge according to the Poole-Frenkel mechanism.

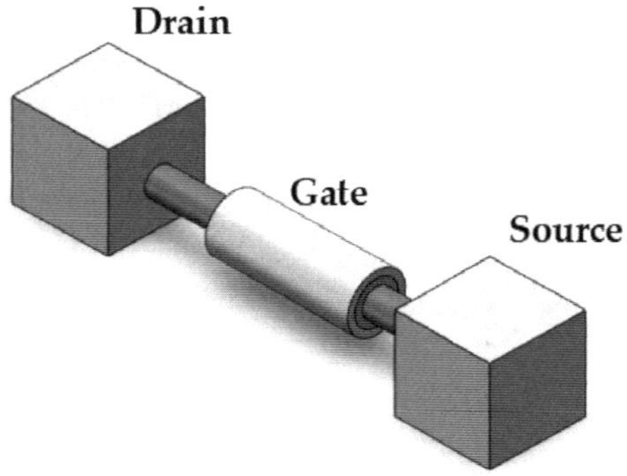

(a)

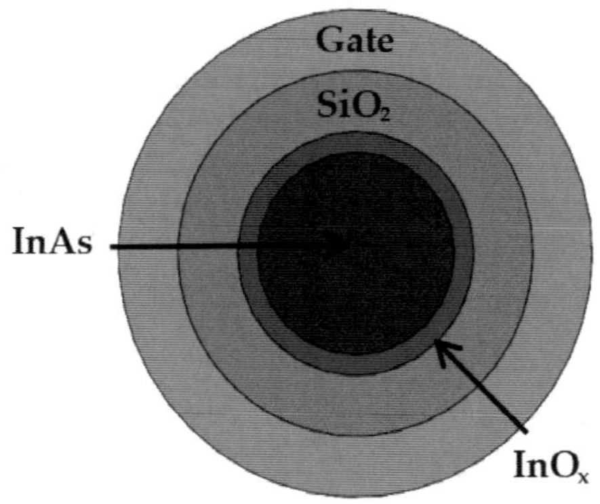

(b)

Figure 1. (a) Schematic diagram of GAA InAs nanowire field effect transistor. (b) Cross section diagram.

Table 1. Main parameters of GAA InAs NW FETs

Device Parameter	Value
Nanowire diameter	20 nm
In_2O_3 thickness	3 nm
Gate length	0.2 um
Gate Oxide thickness	100 nm
Channel doping	10^{19} cm^{-3}
Temperature	300 K
Gate workfunction	5.1 eV

3. RESULTS AND DISCUSSION

To simulate the influence of oxide layer traps on the electrical properties of InAs NW FET, the traps are evenly set at the energy of 0.1 eV under the conduction and a concentration of 10^{19} cm^{-3}. The transmission characteristics of InAs NW FETs under the forward voltage sweep and the reverse voltage sweep with the drain voltage of 0.1V are displayed in Figure 2 (a). The current on/off ratio of the device is calculated to be about 10^4. InAs nanowires with high electron mobility increase the possibilities that hot electrons are trapped by the oxide layer, resulting in an altered threshold voltage [8]. A memory window of over 1 V is obtained due to the native oxide layer of InAs nanowires accommodating charge trapping sites to store information. Therefore, InAs nanowires with native oxides can be used as a neuromorphic device to achieve artificial synaptic function. The curve of device drain current versus time is displayed in Figure 2 (b), where the gate voltage pulse has an amplitude of 2 V and a pulse width of 1 ms at V_{ds}=0.1 V. As shown in Figure 2 (b), when the gate voltage of 2 V is applied at the device, the electrons in the InAs conductive channel are trapped by the native oxide layer, causing that I_{ds} drops continuously. When the gate voltage returns to 0 V, the electrons captured by the trap layer cannot be completely released, making I_{ds} decrease significantly compared to the un-added gate voltage pulse. This is similar to the potentiation behavior of biological synapses, where a pre-synaptic stimulation (2 V, 1ms) causes changes in excitatory postsynaptic currents (I_{ds}), and the involved charges act as neurotransmitters in the synapse.

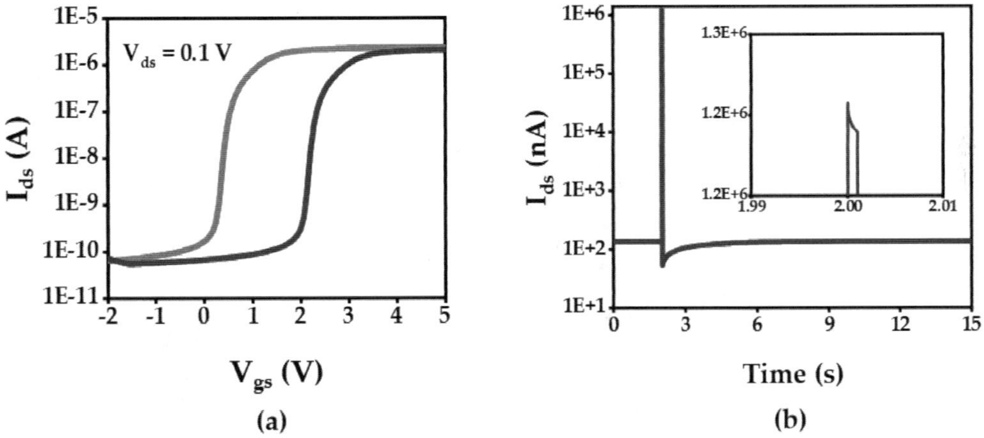

Figure 2. (a) Transfer characteristic of GAA NW FETs at V_{ds}=0.1 V. (b) Change curve of drain current (I_{ds}) versus time after a single pulse of gate voltage. The inset shows the drain current (I_{ds}) changes during the applied pulse.

The simulation of neural networks is achieved by simulating the communication of neurons passing through synapses [9, 10]. Stimulation generated by the pre-synaptic membrane triggers the release of neurotransmitters, as detected by receptors in the postsynaptic membrane, causing changes in the potential and weight of the postsynaptic membrane. Similar to biological synapses, the gate voltage pulse served as stimuli of the pre-synaptic membrane, and the drain current and channel were treated as potentials and weights of the postsynaptic membrane, respectively. As shown in Figure 3 (a), the charge trapping effect is caused by traps. The high mobility of an InAs NW results in an average relatively long free path into the hot electrons, which leads to an increased probability of being trapped. Thus, when the energy of the energetic hot electrons generated in InAs NW by the gate voltage excitation is above the height of the barrier, it is captured by the oxide trap layer before thermalization back to the conduction band. The electrons involved in the combination in the conduction band decrease, the electron density is reduced, which results in the decrease of the drain current. On the basis of retention time, synaptic plasticity is classified as STP and LTP, and the intensity of the stimulation in our experiment can be expressed in terms of the magnitude of the gate voltage. Therefore, simulating the strength of the artificial synaptic stimulus by controlling the size of the gate voltage pulse makes the synapses behave differently. A cross-sectional diagram of the trap charge under applied different gate voltage strengths is shown in Figure 3 (b). At a gate voltage of 2 V, the number of electrons trapped by the native oxide layer is greater than that at the gate voltage of 1.5 V. The larger gate voltage reduces the barrier of the InAs nanowires and the oxide layer, making it easier for thermal electrons to capture across the barrier by the trap layer, causing the decline of the drain current and the rise of the electrons in the oxide layer.

The electron affinity and band gap of zinc blende (ZB) InAs were 0.36 eV and 4.9 eV. The conduction band positions and band gaps of In_2O_3 are available from ref [11]. With the application of a gate voltage, the barriers between conduction band and trap energy level can be lowered, which makes electrons captured more easily in the InAs nanowire channel. The effect of the amplitude of the gating pulse on the simulated synaptic behavior of the device was investigated. The relationship between the applied gate voltage and the excitatory postsynaptic current is shown in Figure 3(c). The excitatory postsynaptic current ranging from 1.5 V to 2 V is 132 nA, 128 nA, 111 nA, 85 nA and 53 nA. The corresponding retention time is shown in Figure 3(d). As the gate voltage enhances, the barrier is lowered, causing the electrons easier cross the barrier to be captured by the trap layer, resulting in a smaller excitatory postsynaptic current and a longer recovery time. When the traps in the oxide layers are not fully occupied, pulse stimulus recovery time was positively correlated with pulse amplitude. Therefore, the transition from STP to LTP can be achieved by adjusting the gate voltage.

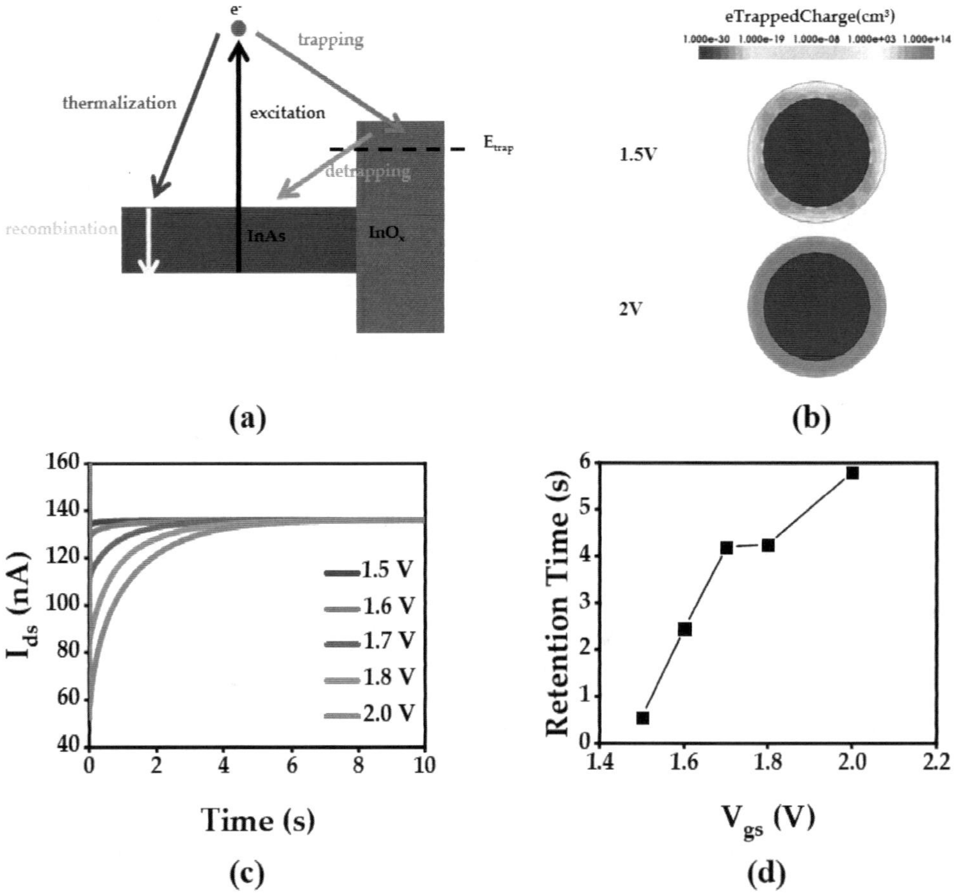

Figure 3. (a) Energy band chart of the processes of excitation, recombination, thermalization, trapping and de-trapping, and E_{trap} refers to the energy level of traps. (b) Distribution diagram of an electron trapped charge in InAs NW FET under different gate voltages. (c) The excitatory postsynaptic current with the different gate voltage (1.5 V ~ 2.0 V). (d) The retention time of the synaptic weight at different gate voltages (1.5 V ~ 2.0 V).

4. CONCLUSIONS

In this paper, we design and simulate a GAA InAs NW FET with biological synaptic behavior. The native In_2O_3 oxide layer on the surface of InAs NW acts as a trap layer capturing the carrier generated in the channel. Synaptic behavior is simulated in the range of low gate voltage, mainly manifested in the synaptic plasticity of STP and LTP. As the gate voltage enhances, the device changes from STP to LTP, which is similar to the characteristics of biological synaptic cooperativity. These works demonstrate the feasibility of GAA InAs NW FETs for emerging neuromorphic networks.

ACKNOWLEDGMENT
This work was supported by the National Natural Science Foundation of China (61935003) and the Fund of State Key Laboratory of Information Photonics and Optical Communications (Beijing University of Posts and Telecommunications), P. R. China (IPOC2020ZZ01).

REFERENCES

[1] Drachman, D.A. "Do we have the brain to spare?" Neurology. Papers 64(12), 2004-2005 (2005).
[2] Indiveri, G. and S.C. Liu. "Memory and Information Processing in Neuromorphic Systems," Proc. IEEE. Papers 103(8), 1379-1397 (2015).
[3] Del Alamo, J.A. "Nanometre-scale electronics with III-V compound semiconductors," Nature. Papers 479(7373), 317-323 (2011).
[4] Fusi, S. et al., "Spike-driven synaptic plasticity: Theory, simulation, VLSI implementation," Neural Comput. Papers 12(10), 2227-2258 (2000).
[5] Li, B. et al., "Mimicking synaptic functionality with an InAs nanowire phototransistor," Nanotechnology. Papers 29(46), 6 (2018).
[6] Fahad, HM et al., "Silicon Nanotube Field Effect Transistor with Core-Shell Gate Stacks for Enhanced High-Performance Operation and Area Scaling Benefits," Nano Lett. Papers 11(10), 4393-4399 (2011).
[7] Yang, Y. et al., "Hot Carrier Trapping Induced Negative Photoconductance in InAs Nanowires toward Novel Nonvolatile Memory," Nano Lett, vol. 15, no. 9, pp. 5875-82, Sep 9 2015.
[8] Li, JS et al., "Anomalous photoconductive behavior of a single InAs nanowire photodetector," Appl. Phys. Lett. Papers 107(26), (2015).
[9] Zhu, X.J. et al., "Emulation of synaptic metaplasticity in memristors," Nanoscale. Papers 9(1), 45-51 (2017).
[10] Zucker, R.S., "Short-term synaptic plasticity," Papers 12, 13-31 (1989).
[11] Pashley, Haberern, Feenstra, and Kirchner. "Different Fermi-Level Pinning Behavior on N- and P-Type Gaas(001)". Phys. Rev. Papers B 48, 4612-4615 (1993).

A low supply voltage sensitivity low power oscillator design

Hao Shi, Zhongjie Guo, Ruokai Bai, Mengli Li, Ziyi Qiu

School of Automation and Information Engineering, Xi'an University of Technology, Xi'an 710048

* Corresponding author: 876954479@qq.com

ABSTRACT: With the rapid development of social economy, the integrated circuit industry has developed rapidly. At the same time, people's requirements for the cost, power consumption, integration, stability and other aspects of integrated circuit chips have increased. Oscillators, as an important module for providing clocks to chips, have higher requirements for on-chip oscillators. The purpose of this paper is to achieve a low supply voltage sensitive and low power oscillator. Based on this research purpose, this paper proposes a power adjustment oscillator circuit of the flip level technique, which realizes that the clock signal duty cycle is 50% under the typical case of a supply voltage of 4 V. The center frequency is 512 KHz, of which the process angle trimming range of the capacitor trim circuit is 52.37%-211.28%. With the supply voltage range of 2.5 V to 5.5 V, the clock signal has a frequency variation of 0.299%/V, has low supply voltage sensitivity, and consumes only 400 nA with low power dissipation.

1. INTRODUCTION

With the development of the times, people's demand for the performance of the on chip integrated oscillator is also getting higher and higher. RC oscillator has been widely used due to its simple structure, low cost, small power consumption, adjustable frequency and other advantages, but RC oscillator is easily affected by the supply voltage and the power supply voltage noise. In practical applications, the supply voltage is often unstable, or will be applied to different supply voltages, so it is necessary to design an oscillator with low supply voltage sensitivity. Most of the low supply voltage sensitivity oscillators under study are charging R C through a constant current source, resulting in low sensitivity to the supply voltage. In this paper, a power supply adjustment circuit is proposed, using a diode-connected MOS to generate a fixed clamp voltage and a fixed voltage, thereby achieving low sensitivity to the supply voltage. In order to reduce the short-circuit current of the inverter, the short-circuit current control technology is used to realize that the current flows through the flip level circuit only when the level is flipped, which achieves the low power consumption of the oscillator [1].

In order to achieve the low voltage sensitivity and low power consumption oscillator design, this paper proposes a power supply adjustment circuit based on the flip level technology, realizing frequency stability in the range of 2.5 V-5.5 V supply voltage, considers the influence of the process angle on the frequency, and uses the trimming capacitor to compensate for the influence of the process angle. The stability of its frequency at different process angles is guaranteed.

2. OSCILLATOR MAIN CIRCUIT DESIGN

2.1 Oscillator main circuit design

The oscillator technique based on the level flipping technique proposed in this paper is shown in Figure 1. The main principle of this oscillator is the use of level flipping technology to convert the time required to charge and discharge the capacitor, where V_R flip levels from high to low are V_{DD}-V_{SGMP1}, V_R flip levels from low to high are V_{GSMN1}. Since the resistance of the resistor R is large, the charging current is basically the same as the discharge current, and the main function of the oscillator is realized through the shaping circuit. The oscillator frequency is:

$$f = \frac{I_{AVG}}{2C_1(LV_{DD} - V_{SGMP1} - V_{GSMN1})} \tag{1}$$

In the above equation, I_{AVG} is the average charge and discharge current through R and C:

$$I_{AVG} = \int_0^{0.5T} \frac{LV_{DD}}{R} e^{\frac{t}{\tau}} dt \tag{2}$$

In Figure1, MP_2 is 18 inverted tubes in the linear region, and MN_2 is 35 Inverted tubes in the linear region are used to make large resistors. The digital trim circuit reserves different capacitance values of capacitors that can be integrated into the circuit, thereby causing a trimming effect on the frequency.

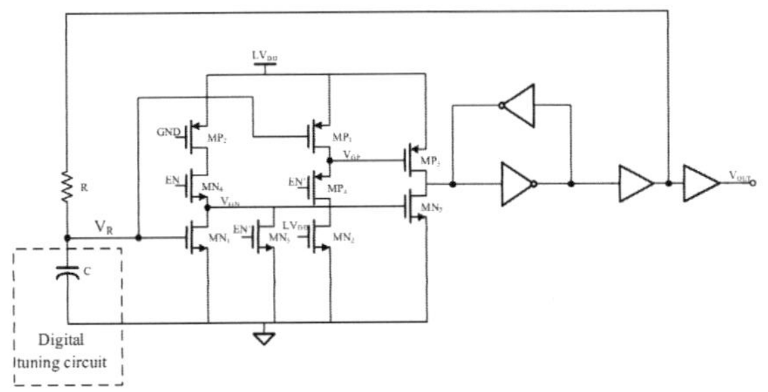

Figure 1 System circuit design block diagram

2.2 Power supply adjustment circuit

As shown in Equations (1) and (2), the frequency of the oscillator will vary greatly with the change of the supply voltage. In order to reduce the sensitivity of supply voltage, this paper designs a power supply voltage adjustment circuit to decrease the influence of the supply voltage on the frequency accuracy of the oscillator in the range of 2.5 V-5.5 V, and uses the supply voltage adjustment circuit to stabilize V_{DD} to LV_{DD}, where the voltage value of LV_{DD} is stabilized at 2.1 V. The structure of the power supply adjustment circuit is very simple: a diode is used to clamp circuit, and the output voltage is stabilized on a certain voltage value, and the negative feedback is connected to improve the load capacity of the clamping voltage.

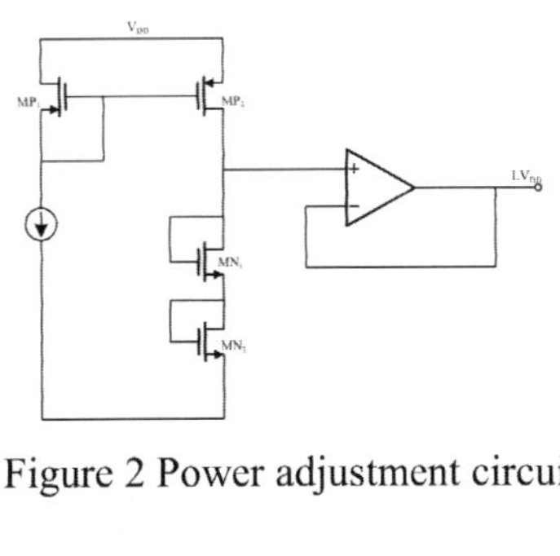

Figure 2 Power adjustment circuit

Figure 2 shows the supply voltage regulation circuit, which consists of a diode clamp circuit, an op amp circuit, and a current source mirror circuit. The current source circuit consists of MP_1 and MP_2, and the MOS tube width ratio is consistent. NMOS tube saturation region current formula:

$$I = \frac{1}{2}\mu_n C_{ox}\frac{W}{L}(V_{GS} - V_{TH})^2 \quad (3)$$

As shown in Equation (3), when certain current flows through the branches of MN_1 and MN_2, its:

$$V_{GS} = \sqrt{\frac{2I}{\mu_n C_{ox} W/L}} + V_{TH} \quad (4)$$

Available by Formula (4), when the current flows for a certain amount of time, the MOS tube V_{GS} is certain. Due to the diode-connected MOS tube $V_{GS} = V_{DS}$, the voltage drops of two fixed V_{GS} will be generated on MN_1 and MN_2, s9o adjust the current and aspect ratio can obtain a certain output voltage. Re-accessing the negative feedback op amp increases the load capability of the clamping voltage. A fixed LV_{DD} is created after the supply voltage regulation circuitry reduces the sensitivity of the oscillator to the supply voltage.

2.3 Capacitor trim circuit design

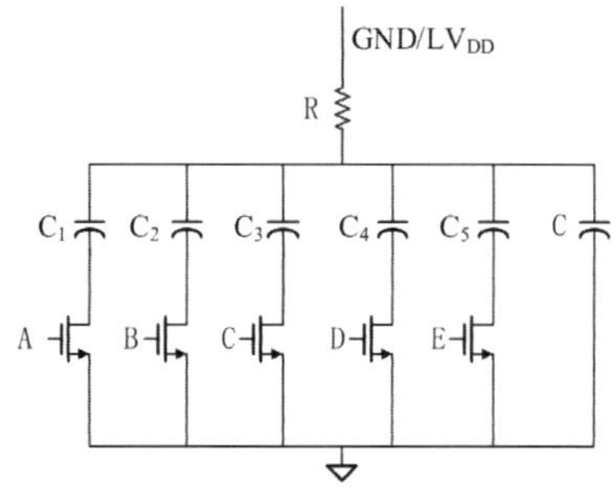

Figure 3 Capacitor repair circuit

Figure 3 shows the oscillator's digital trim circuit. During the interaction between the chip and the IIC, the oscillator frequency at different process angles and at different temperatures is usually trimmed by the IIC. This capacitor trimming through the NMOS tube as a switch tube, controls different capacitor access circuit, to achieve the oscillator digital trimming function, wherein the capacitance ratio of the capacitor is: C_1: C_2: C_3: C_4: C_5 = 16: 8: 4: 2: 1. Under normal circumstances, it is for high levels, which is connected to the circuit, so that the frequency of the oscillator can be adjusted upwards and downwards, generally in the range of $0.5f_0$-$2f_0$. In fact, the revision range implemented in this paper is 52.37%-211.28%.

3. CIRCUIT SIMULATION RESULTS AND DATA ANALYSIS

The low voltage sensitive low power oscillator in this article is designed based on a 0.18 μm BCD process. At room temperature, when the supply voltage is 4 V, the simulated circuit is shown in Figure 1, and the voltages V_R, V_{GN}, and across the charge and discharge capacitor are simulated the waveforms of V_{GP}, circuit power consumption I_R, and oscillator output are shown in Figure 4.

As shown in Figure 4, the level reversal voltage V_R varies from 0.3 V to 1.7 V, and when the oscillator frequency reaches 512 KHz, the maximum charging current is about 960 nA, but its average power consumption is only 400 nA.

Dynamic current is generated only when the level of the oscillator is flipped, this is because when the level is flipped, the level reversal circuit will produce dynamic current to the ground of the path. V_{GN} is near ground level most of the time during the oscillation period, and V_{GP} is near V_{DD} mostly. MP_3 and MN_3 tubes are not turned on at the same time, and there is no quiescent current but dynamic current when the level is reversed. When the V_R level is reversed, V_{GN} and V_{GP} make MN_3 And MP_3 tubes are turned on separately, eliminating the short-circuit current. Therefore, the circuit consumes less current. In order to reduce the short-circuit current of the inverter, the short-circuit current control technology is used to drive the NMOS and PMOS respectively, which effectively controls the short-circuit current and realizes the low temperature drift and low power consumption of the oscillator.

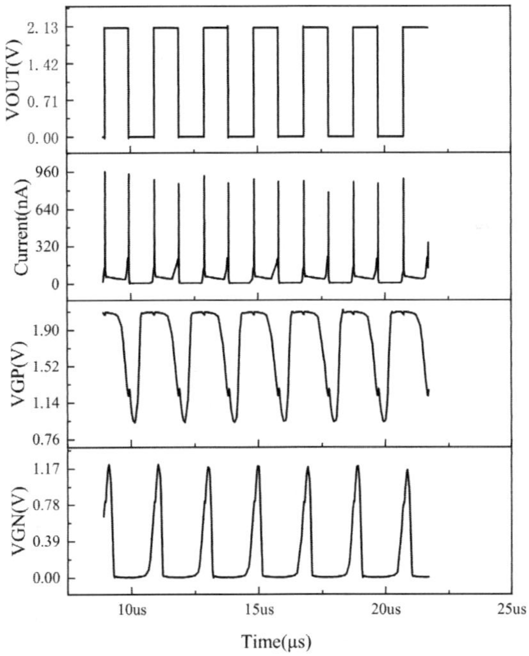

Figure 4 Transient simulation waveform of oscillator

Figure 5 below shows the oscillator frequency with the voltage variation, the supply voltage from 2.5 V to 5.5 V, the oscillator frequency changes from 515.6 KHz to 511.0 KHz, 0.299%/1V compared to typical 4 V, 27 °C.

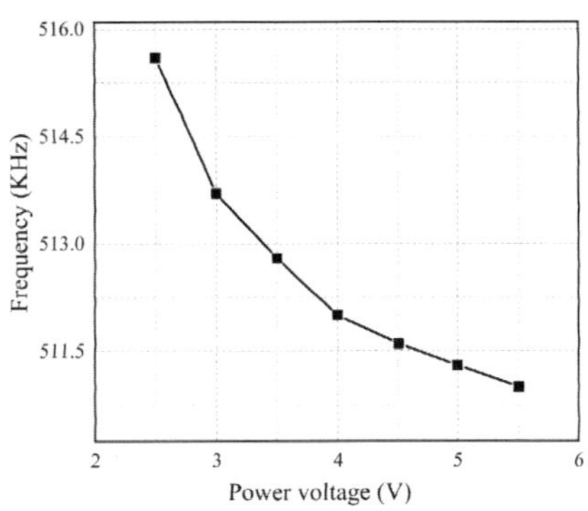

Figure 5 Oscillator frequency versus supply voltage

Table 1 below lists the oscillator circuits and literature [2], literature [3], literature [4], literature [5], literature [6] comparison of various parameters. This design uses a 4 V supply voltage as a reference to achieve an oscillator frequency of 512 KHz over the supply voltage range of 2.5 V to 5.5 V. The oscillator frequency varies evidently by only 0.299% per 1 V of supply voltage. Due to the frequency supply voltage sensitivity of the compared literature, this design can also operate over a wider supply voltage range. Its power consumption is only 1 μW, which is only higher than the power consumption of the literature [3]. But the design of the literature [3] is sensitive to its supply voltage 116 times, so relatively speaking, this design achieves low power consumption in the case of low supply voltage sensitivity.

Table 1 Comparison of literature parameters

	Ref [2]	Ref [3]	Ref [4]	Ref [5]	Ref [6]	This work
Technology	0.18μm	0.18μm	0.13μm	0.18μm	0.13μm	0.18μm
Frequency (MHz)	0.122	3.2	14	1.2	3.01	0.512
Power(μW)	40	0.0144	38.4	45	5.8	1
Power voltage (V)	1.4 to 2.0	0.6 to 1.8	1.4 to 1.6	1.7 to 1.9	0.99 to 1.01	2.5 to 5.5
Frequency variation%/1V	3.98	5	2	1.6	180	0.299

4. CONCLUSION

In this paper, the advantages and disadvantages of the traditional RC oscillator structure, the oscillator structure based on the level flipping technique, and the relationship between its output frequency and supply voltage are analyzed. On this basis, a supply voltage clamp circuit is designed, which realizes the oscillator duty cycle of 50% at a supply voltage of 4 V at room temperature with a frequency of 512 KHz. The oscillator's supply voltage sensitivity is reduced, enabling a 1 V oscillator frequency change for every 1 V change in the supply voltage between 2.5 V and 5.5 V 0.299%. Under other design requirements, its frequency can be changed flexibly. At the same time, the process angle trimming circuit is designed, and the frequency range of trimming is 52.37%-211.28%. In terms of the future work, this paper realizes the low power supply voltage sensitivity in the voltage range of 2.5 V-5.5 V and low power consumption of the oscillator. There is no research on lower low power supply voltage. With the progress of technology, the power supply voltage is also reduced, and this part of the research can be added in the future. In order to reduce power consumption, the value of the resistance in the circuit is relatively large, which wastes the area. In the subsequent improvement, appropriate consideration can be given to reduce the resistance to reduce the waste of the area.

REFERENCES

[1] Paidimarri, D. Griffith, A. Wang, A. P. Chandrak Asan and G. Burra, "A 120 nW 18.5 kHz RC oscillator with comparator offset cancellation for ±0.25% temperature stability,"2013 IEEE International Solid -State Circuits Conference Digest of Technical Papers, 2013, pp.184-185.

[2] H. V. Nguyen, J. Kim and S. Kim, "Temperature and Supply Voltage Insensitive Relaxation Oscillator Using Voltage Reference," TENCON 2018 - 2018 IEEE Region 10 Conference, 2018, pp. 0555-0558.

[3] S. Dai and J. K. Rosenstein, "A 14.4 nW 122 KHz dual-phase current-mode relaxation oscillator for near-zero-power sensors,"2015 IEEE Custom Integrated Circuits Conference (CICC), 2015, pp. 1-4.

[4] K. Choe, O. D. Bernal, D. Nuttman and M. Je, "A precision relaxation oscillator with a self-clocked offset-cancellation scheme for implantable biomedical SoCs,"2009 IEEE International Solid-State Circuits Conference - Digest of Technical Papers, 2009, pp. 402-403,403a.

[5] Y. Tokunaga, S. Sakiyama, A. Matsumoto and S. Dosho, "An On-Chip CMOS Relaxation Oscillator with Voltage Averaging Feedback," in IEEE Journal of Solid-State Circuits, vol. 45, no. 6, pp. 1150-1158, June 2010.

[6] K. Huang and D. D. Wentzloff, "A 1.2-MHz 5.8- μW Temperature-Compensated Relaxation Oscillator in 130-nm CMOS," in IEEE Transactions on Circuits and Systems II: Express Briefs, vol. 61, no. 5, pp. 334-338, May 2014.

Research on the Mechanism of Sensing Signal Processing

Chuanpeng Ning[1] and Xin Yu[1]

School of Electrical and Electronic Engineering, Changchun University of Technology, Changchun, Jilin 130012, China

ningchuanpeng@163.com

ABSTRACT: The speed sensor is an important support for the development of modern industrial automation. It plays a great part in automobiles, machinery manufacturing, and aerospace, and is widely used in CNC machine tools, cutting laser machines, industrial transmission, and other mechanical equipment. In this paper, a sensor is designed, which uses the signal generated by the body's rotation as a sine wave signal to transmit frequency information. The sinusoidal signal is transmitted to the signal processing circuit, and two voltage dividers are used to reduce the signal amplitude to the applied voltage of the MCU below 5V. After processing the circuit, the sinusoidal signal can be converted into the square wave signal with the same frequency. The square wave signal is transmitted to the microcontroller to read the square wave's pulse frequency, and the body's rotational speed can be output after the formula is calculated in the program read by the microcontroller in advance.

1. Introduction

Sensors are the bridge between the physical and digital worlds and have a wide range of uses in today's digital age. As one of the three technical pillars of modern information technology, the development of sensors is an important research pillar for promoting digital industrialization [3] [5] and is widely used in the fields of industrial automation and communication [1] [2] [8]. Among various mechanical motions, rotation is one of the most basic motions in industry, so the measurement of rotational speed has always been a research hotspot in the field of engineering [6-7]. The types of traditional rotary sensors mainly include optoelectronic transformation [9], electromagnetic sensors [10], and electrical effect sensors [4].

This paper simulates the sine wave signal generated in the rotation, through the signal processing circuit composed of the LM324 chip as the core, the sine wave signal is converted into a square wave signal, and the duty cycle of the square wave is adjusted by the pre-resistor, which improves the MCU's performance. Accuracy of signal processing. Due to the different amplitudes of the sine wave signals generated by various rotation speeds, the output waveform voltage amplitude is too high when the rotation speed is too high, which is easy to cause damage to the MCU and limits the use range of the sensor. In this study, a voltage divider resistor is added at the back end of the square wave output. The voltage value of the square wave can be adjusted by setting different resistance values, and the amplitude of the square wave can be adjusted to the applied voltage range of the MCU. In this study, the frequency values of 10-90 Hz were measured with an error value of 0.11%.

2. System and Working Principle

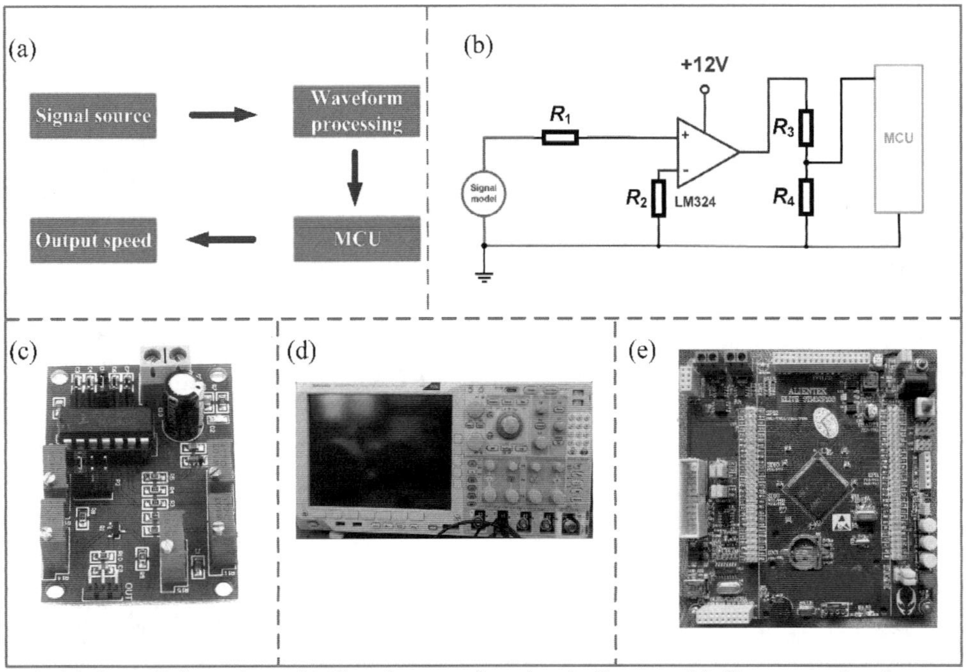

Figure 1. (a) Overall flow chart. (b) Signal processing circuit diagram. (c) Sine wave signal generator. (d) Digital oscilloscope. (e) MCU.

Figure 1(a) is the overall block diagram of the experiment. The specific principle is that the sine wave signal output by the signal generator is converted into a square wave signal after passing through the signal processing circuit. Then input the square wave signal into the MCU for pulse counting, output the frequency value, calculate the speed according to the set frequency and speed conversion formula, and output it to the serial port to display the value.

Figure 1(b) shows the signal processing circuit, which uses the LM324 chip to convert sine wave and square wave. The sine wave signal from the signal source is transmitted to the LM324 chip through the two resistors R_1 and R_2 for conversion. Figure 1(c) shows a function generator, which uses an integrated circuit to simulate the signal value generated at the rotational speed, and transmits the sine wave signal to the signal processing circuit. Figure 1(d) shows the digital oscilloscope used in the experiment to display sine wave and square wave data. Figure 1(e) shows the actual picture of the MCU used in the experiment. Since the input voltage of the MCU is below 5V, the voltage divider resistors R_3 and R_4 are added after the square wave output by the LM324 chip to reduce the voltage of the square wave signal. to the applied voltage range of the MCU, and transmit the signal to the MCU for processing.

When the square wave signal is transmitted to the MCU, the MCU triggers an interrupt through two rising edges of a square wave, and then the timer counts. The total number of counts multiplied by the counting unit time is the period of the square wave. According to the speed measurement formula:

$$n = \frac{f}{p} \tag{1}$$

In the formula n is the speed, the unit is r/s; f is the frequency, the unit is Hz; p is the number of pole pairs of the motor that sends out the signal source. Here we take $p=1$ as an example to analyze, and finally output the measured speed through the serial port of the microcontroller value.

3. Experiment and Comparison

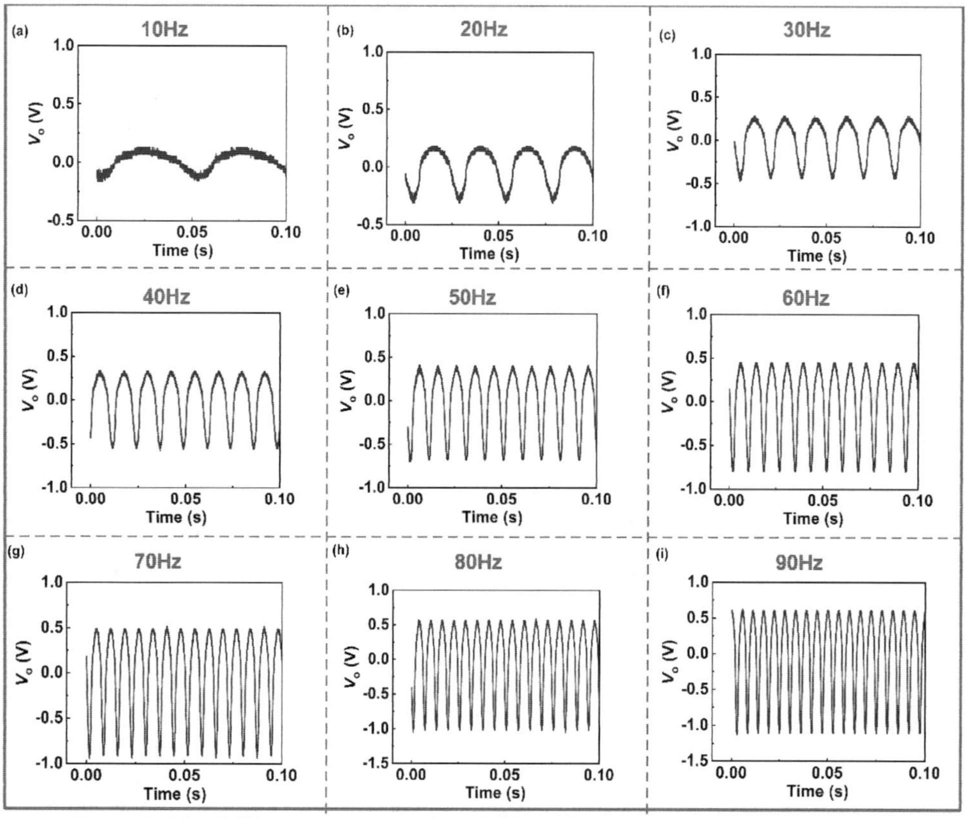

Figure 2. (a)~(i) Sine waves output by 10-90Hz sine wave generator respectively.

Figure 2 is the sine wave diagram of different frequencies of 10-90Hz output by the signal transmission device displayed by the digital oscilloscope. Since the sine wave signals sent by the rotational speed source are pulse signals in opposite directions in most cases, positive and negative are used in the output of the simulator in this paper. From the picture data, it can be concluded that the voltage of the output waveform is different in the state of different frequencies. As the frequency increases, the amplitude of the output sine wave also increases.

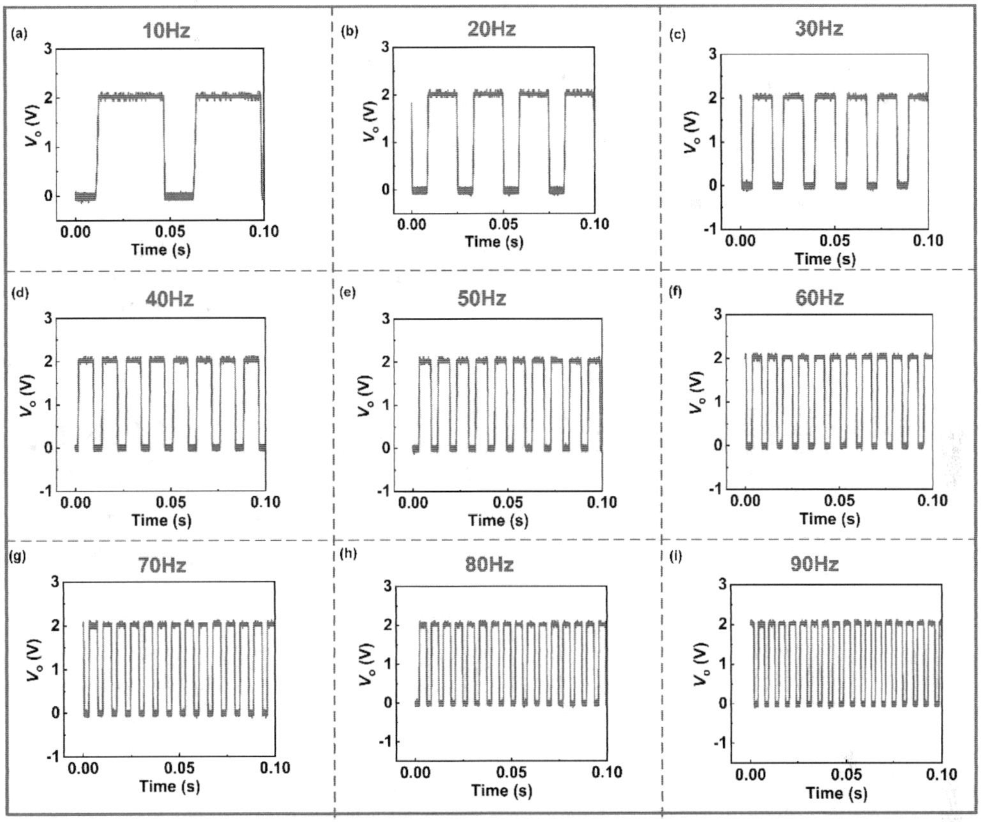

Figure 3. (a)-(i) Respectively are the square waves output by 10-90Hz sine waves after signal processing by the circuit.

Figure 3 shows the square waves of different frequencies after signal processing displayed by the digital oscilloscope. The signal processing circuit adopts the LM324 chip and uses a 12V DC power supply to power it, converting the sine wave into a square wave. By adjusting the resistors R1 and R2 in the circuit, the duty cycle of the output square wave is adjusted between 30% and 70%, which reduces the error caused by missing or over-reading due to too small a duty cycle when the microcontroller reads. This improves the accuracy of the speed measurement. Since the amplitude of the sine wave signal output by different signal sources is different, if the amplitude of the signal source is too large, it is easy to cause damage to the MCU. Therefore, to suit sine wave signal sources of different amplitudes, divider resistors R3 and R4 are added at the back end of the output of this experimental circuit. Signal reading and processing at the terminal.

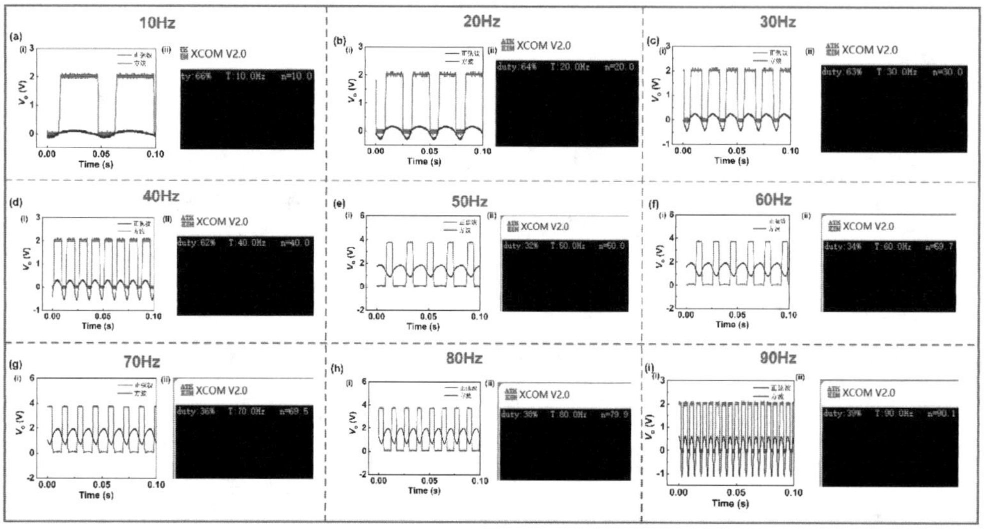

Figure 4. (a)-(i) (i) Comparison of 10-90Hz sine wave and square wave respectively; (ii) MCU output speed display value.

Figure 4(i) is a comparison between the sine wave signal and the square wave signal displayed by a digital oscilloscope. From (i), it can be seen that the square wave signal after signal processing and the sine wave output by the original signal source, signals are of the same frequency, and there is no deviation in the measurement of the rotational speed value of the original rotational speed. Figure 4(ii) shows the speed value and frequency information output by the serial port of the microcontroller in the computer. According to the error rate formula:

$$\lambda = \frac{n - n_1}{n} \times 100\% \tag{2}$$

In the formula λ is the error rate of the rotational speed, n is the actual rotational speed value, and n_1 is the measured rotational speed value. From this, it can be obtained that the error rate of the rotational speed sensor is 0.11%.

4. Conclusion

To sum up, this paper designs a sine wave to square wave circuit, which can adjust the duty cycle of the square wave by adjusting different resistors, and using the voltage divider resistor to control the signal amplitude within the range that can be read by the microcontroller. The frequency signal in the sine wave is converted into a square wave through circuit processing, and finally, the single-chip microcomputer reads and processes the square wave signal and outputs the rotational speed value of the rotational speed. This study can accurately measure the rotational speed at a frequency of 10-90 Hz, and control the rotational speed error to 0.11%. This research provides a feasible solution for rotational speed measurement and can be used to measure the occasions where various rotational speed sources emit sine waves.

Acknowledgment

The authors are grateful for the Jilin Province Science and Technology Department Project (YDZJ202201ZYTS374).

References

[1] Cheng, L.L.; Cigada, A.; Lang, Z.Q.; Zappa, E.; Zhu, Y . P . An output-only ARX model-based

sensor fusion framework on structural dynamic measurements using distributed optical fiber sensors and fiber Bragg grating sensors. *Mechanical Systems and Signal Processing* 2021, 152, 107439.

[2] Erdogan, Y .S.; Ada, M. A computer-vision based vibration transducer scheme for structural health monitoring applications. *Smart Materials and Structures.* 2020, 29, 085007.

[3] Fleming, A.J. A review of nanometer resolution position sensors: Operation and performance. *Sensors and Actuators, A: Physical* 2013, 190, 106–126.

[4] Ferrari, M.; Ferrari, V .; Mariola, D.; Taroni, A. Modeling, fabrication and performance measurements of a piezoelectric energy converter for power harvesting in autonomous microsystems. *IEEE Transactions on Instrumentation and Measurement.* 2006, 55, 2096–2101.

[5] George, B.; Tan, Z.; Nihtianov, S. Advances in capacitive, eddy current, and magnetic displacement sensors and corresponding interfaces. *IEEE TRANSACTIONS ON INDUSTRIAL ELECTRONICS.* 2017, 64, 9595–9607.

[6] Hagiwara, N.; Suzuki, Y ; Murase, H. A method of improving the resolution and accuracy of rotary encoders using a code compensation technique. *IEEE Transactions on Instrumentation and Measurement.* 1992, 41, 98–101.

[7] Liu, X.K.; Zhan, B.N.; Yu, Z.C.; Peng, K.; Pu, H.J. A miniaturized capacitive absolute angular positioning sensor based on a dual two-stage secondary re-modulation scheme with time-division multiplexing. *Sensors and Actuators, A: Physical* 2020, 310, 112043.

[8] Miedl, F.; Tille, T. 3-D surface-integrated touch-sensor system for automotive HMI applications. *IEEE/ASME Transactions on Mechatronics.* 2016, 21, 787–794.

[9] Peters K J H, Rodriguez S R K. Exceptional precision of a nonlinear optical sensor at a square-root singularity. *Physical Review Letters*, 2022, 129: 013901.

[10] Zhou L, Wu F, Hall R, et al. Electromagnetic sensors for in-situ dynamic microstructure monitoring of recovery and recrystallization in interstitial free steels. *Journal of Magnetism and Magnetic Materials*, 2022, 551: 169187.

A Fast Transient Low-Dropout Regulator with High-Power Supply Rejection

Ruokai Bai, Zhongjie Guo, Jiale Yang, Hao Shi, Xiao Han

School of Automation and Information Engineering, Xi'an University of Technology, Xi'an 710048

* Corresponding author: 1125431401@qq.com

Abstract: With the continuous popularization of portable electronic products in electronic information, aerospace, biomedicine and other fields, low-dropout regulator (LDO) has become an important part of modern power management chips. Compared with other power management modules, it has the advantages of low power consumption, high power supply rejection, and strong anti-noise capability. However, the LDO power tube has a large width-to-length ratio, so when the load changes transiently, the gate of the power tube cannot respond in time, resulting in an overshoot and undershoot of the output voltage and a long recovery time. In this paper, the LDO transient enhancement circuit is designed. When the load changes transiently, the power tube gate capacitor is quickly charged and discharged, which overcomes the overshoot voltage and undershoot voltage. While improving the transient response of traditional LDO, the structure ensures the power supply rejection (PSR) and other performance of the circuit. According to the results, with an input of 2.5 V, the output voltage is regulated to 1.2 V, and PSR is -80 dB. When the load changes by 20 mA in 1 μs, it is stable within 1.2 μs while the undershoot voltage is controlled at 51 mV and the overshoot voltage is controlled at 65 mV.

1. INTRODUCTION

With the popularity of portable electronic products such as mobile phones and computers, the output accuracy and load transient capability of power management chips are required. Compared with other power management modules, a low-dropout regulator (LDO) has the characteristics of simple structure, small scale, low cost and strong noise resistance in power management chips, and has been widely used in small electronic products, which can provide stable voltage to other modules in the product [1-2].

The traditional LDO, which is composed of a voltage reference, an error amplifier, a power tube and a feedback circuit, when the load changes, adjusts the power tube gate voltage through a negative feedback loop to provide the corresponding current for the load while ensuring the stability of the output voltage. However, in the traditional circuit, when the load changes transiently, due to the limited load adjustment ability of the circuit, it cannot respond in time, resulting in the overshoot or undershoot clock, and the output value can be restored after a period of time [3].

In this paper, a transient response enhancement circuit is proposed. When the load changes, the power tube gate is quickly charged or discharged through the transient enhancement circuit, which optimizes the overshoot or downshoot and reduces the stabilization time.

2. LDO TRANSIENT ENHANCEMENT CIRCUIT DESIGN

2.1 Transient response analysis

When the load current changes transiently in a very short period of time, from light load to heavy load, because the gate of the power tube cannot immediately respond to the voltage adjustment, at this time the corresponding current for the load is provided by the load capacitor, so the output voltage is reduced. After the feedback loop, the power tube gate voltage is reduced; the corresponding load current is provided; and the output voltage increases to the stable value.

Correspondingly, when changing from heavy load to light load, since the gate of the power tube cannot respond to the increase immediately, the output voltage will first increase and then decrease, and finally stabilize. As shown in Figure 1, since the power tube is used to drive the post-stage circuit, the aspect ratio is very large, and the gate capacitance is also increased. Because the gate capacitance is too large, the time for charging and discharge will increase accordingly, which will affect the transient response of LDO. The time from load change to output voltage recovery is divided into transient response time (t_R) and settling time (t_S) [4].

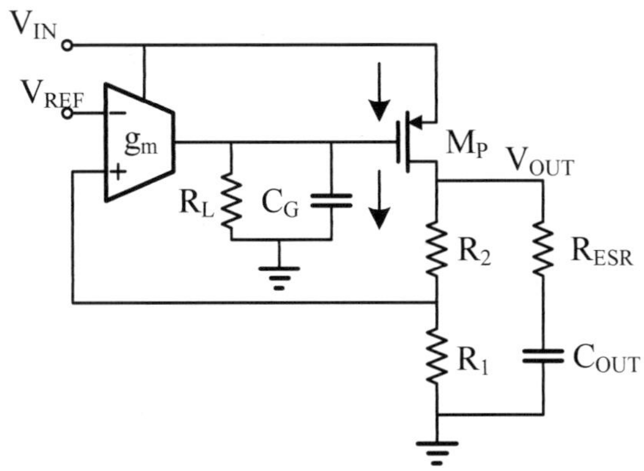

Figure 1. LDO Circuit Analysis

$$C_G = C_{ox}WL \tag{1}$$

$$t_R \approx \frac{1}{BW} + C_G \frac{\Delta V}{I_{SR}} \tag{2}$$

$$\Delta V \approx \frac{\Delta I}{C_{OUT}} t_r + \Delta I R_{ESR} \tag{3}$$

C_{ox} is the gate oxide layer capacitor. W and L are the power tube width and length, respectively. BW is the closed-loop bandwidth of the system. ΔV is the output voltage change value. I_{SR} is the power tube charge and discharge current. C_{OUT} is the load capacitance. R_{ESR} is the equivalent series resistance of the output capacitor. ΔI is the change in output current. Therefore, the main factors affecting t_R are bandwidth and gate charge-discharge current, while t_S is determined by open-loop stability. So this paper adds additional charge-discharge current to improve transient response capability while improving bandwidth and stability [5].

This paper designs two large currents to charge and discharge the power tube gate capacitor. When the light load becomes a heavy load, the output voltage downshoot is suppressed by discharging the power tube gate capacitor; when a heavy load becomes a light load, the upcharge voltage is suppressed by charging the power tube gate capacitor.

2.2 Transient reinforcement circuitry

The block diagram of the LDO transient enhancement circuit proposed in this paper is shown in Figure 2, which is mainly composed of an operational transconductance amplifier (OTA), a common-emission amplification circuit, and a charge and discharge branch. V_G is connected to the gate end of the power tube in LDO. As the load increases, V_{FB} decreases and the base current of the triode T_3 increases, passing through T_3 and R_5. The common-beam amplification circuit is formed so that the gate input of M_N is increased and the M_N tube is turned on to LDO. The gate of the power tube is discharged to increase the V_{OUT} voltage.

When the load decreases, the V_{FB} increases, that is, the base current of the triode T_1 increases. T_1, T_2, the common-beam amplification circuits composed of R_3 and R_4 reduce the gate input of M_P so that the M_P tube is turned on to LDO. The power tube gate is charged to reduce the V_{OUT} voltage. At the same time, to prevent the small fluctuation of V_{OUT} from causing the transient enhancement circuit to continue to work, resulting in overcompensation, a current source is introduced at the output of the OTA, when the output voltage fluctuates in a small range, The transient reinforcement circuit is not operational.

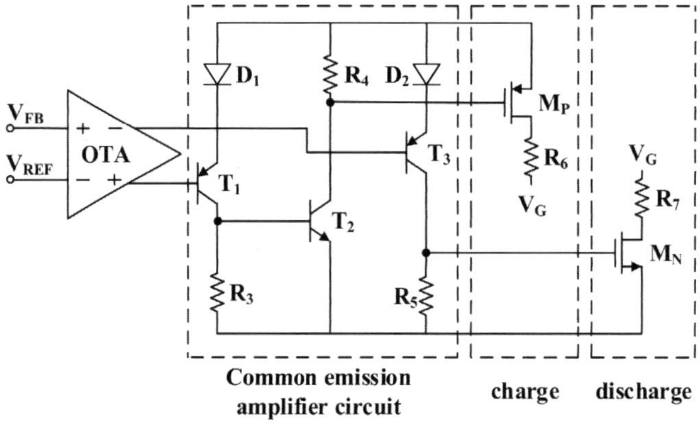

Figure 2. Transient Enhancement Circuit

2.3 LDO circuit design

The design in this paper takes into account the power supply rejection ratio, load regulation, linear regulation and stability of the LDO circuit. The error amplifier uses a folded source common gate amplifier with high or low frequency gain. The stability of the circuit does not depend on large load capacitors and makes use of the characteristics of the large output impedance of the cascode. The main pole is designed to be located at the output node of the error amplifier. For the stability of the system, a compensation capacitor C_M is added [6].

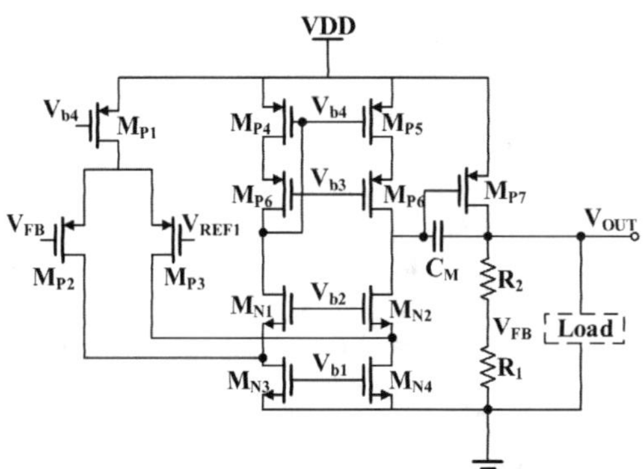

Figure 3 LDO Design Circuit

$$P_{EA} = \frac{1}{R_{EA} \times C_G} \tag{4}$$

$$R_{EA} = g_{mn2} r_2 r_4 \parallel g_{mp6} r_5 r_6 \tag{5}$$

R_{EA} is the output impedance of the cascode amplifier. C_G is the grid capacitance of the power tube. g_{mp7} is the transconductance of the power tube. R_1 and R_2 are the sums of the resistance values below the power tube. C_M is the compensation capacitor. g_{mn2} and g_{mn6} are the transconductances of M_{N2} and M_{P6} respectively. $r_2r_4r_5r_6$ are the saturation region impedance of $M_{N2}M_{N4}M_{P5}M_{P6}$.

3. SIMULATION RESULTS AND ANALYSIS

Under the 18 μm CSMC BCD process, the circuit was simulated in Spectre using cadence software, which mainly demonstrated the transient response and power supply rejection. 40 us is simulated in a transient state. The load current changes from 0 mA to 20 mA at 10 us to 11 us, and the load current changes from 20 mA to 0 at 30 us to 31 us to get the transient simulation results. In addition, a 1 V AC signal is added to the VCC signal to get PSR at the output.

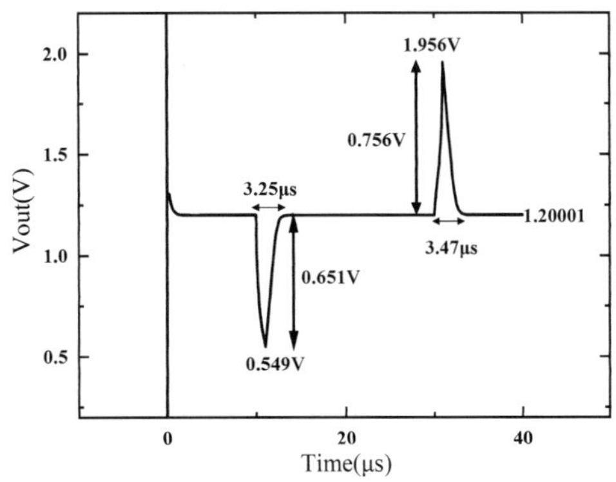

Figure 4. Before Transient Response Enhancement

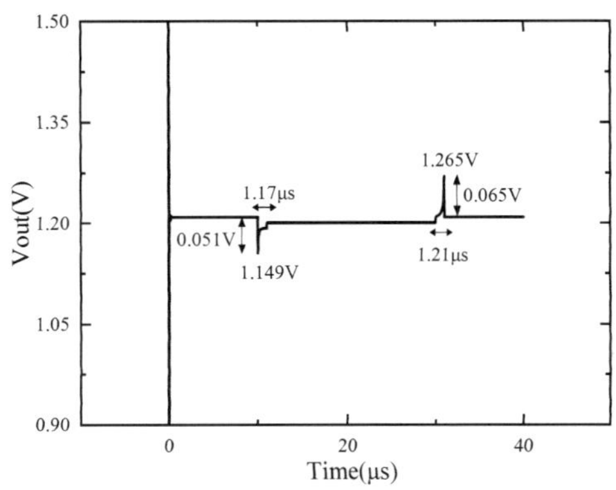

Figure 5. After Transient Response Enhancement

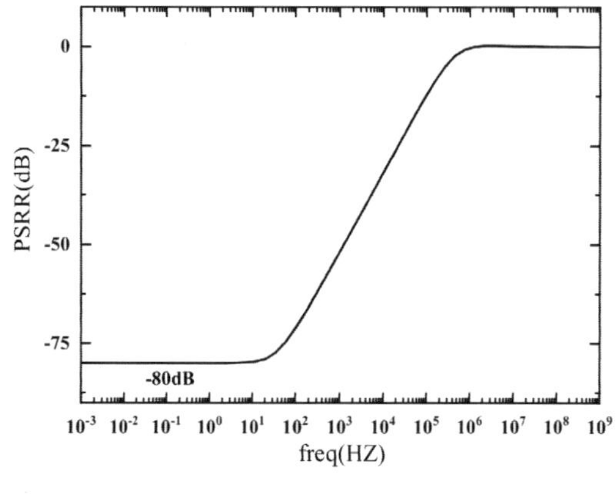

Figure 6. PSR Simulation Results

According to the simulation results that after adding the transient enhancement circuit, the transient response of the LDO is significantly improved; the overshoot voltage is improved by 691 mV; the undershoot voltage is improved by 600 mV; the transient response time is improved by about 2 μs. In addition, since the transient enhancement circuit does not work when the LDO does not have a transient response, the PSR of the LDO is hardly affected under static operation, reaching -80 dB.

4. CONCLUSION

In this paper, the LDO transient response enhancement circuit is designed based on the 0 18 μm CSMC BCD process. When the load changes transiently, the transient response is suppressed by rapidly charging and discharging the gate capacitance of the power tube. The important conclusions are summarized as follows: (1) the voltage is 1.2 V at the supply voltage of 2.5 V, and PSR is -80 dB; (2) when the load changes by 20 mA in 1 μs, it is stable within 1.2 μs while the undershoot voltage is controlled at 51 mV and the overshoot voltage is controlled at 65 mV. This technology achieves transient corresponding enhancements while ensuring other properties such as power supply rejection. In future studies, LDO should be able to cope with more drastic transient changes in a shorter period of time, further enhancing its ability to cope with transient responses and output a stable voltage. At the same time, LDO's power supply rejection, anti-noise ability, and output accuracy are also very important indicators. While improving the transient state, these indicators must also be guaranteed.

REFERENCES

[1] M. Khan and M. H. Chowdhury, "Capacitor-less Low-Dropout Regulator (LDO) with Improved PSRR and Enhanced Slew-Rate". 2018 IEEE International Symposium on Circuits and Systems (ISCAS), 2018, pp. 1-5.

[2] Y. Peng, S. Huang and Q. Duan, "A Fast Transient Response LDO without on-chip Compensated Capacitor". 2021 4th International Conference on Electron Device and Mechanical Engineering (ICEDME), 2021, pp. 163-166.

[3] F. Lavalle-Aviles, J. Torres and E. Sánchez-Sinencio, "A High-Power Supply Rejection and Fast Settling Time Capacitor-Less LDO". in IEEE Transactions on Power Electronics, vol. 34, no. 1, pp. 474-484, Jan. 2019.

[4] N. Adorni, S. Stanzione and A. Boni, "A 10-mA LDO With 16-nA IQ and Operating From 800-mV Supply". in IEEE Journal of Solid-State Circuits, vol. 55, no. 2, pp. 404-413, Feb. 2020, doi: 10.1109/JSSC.2019.2948820.

[5] Y. Zhou, Y. Zheng and K. N. Leung, "An Output-Capacitorless Low-Dropout Regulator with High Slew Rate and Unity-Gain Bandwidth". 2020 IEEE International Symposium on Circuits and Systems (ISCAS), 2020, pp. 1-5.

[6] F. Montalvo-Galicia, G. Diaz-Arango, C. Ventura-Arizmendi, B. Calvo and M. T. Sanz-Pascual, "Comparison of Two Internal Miller Compensation Techniques for LDO Regulators". 2019 16th International Conference on Electrical Engineering, Computing Science and Automatic Control (CCE), 2019, pp. 1-4.

Optimization of Planar Array Electrostatic Sensor for Metal Surface Defect Detection

Qian Xue[1, a], Xijuan Chen[2*]

[1]College of Electronic Information and Automation, Civil Aviation University of China, Tianjin, China

[2]College of Electronic Information and Automation, Civil Aviation University of China, Tianjin, China

[a]qxue@cauc.edu.cn

*Corresponding author: 1196036697@qq.com

Abstract—Static electricity is usually generated in the damage area of the metal surface due to contact friction, and the charge distribution density can reflect the size, shape, and relative position of the damage area. Based on this, a planar array electrostatic sensor is designed to detect metal surface defect in this paper, and the shielding method, number of electrodes, electrode shape, and arrangement of the sensor are optimized taking account of the induced charge value, the uniformity of sensitivity and the image correlation coefficient. Different image reconstruction algorithms (e.g. Landweber algorithm, conjugate Gradient algorithm, Tikhonov regularization and primary dual interior point method) are utilized to evaluate the performance of the designed electrostatic sensor. The results demonstrated that the sensor with hexagonal electrode shape, integrated shielding, a new arrangement, a duty cycle of 80%, and a peripheral shielding electrode, has better image quality for all the tested damage models. When using the PDIPA algorithm for image reconstruction, the image correlation coefficient can exceed 0.9.

1. INTRODUCTION

Some uncontrollable defects often occur on the surface of sheet metal in the production process, such as scratches, cracks, dirty spots, metal surface peeling, scarring, peeling, defects [1], etc. The surface contact damage of these parts often leads to mechanism failure or even complete failure. There are many detection methods for metal surface contact damage. Manual visual inspection [2] is the initial method used. This method is prone to misjudgment and error, and the efficiency is low. Eddy current testing [3] mainly uses the eddy current phenomenon to detect defects, which -is greatly affected by the shape of the test pieces, inefficient and difficult to identify the types of defects. Magnetic flux leakage detection [4], is low-cost, and can detect surface defects and small internal defects. Although it can only detect metal materials, it's unable to identify the types of defects. Ultrasonic flaw detection [5] requires couplant to detect internal damage, but there is a near-field blind area, which has poor detection effect on the near surface with additive flaws of expensive equipment and high detection costs. Infrared detection technology [6] estimates the depth of the defect by calculating the local heating value, but only limited types of defects can be identified. Magnetic particle inspection [7] cannot detect non-magnetic materials, and it is not easy to detect surface defects such as scratches. The entire inspection process is more complicated and poor in real-time. Penetration inspection [8] can only detect defects with loopholes on the workpiece surface. Traditional image processing technology [9] has poor adaptability and strict requirements for polishing, and it is difficult to meet the demands of general detection of surface defects.

There is no unified theoretical explanation for the cause of static electricity caused by metal contact damage at present. The generation of static electricity is caused by many factors. When different metal surfaces contact each other, plastic deformation occurs at the top of the contacted micro convex body, which leads to the formation of phase transition zone. The surface phase change regions of different metals have different contact potential differences, and mutual contact leads to static electricity on the metal surface. The fluctuation of potential can reflect the corrosion process of metal material surface [10], [11]. Even if there is no obvious damage on the metal surface, the metal surface will have weak static charge for various reasons, and this weak static charge is easy to lose. Zuo Hongfu team of Nanjing University of Aeronautics and Astronautics carried out the experiment of detecting two damages on the metal surface through a single electrostatic sensor. the team's research results show that the signal measured by the sensor is easily disturbed by environmental noise, and the amplitude of the charge signal is too small. Through the signal processing device, it is obtained that there is an obvious time interval between the signal peaks corresponding to the two damages, which verifies the feasibility of using multiple electrostatic sensors to locate the damage on the metal surface. Metal surface defect detection technology based on electrostatic induction data [13] has the advantages of real-time monitoring and direct measurement. However, the electrostatic signal is generally too weak for the sensor to detect, so it needs to be further optimized.

2. OF PLANAR ARRAY ELECTROSTATIC SENSOR

The geometric size and material parameters of the sensor determine the induced charge and the corresponding sensitive field distribution on the electrode. The characteristics of the sensitive field distribution of electrostatic sensors are important factors that affect the quality of image reconstruction. Due to the complexity of the sensitive field of the electrode array, it is hard to solve it by a specific mathematical model, and it is generally solved by a numerical method. In this paper, the sensitive field of the sensor is analyzed and solved by finite element method. The sensitivity distribution characteristics of the sensor are affected by the structure of the sensor and the size parameters of the electrodes, such as the number, radius, shape, and arrangement of the electrodes.

2.1 Model of Planar Array Electrostatic Sensor

The model is 5×5, a total of 25 sensing probes are arranged in a planar array, and the signals measured by the probes are used to reconstruct the image of metal plane electrostatic charge distribution [15]. As shown in Figure 1, the center distance between two probes is 7mm The label of each probe is arranged as shown in Figure 1, and the probe material is copper. The effective measurement plane is a 28mm×28mm square sheet with a thickness of 0.1mm, which is divided into 256 areas of 16×16, and each area is a square with a side length of 1.75mm×1.75mm. The shell is formed by the difference set of two cylinders with an inner circle radius of 23mm and an outer circle radius of 25mm. The thickness is 2mm and filled with copper. The three-dimensional model is shown in Figure 2.

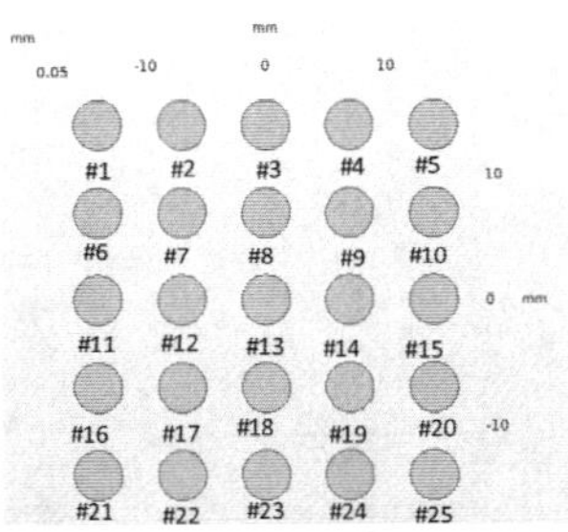

Figure 1 Electrode label

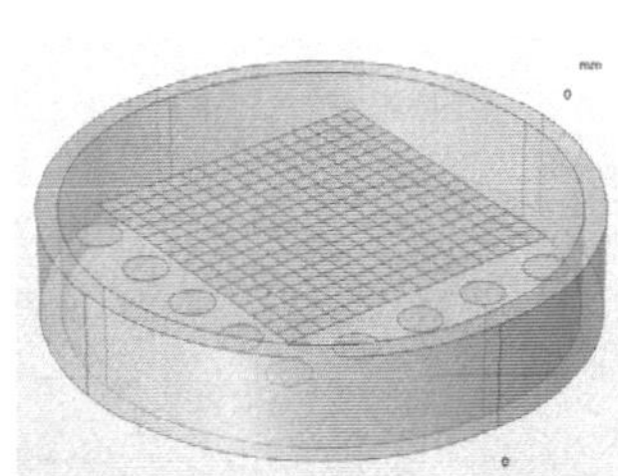

Figure 2 Three-dimensional model

In electrostatic imaging system, there are many image reconstruction algorithms used to solve inverse problems. Including Landweber algorithm, CGLS (conjugate gradient algorithm), L1 norm regularization algorithm, Tikhonov regularization algorithm, PDIPA (primitive dual interior point algorithm) and so on. this paper makes a comparative analysis of imaging through these five reconstruction algorithms.

3. ELECTROSTATIC SENSOR STRUCTURE OPTIMIZATION

The role of the sensor is indispensable in the metal surface defect detection technology. In this paper, aiming at the uniformity of sensitivity distribution, the magnitude of induced charge and image correlation coefficient. And the shielding mode, the number of electrodes, electrode shape and electrode arrangement of the sensor are compared and optimized.

3.1 Optimization of shielding method

To weaken the influence of the external electric field on the measured value of the sensor, different electrode shielding methods are designed and compared. A fence-type shield is shown in Figure 4, the height of the fence is slightly higher than that of the electrode sheet, each fence is filled with copper and grounded to shield the interference between the electrodes. Figure 5 shows the ring shield for each electrode. The thickness of the ring is 0.2mm and is slightly higher than the electrode sheet. It is filled with copper and grounded. There is a gap between the ring and the electrode sheet. The gap is 0.3mm and is filled with air. Figure 7 is designed as an integrated shielding base. When making the base, 25 round holes are reserved. The size is 2.5mm, which is 0.5mm larger than the size of the electrode sheet. It is used to place the electrode sheet. There is no direct contact between the electrode sheet and the base, and the gap is filled with air.

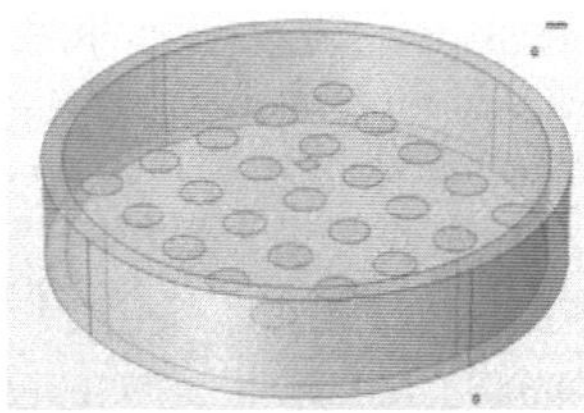

Figure 3 Initial model

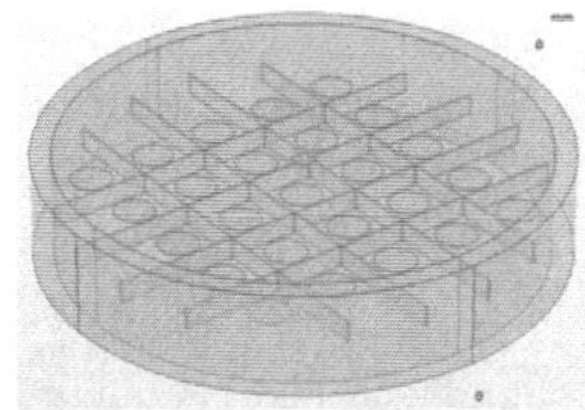

Figure 4 Barrier shield model

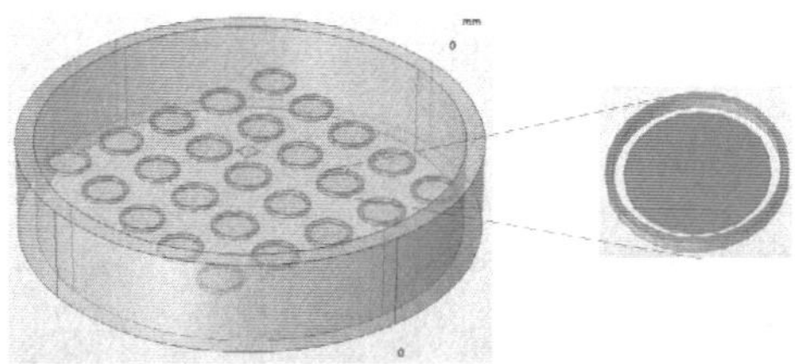

Figure 5 Each electrode plus ring shield model

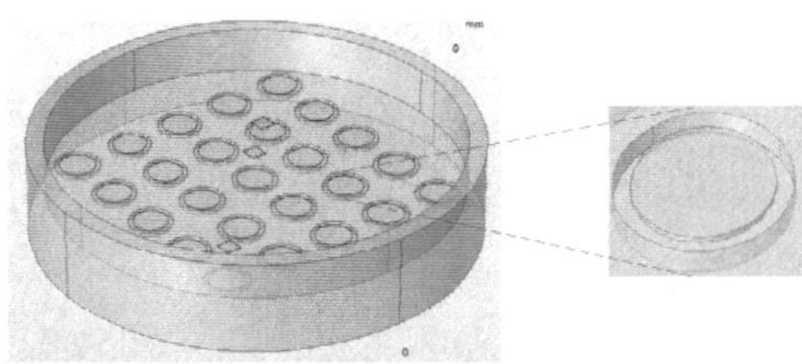

Figure 6 Shield and base integrated model

Taking the center of the effective measurement plane as the origin of coordinates, the coordinates of the electrodes are shown in Table 1:

Table 1 Electrode coordinates

#1	#2	#3	#4	#5
(-14,-14)	(-7,14)	(0,14)	(7,14)	(14,14)
#6	#7	#8	#9	#10
(-14,7)	(-7,7)	(0,7)	(7,7)	(14,7)
#11	#12	#13	#14	#15
(-14,0)	(-7,0)	(0,0)	(7,0)	(14,0)
#16	#17	#18	#19	#20
(-14,-7)	(-7,-7)	(0,-7)	(7.-7)	(14,-7)
#21	#22	#23	#24	#25
(-14,-14)	(-7,-14)	(0,-14)	(7.-14)	(14,-14)

The distribution density of the electrostatic charge on the metal surface reflects the damage. There are four positional relationships between the electrode sheet and the space charge: in the center of the sensor; between the electrode and the electrode; partly above the electrode and partly in the gap position; at the edge of the effective measurement plane. In this paper, the charge coordinates corresponding to the four positional relationships are selected, as shown in figure 7. Four kinds of shielding sensors are used to measure the induced charge value, and the sensitivity matrix is calculated to solve the inverse problem, and imaging is carried out through the Landweber algorithm. The results are shown in Table 2.

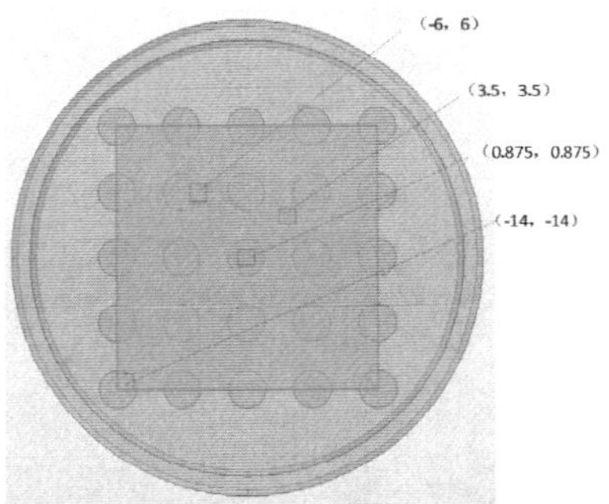

Figure 7 Coordinates of charge position

Table 2 The imaging comparison of each sensor when a charge is placed

Charge coordinate	Original model	Ring shield	Fence shield	Integrated shield base
Electrode distribution				
Sensitivity of No. 1 electrode				
Sensitivity distribution	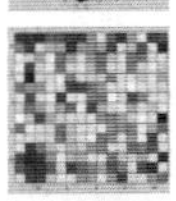	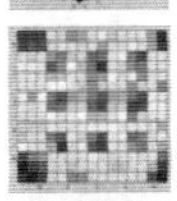		
(-0.875, -0.875)				
(3.5, 3.5)		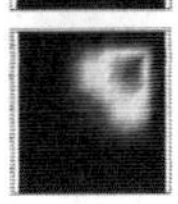	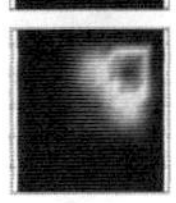	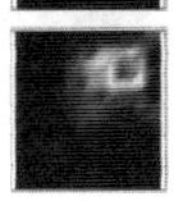

(-6, 6)	
(-14, -14)	

After shielding between the electrodes, the imaging effect has been significantly improved and many artifacts have been reduced. From the sensitivity distribution diagram, the sensitivity distribution of the three types of sensors with ring-shaped shielding, fence-type shielding, and shielding base integrated for each electrode is relatively uniform; when the charge is located directly above the electrode sheet, the imaging effect of the three shielding methods are similar; when the charge is located between the poles, from the image point of view, the integrated sensor with shielding and base has the best imaging effect; when the charge is partly above the electrode and part is in the empty position, the barrier-type shielding effect is the best; when the charge is located at the edge of the effective measurement plane, the imaging effects of the three sensors are similar.

Due to the inevitable noise interference in the actual measurement, the measurement accuracy will be seriously affected if the induced charge value is too small. The magnitude of electrode induced charge value at different relative positions should be considered in the sensor design.

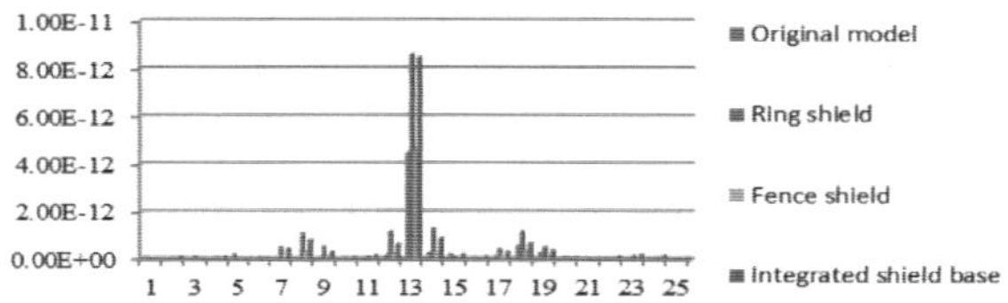

Figure 8 The induced charge value when the charge is directly above the electrode sheet

When the charge is located in the center of the sensor, by comparing the induced charge value on the 13th electrode, it can be seen that the sensor with a ring-shaped shield on each electrode and the sensor with an integrated shield base are close to each other.

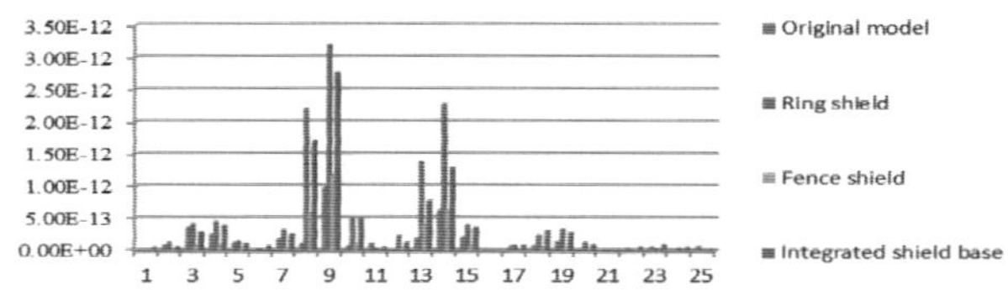

Figure 9 When the charge is located between the electrode and the electrode, the measured value of the charge on each electrode

When the electrode is located between the poles, each electrode plus the ring-shaped shielded sensor can sense the largest amount of charge, followed by the sensor with an integrated shielding base.

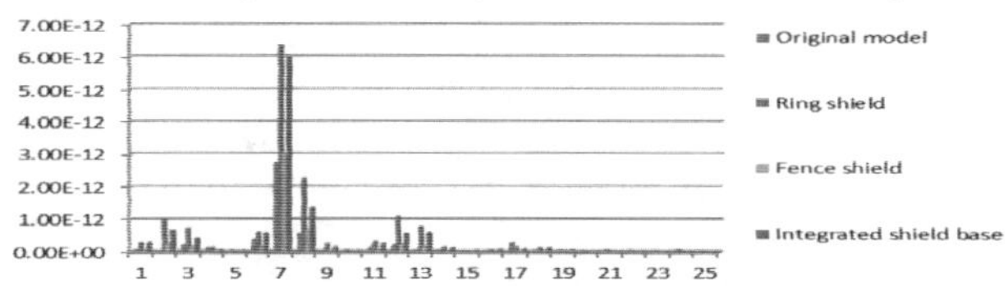

Figure 10 The measured value of the amount of charge on each electrode when the charge part is above the electrode sheet and the part is in the empty position

When the charge is located above the electrode sheet and the part is in an empty position, compare the induced charge value on the No. 7 electrode. The sensor with a ring shield on each electrode has the largest induction value, followed by the sensor with an integrated shield base. Through the comprehensive comparison of the induced charge value and the image reconstruction effect, the integrated sensor with the shield base is significantly better than the other three types.

3.2 Optimization of the number of electrodes

Compared with other electrical tomography technologies, electrostatic sensors are more ill-qualified, and the independent measurement values obtained are equal to the number of electrodes in the sensor. Increasing the number of electrodes can increase the independent measurement value, but the electrode area is correspondingly reduced, which reduces the accuracy of the induced charge measurement. Therefore, the design of the number of electrodes and geometric dimensions should consider the number of measured values and accuracy.

As shown in Figure 11, the electrode radius is set to x, the electrode spacing is set to y, the center distance is $2x+y$, and the unit is millimeter (mm), then the duty cycle is expressed as $a=\frac{2x}{2x+y}$.

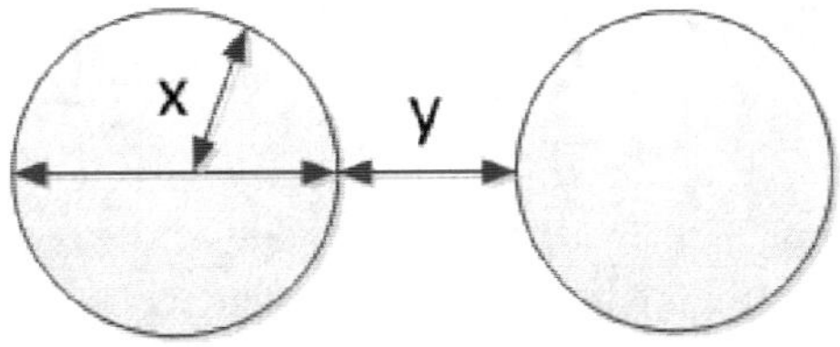

Figure 11 Schematic diagram of duty cycle

The 5x5 array sensor has an electrode radius of 2mm, an electrode spacing of 3mm, and a center-to-center distance of 7mm, which means the duty cycle is 4/7. When optimizing the number of electrodes, keep the duty cycle and the effective measurement plane unchanged. As the number of electrodes gradually increases, the electrode radius gradually decreases. The sensitivity distribution of sensors with different numbers of electrodes and the effect of Landweber image reconstruction are shown in Table 3.

Table 3 The imaging effect of sensors with different numbers of electrodes

Charge coordinate	4x4	5x5	6x6	7x7	8x8	9x9
Electrode distribution						
Sensitivity of No. 1 electrode						
Sensitivity distribution	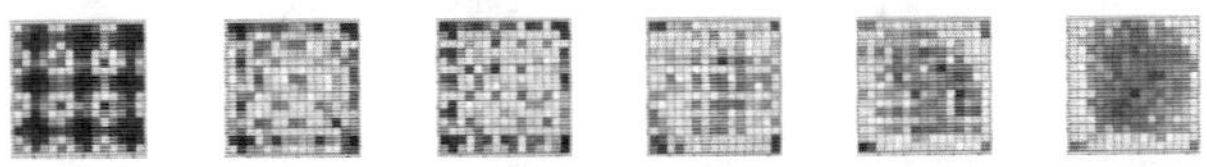					
(0, 0)						
(7, 7)	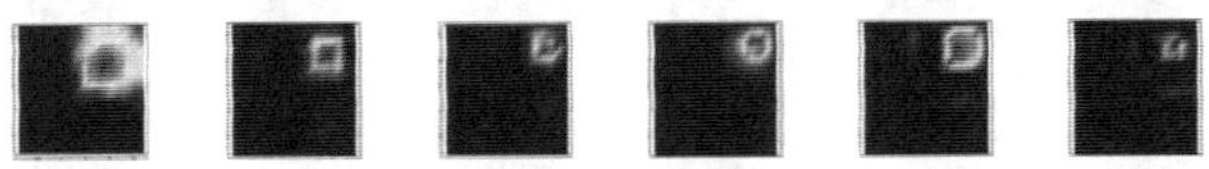					
(-7, 7)						
(-14, -14)						

The greater the number of electrodes, the more uniform the sensitivity distribution, the better the image reconstruction effect, and the higher the accuracy. It can be seen from Table 3 that the number of electrodes from 25 to 81 has better imaging quality.

Table 4 Comparison of the image reconstruction of the two regions charged with static charges by the sensors with different numbers of electrodes

Charge coordinate	5x5	6x6	7x7	8x8	9x9
(7, 7) (-7, -7)	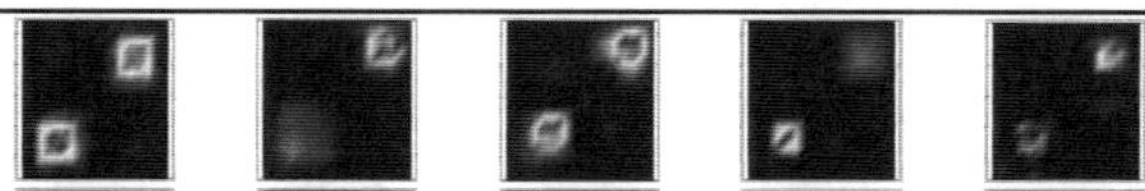				
(7, 12.25) (7, 0)	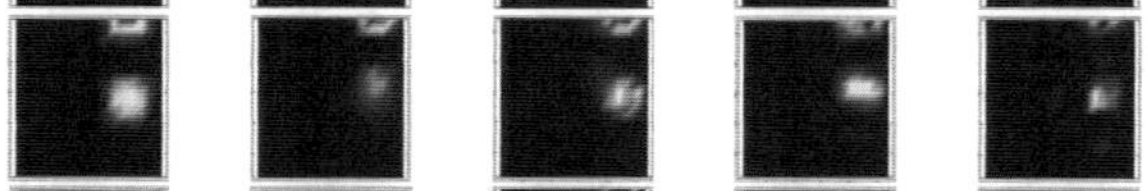				
(7, 12.25) (7, 3.5)	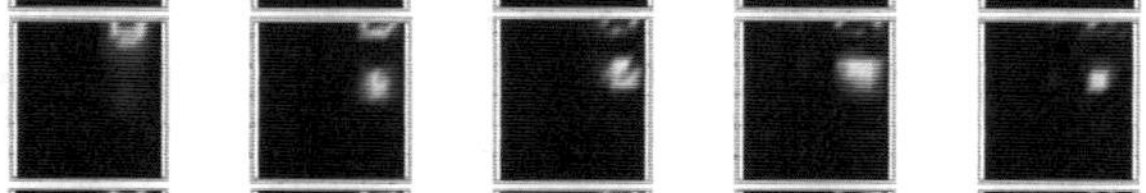				
(7, 12.25) (7, 7)	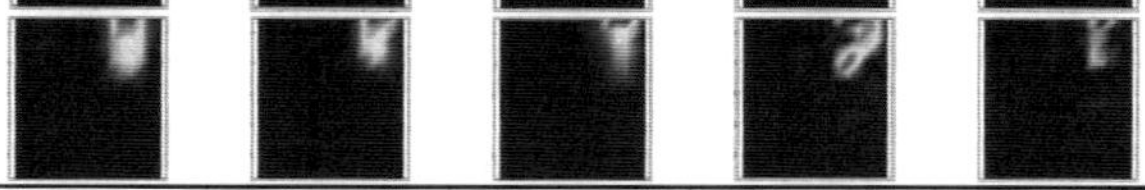				

For two areas with an electrostatic charge, when the distance is large, the first and second rows in Table 4, the above sensors can distinguish the two areas charged. When the distance is close, the charge distance between the two regions is 8.75mm (the third row in the table), the 5×5 array cannot distinguish the charged area. When the distance is closer to 5.25mm (the fourth row in the table), the 5×5 array, 6×6 array and 7×7 array cannot be distinguished. Only 8×8 array and 9×9 array sensors can be distinguished, and the imaging effect of the 9×9 array sensor is the best.

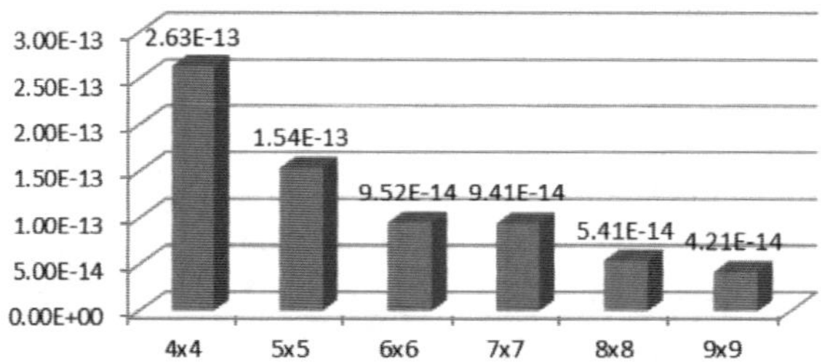

Figure 12 The amount of induced charge of sensors with different numbers of electrodes (take the average value)

3.3 Optimization of electrode shape

Under the condition that the area of the electrode sheet is kept constant, electrodes of different shapes are designed. As shown in Figure 13, the radius of the circular electrode is 2mm, and the area of the electrode sheet is 12.56mm^2. Based on the area of the circular electrode sheet, the relevant parameters of the other four shapes are calculated. The side length of the square electrode is 3.544mm; the length of the rectangular electrode is 4mm and the width is 3.14mm; the inner radius of the ring electrode is 0.64mm and the outer radius is 2.1mm; the side length of the regular hexagon is 2.2mm.

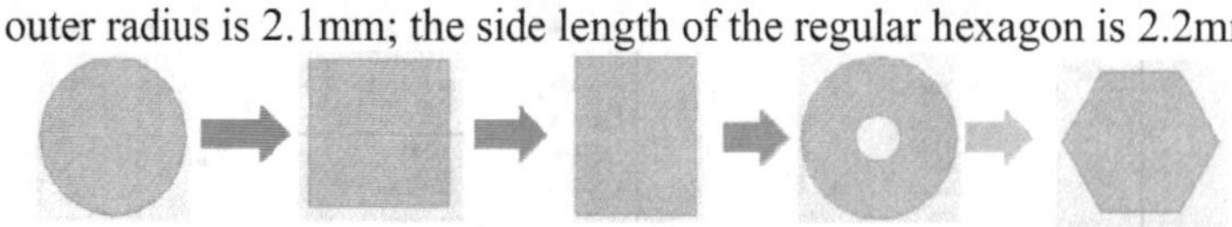

Figure 13 Electrode shape

When the effective measurement plane is locally charged with an electrostatic charge, set a region with an electrostatic charge, and use sensors with different electrode shapes to perform image reconstruction. The results are shown in Table 5.

Table 5 Sensor imaging effects of different electrode shapes

Charge coordinate	Round	Square	Rectangu lar	Ring	Hexagonal
Electrode distribution					
Sensitivity of No. 1 electrode					
Sensitivity distribution	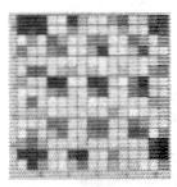	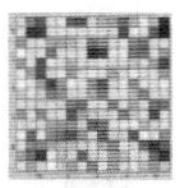		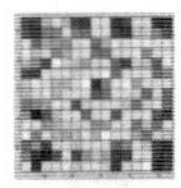	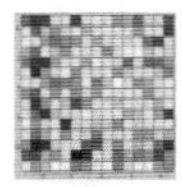
(-0.875, -0.875)					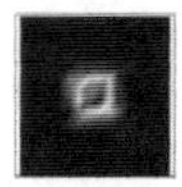
(3.5, 3.5)	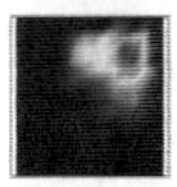			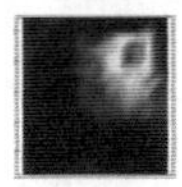	
(-14, -14)					

Through the comparative analysis in Table 5, the sensitivity distribution of sensors with regular hexagon electrodes is more uniform and image artifacts are less.

3.4 Optimization of the electrode arrangement

The electrode arrangement affects the performance of the sensor. Generally, the sensors are arranged in a square array, and the utilization rate of the electrodes at the edge of the effective measurement is not high, and the effective part of the electrode sheet is only 1/6 or 1/2.

In this section, according to the characteristics of the regular hexagon, the regular hexagon electrode at the center of the sensor is taken as the center and diverges to the six sides, and the duty cycle remains at 4/7. Due to the characteristics of the regular hexagon, the duty cycle of the adjacent electrode pieces is 4/7,and the number of electrodes is 23, which effectively improves the utilization of edge electrodes, as shown in Figure 14(b). Figure 14(c) shows that a circle of shielding electrodes is set around the measuring electrode sheet, which can improve the ability of the edge electrode to acquire the amount of charge and improve the sensitivity and anti-interference ability of the edge electrode. Figure 14(d) shows that the duty cycle is adjusted to 80%, 45 electrodes, and a circle of shield electrodes on the periphery.

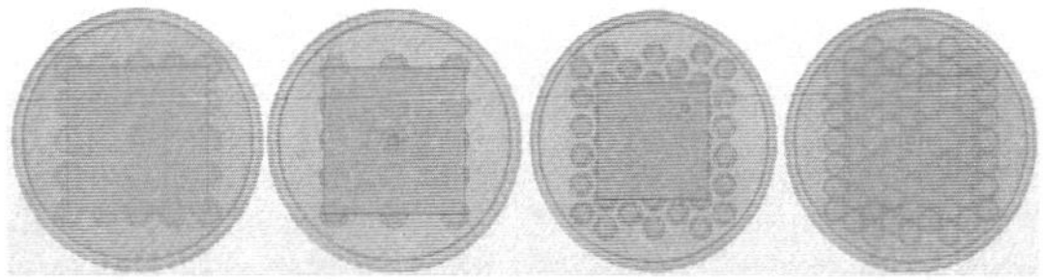

(a) (b) (c) (d)

Figure 14 (a) Ordinary hexagonal array (b) After rearrangement (c) The outer circle of shielding electrode (d) The duty cycle is 80% The outer circle of the shielding electrode

Table 6 Landweber imaging comparison of different electrode arrangements

Electrode arrangement	Sensitivity distribution	(-0.875, -0.875)	(3.5, 3.5)	(-14, -14)	(7, 7) (-7, -7)	(7,12.25) (7,3.5)

form	
Ordinary hexagonal array	
After rearrangement	
Add shield electrode	
Duty ratio 80%, plus shielding electrode	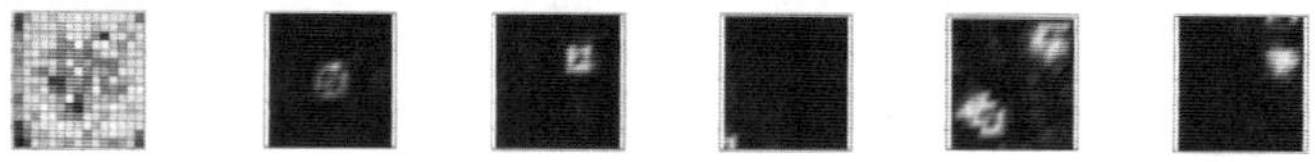

As shown in Table 6, when a local area is charged or two areas are charged, the sensor with a rotating hexagonal arrangement and a shielding electrode added to the periphery produces fewer image artifacts (from the third row in the table). It is obvious from the last column that the image quality and resolution are improved compared with conventional array sensors. Subsequently, the duty cycle was adjusted to 80%, and the number of electrodes was 45. From the comparison in Table 6, the image reconstruction quality of the adjusted sensor was improved compared with the first three.

4. SIMULATION EXPERIMENT RESULTS AND ANALYSIS

Through the previous analysis, a sensor with a hexagonal electrode shape, an integrated shielding base, a new arrangement, a duty cycle of 80%, and a circle of shielded electrodes around the sensor has a better imaging effect. Now, with this sensor, some common algorithms are used to perform an imaging comparison between a local area of a metal plane and two areas of a metal plane. The imaging results are shown in Table 7. It can be seen from the reconstructed image that the image reconstructed by PDIPA has the best effect.

Table 7 Comparison of image reconstruction with different algorithms

	landweber	cgls	L1	Tikhonov	PDIPA
(7, 7) (-7, -7)	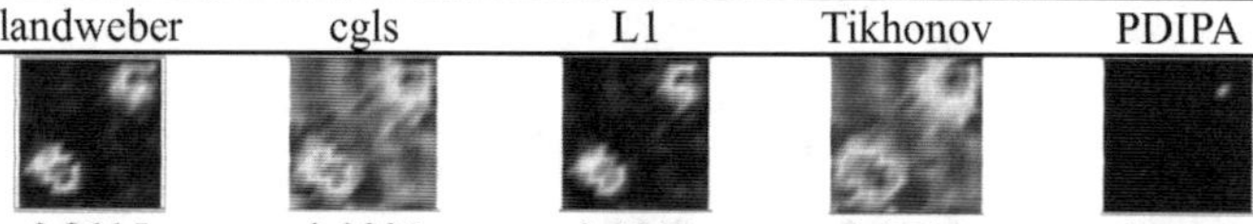				
Corr	0.3115	0.2897	0.3210	0.2793	0.9011
(7, 12.25) (7, 7)					
Corr	0.4746	0.4576	0.4720	0.4595	0.9146
(7, 12.25 (7, 3.5)	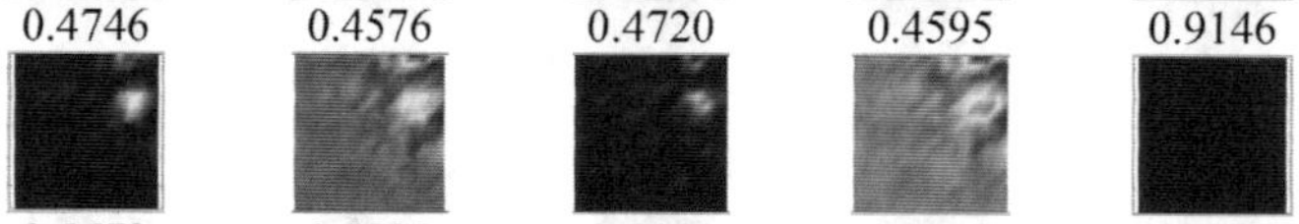				
Corr	0.5672	0.5585	0.5445	0.5341	0.9995
(7, 12.25) (7, 0)					
Corr	0.5608	0.5260	0.5365	0.5188	0.6597

Taking the image correlation coefficient as the evaluation index, the calculation formula of the image correlation coefficient is:

$$Corr = \frac{\sum_{i=1}^{M}\left(g_1 - \overline{g_1}\right)\left(g - \overline{g}\right)}{\sqrt{\sum_{i=1}^{M}\left(g_1 - \overline{g_1}\right)^2 \sum_{i=1}^{M}\left(g - \overline{g}\right)^2}} \tag{1}$$

Where g represents the ideal charge distribution matrix; $\overline{g}$ represents the average value of g; g_1 represents the gray value matrix of the reconstructed image based on different algorithms, and $\overline{g_1}$ represents the average value of g_1.

The calculated image correlation coefficient is shown in the Corr value in Table 7, and it is more intuitive to express it in the form of a histogram, as shown in Figure 15. It can be seen more intuitively from the comparison graph of image correlation coefficients that the image correlation coefficient of the PDIPA algorithm is the highest.

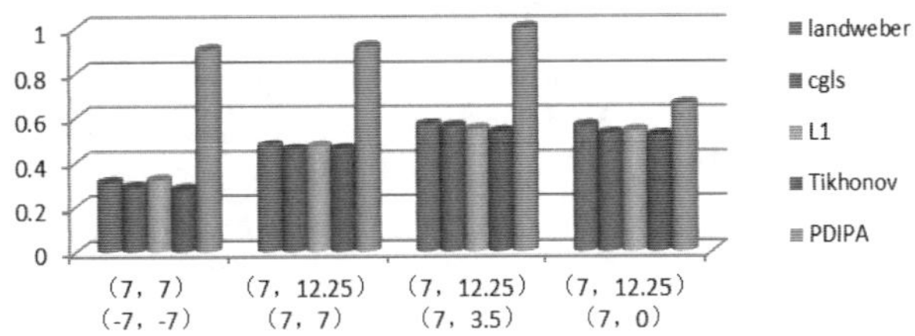

Figure 15 Correlation coefficients of image reconstruction with different algorithms

5. CONCLUSION

The main work of this paper is to optimize a planar array sensor for metal surface defect detection. Several typical cases of local electrostatic charge on the metal surface are set up, and the designed sensors are used to induct and identify the typical charge types and image reconstruction. Through the optimization design of the shielding method, the number of electrodes, electrode shape, electrode arrangement method, etc. The simulation experiment results show that the shielding method is an integrated shielding base, the electrode shape is hexagonal, the arrangement is in a new way, and the duty cycle is 80% and the sensor with a shielding electrode on the periphery has better identification and induction of electrostatic charges on the metal surface. At the same time, five commonly used algorithms such as Landweber, CGLS, L1 norm regularization, Tikhonov regularization, and PDIPA are used for image reconstruction. The results show that the PDIPA algorithm is the best, the image reconstruction effect is the best, and the image correlation coefficient is the highest.

REFERENCES

[1] Liu M. (2020) Research on Metal Sheet Surface Defect Detection System Based on Deep Learning. D. Qingdao University of Science and Technology, Science,

[2] Tang B, Wang X D, Liu Z, Liu Y J. (2017) Machine Vision Inspection and Recognition of Steel Plate Surface Defects. C. Proceedings of 2017 6th International Conference on Manufacture Engineering, Quality and Production.

[3] Han N, Zhang Z J, Yin W L, Zhao C Y. (2019) Research on metal surface defect detection method based on eddy current sensor. J. Journal of Transducer Technology, 32(05):704-710.

[4] Li C, Zhang Z J, Han N, Zhao C Y. (2021) Design of metal surface defect detection system based on GMR sensor. J. Computer Measurement and Control, 29(01):59-65+70.

[5] Zhou X T, Huang X Y. (2021) Research on Visual Inspection of Metal Material Surface Defects Based on Laser Ultrasonic. J. Industrial Control Computer, 34(04):35-36.

[6] Xie J, Xu C H, Chen G M. (2012) Infrared Thermal Image Surface Defect Detection of Metal Workpiece Based on MRF Model. J. Petroleum Field Machinery, 41(11):20-24.
[7] Liu Y Z. (2016) Research on Defect Detection Technology of Steel Structure Based on Electromagnetic Ultrasonic. D. Hebei University of Technology.
[8] Liu H J. (2020) Research on the Detection and Recognition Method of Metal Workpiece Surface Defects, Xiangtan University. 65
[9] Zhou C. (2013) Research on steel plate surface defect detection system based on image processing. D. Liaoning University of Technology.
[10] He Y D, Qi H B. (2005) Introduction to Material Corrosion and Protection. M. Beijing: Mechanical Industry Press, 96-101.
[11] Wang M F, Wei H Y, Chen D C, Wan B, Liu G H, Li G D. (2016) Kelvin probe technology to detect early pitting sensitivity of stainless steel with different surface roughness. J. Corrosion and Protection, 37(11):896-899.
[12] Zhong Z R, Zuo H F, Guo J C, Jiang H. (2020) An estimation method of particle charge based on an array electrostatic sensor. J. Chinese Journal of Scientific Instrument, 41(07):80-90.
[13] Jiang H, Zuo H F, Guo J C, Zhong Z R. (2021) An electrostatic monitoring method for aero-engine blade tip radial clearance. J. Journal of Aeronautical Dynamics, 36(03):466-476.
[14] Liu R C, Xu C, Sun J Z. (2021) Simulation analysis of structural parameters and spatial characteristics of electrostatic sensors. J. Computer Simulation, 38(06):184-189+194.
[15] Xu Y M. (2016) Research on the health monitoring and maintenance strategy of the flap mechanism of civil aircraft. D. Nanjing University of Aeronautics and Astronautics.
[16] Liu J. (2020) Research on EST image reconstruction algorithm based on machine learning. D. Civil Aviation University of China.

Research on the electronic switch of power management circuits for triboelectric nanogenerator

Mingqiang Lv, Guan Liu, Yi Zhang, Linchong Han and Jing Zhao*

School of Electrical and Electronic Engineering, Changchun University of Technology, Changchun, Jilin Province 130012, China.

*Corresponding authors' e-mail address: zhaojing@ccut.edu.cn

Abstract. Nowadays, people are increasingly interested in how to collect clean energy from nature more effectively. As an emerging harvest, triboelectric nanogenerators (TENGs) can collect a wide range of clean energy from nature. However, the energy generated by TENG can not be used directly for driving various miniature electronics. In this study, we have designed a triboelectric nanogenerator for wind energy harvesting, which consists of two simple generator sets. In addition, a power management circuit based on this triboelectric nanogenerator has been designed, which consists of a rectifier section, an electronic switch section, and a filter section. Among them, the most important part is the electronic switch section, and the electronic switch is based on triode characteristics. The experiment shows that the output voltage has been to stabilize at 3.89V when the triboelectric nanogenerator structure is in operation, and it can sustainably drive the LED in a stable working condition.

1. Introduction

With the rapid development of the world economy, the demand for oil, coal, and various other natural resources is constantly increasing [1]. This has led to increased levels of environmental pollution and an energy crisis. How to use mechanical energy in the environment has become a major issue for researchers [2]. In 2012, Wang's group invented triboelectric nanogenerators (TENGs), which can collect various mechanical energies in real life [3-4]. The basic principle of triboelectric nanogenerators is the coupling of electrostatic induction effects and triboelectrification, which allows them to collect and convert small amounts of energy from the natural environment into electrical signals. Moreover, the simple mechanical structure of the triboelectric nanogenerator, with periodic charge transfer between the two electrodes, allows the output signal to be alternating current (AC) [5-6]. Such a simple structure also makes it possible for all kinds of mechanical energy in the environment to be harvested, such as wind, water, raindrops, bioenergy, vibration energy, and other energy [7-8]. However, as the output of the triboelectric nanogenerator is an AC signal, there is an urgent need for a power management circuit to convert the generated AC signal into a stable electrical signal to supply various electronic components [9-10].

To this end, we have developed a triboelectric nanogenerator for wind energy harvesting and designed a power management circuit for this purpose. The designed triboelectric nanogenerator consists of a cap, a wind scoop, a rotor, a stator, and a rotating shaft. When the wind scoop receives wind energy to make it rotate, the rotation shaft and rotor rotate with the rotation of the scoop. The power management circuit mainly consists of a double voltage circuit to rectify the AC signal generated by the TENG, a buck circuit is used to reduce voltage and reduce voltage fluctuation at the same time. In addition, the main purpose of this paper is to design an electronic switch for the buck circuit. The experiment shows that the designed electronic switch can be used for buck circuits, and the voltage can be stabilized at around 3.89V, which is good for powering small, low-power devices.

2. System and working principle

2.1 Structure of the triboelectric nanogenerator

The overall structure of the triboelectric nanogenerator mainly consists of a wind scoop, a drive shaft, a stator, and a rotor generating unit. When the wind scoop receives the wind energy from the environment, the rotor and shaft rotate in response to the rotation of the wind scoop. At the same time, the FEP film glued to the rotor and the copper foil glued to the stator contact friction with each other generating electrical signals. Figure 1a shows the signal generated by the prototype. The open circuit voltage of the prepared prototype is 210 V, the short circuit current is 7 μA and the transferred charge is 70 nC. The charge transfer process is shown in Figure 1b. At the beginning of the rotation, the FEP film is only in contact with copper-1 and the electrons in copper-1 are transferred to the FEP film. At this moment, the copper and the FEP film accumulate a certain positive and negative charge respectively. When the rotor turns, the FEP film slides from copper-1 to copper-2, at which point the electrons in copper-2 flow into copper-1 via an external circuit. A positive half-cycle charge transfer is generated. When FEP rotates to contact

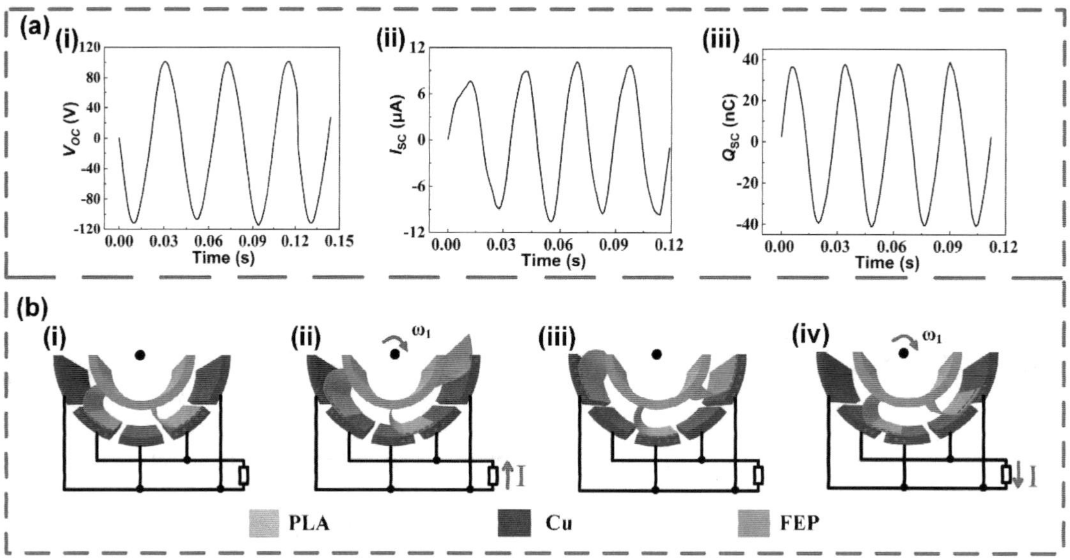

Figure 1. Prototype original signal and large transfer flow chart.

copper-2 only. Similarly, both copper-2 and the surface of the FEP film accumulate an equal number of positive and negative charges. As the rotor continues to rotate and the FEP film again turns to contact with copper-1 and copper-2, the negative charge in copper-1 again flows backward through the external circuit into copper-2. A negative half-cycle charge transfer is generated. A charge cycle is completed.

2.2 Operating principle of the power management circuit

When the triboelectric nanogenerator receives wind power from the environment, the AC voltage generated by the TENG is transformed into a direct current (DC) signal by a double voltage rectification module. However, due to the large voltage, it is not possible to supply the load directly. Therefore, a buck chopper circuit is used. The stable voltage can be received with a small ripple that can supply the load. Furthermore, this circuit needs to be combined with the use of a switch. It is shown in Figure 2a. In this paper, an electronic switch is designed which can be used alone for the buck chopper circuit. The electronic switch consists mainly of a peak detection circuit and a triode. The peak detection circuit detects the highest value of the voltage and opens the triode to supply the back end.

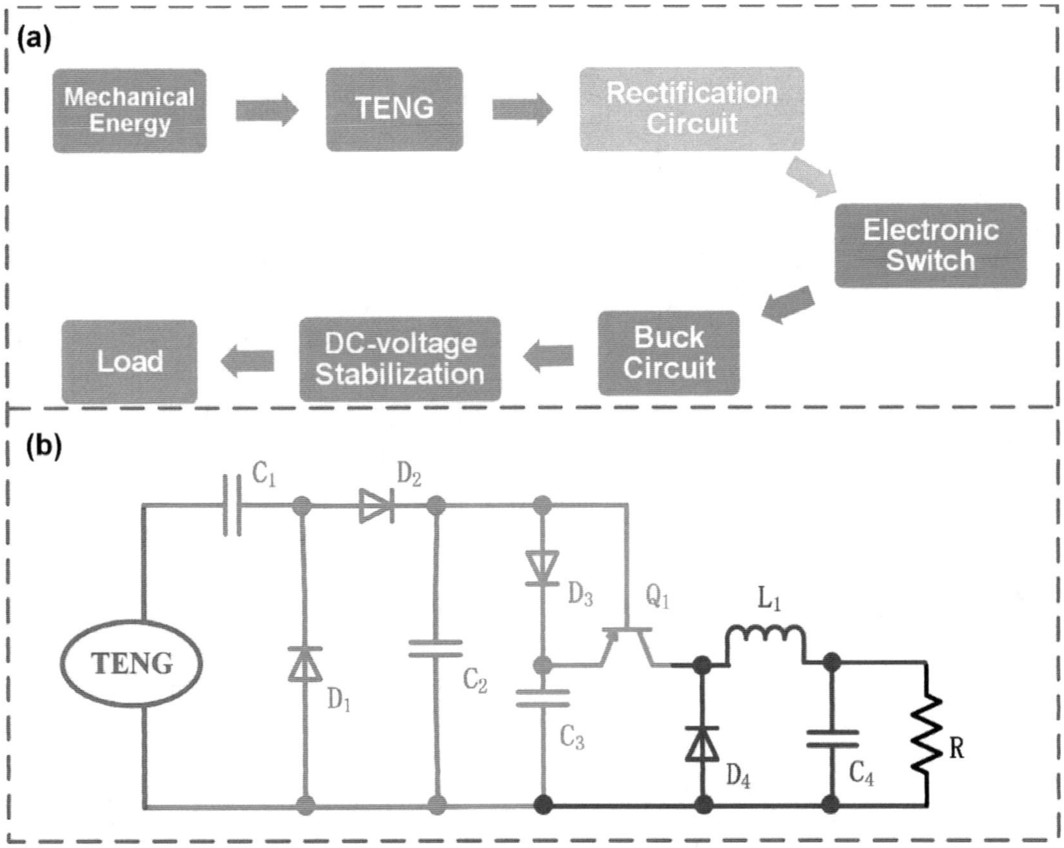

Figure 2. Flow chart of the energy collection system and power management circuit diagram.

The power management circuit is shown in Figure 2b. The AC signal generated by the triboelectric nanogenerator is stored in C_2 via two diodes. At the positive half of the AC signal, the rectifier circuit D_1 stops and D_2 conducts, supplying the AC signal to C_2 via C_1 and D_2. At the negative half of the AC signal, the rectifier circuit D_2 stops and D_1 conducts to supply C_1. C_1 collects the negative half of the AC signal and supplies C_2 at the positive half of the AC signal to achieve the effect of rectifying and storing energy at the same time. When C_3 is stored to maximum energy, the triode is turned on, so that the triode starts to work to supply the back end. The peak detection circuit is mainly composed of D_3 and C_3. Due to the unidirectional conductivity of D_3, the voltage of C_3 can not be released to achieve maximum value. When the emitter of the PNP transistor detects that the voltage of C_3 has reached its highest value, it is higher than the base voltage of the PNP transistor. To meet the conduction condition of the PNP transistor as a switch, the switch opens at this time. When the voltage of C_3 is released, the transistor does not meet conduction conditions, the switch is turned off. The buck circuit is mainly composed of inductor L_1, capacitor C_4, and diode D_4. When the switch is turned on, energy is supplied to the inductor, capacitor, and load. And when switched off, the energy in the capacitor and inductor continues to supply the load through the renewal diode D_4. The inductor cannot make a sudden change in current, ensuring a stable energy supply for the load.

3. Experimental comparison

Figure 3 shows the experimental verification of the overall circuit. Figure 3a shows the verification of the double voltage circuit and the switching circuit. Figure 3b(i) shows the verification of the output voltage of the double voltage circuit, and C_2 eventually stabilizes at a voltage value of nearly 400V. Figure 3b(ii) shows the voltage value of the peak circuit capacitor C_3 and the voltage value of the load R in the switching circuit. When the PNP triode is turned on, the load receives energy. Otherwise, the load voltage value is zero. Thus, the above experiments prove the availability of electronic switches and the ability to stabilize load output.

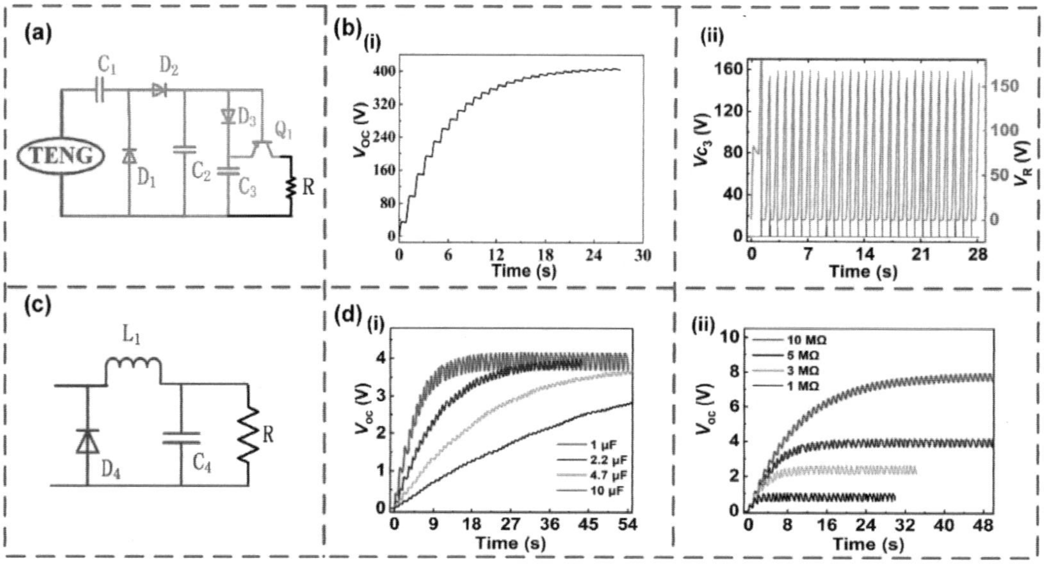

Figure 3. Experimental analysis of power management circuits.

Figure 3c shows the verification of the buck chopper circuit. When the switch is on and off resulting in the load-receiving energy as a pulse signal, it does not provide energy directly to the load. So, a filter circuit is needed to filter the pulsed signal into a stable DC signal. From this starting point, an inductor of 1mH, different resistance values, and different capacitance values were used to verify the filtering ability of the pulse. Figure 3d(i) shows the output performance with different capacitance C_4 driving a 5 MΩ load. The capacitors of 1 μF, 2.2 μF, 4.7 μF, and 10 μF were verified to drive the load, and all eventually stabilized at around 3.89 V. Because it could raise the voltage to 3.89 V in a shorter time with less fluctuation, the 2.2 μF capacitor was chosen. In Figure 3d(ii), resistors of resistance values of 1 MΩ, 3 MΩ, 5 MΩ, and 10 MΩ were chosen to receive the energy from the buck circuit. The larger the resistance value, the larger the voltage that can be received. As most of the small power electronic components are used at voltages from 3 V to 5 V, a load with a resistance of 5 MΩ is most suitable.

4. Summary

In summary, the triboelectric nanogenerator is designed for wind energy harvesting in the natural environment. In order to achieve a stable voltage output from the triboelectric nanogenerator, an electronic switch is designed to provide a new approach to buck chopper circuits. Experiment shows the electronic switch is a universal electronic switch and the electronic switch reduces a part of space for the mechanical structure. At the same time, this study provides a viable solution for stabilizing voltages in TENGs. Based on the stabilized voltage values, it can be used for LED lighting. This work provides a universal electronic switch, which provides a variety of possibilities for power management circuits. There is no doubt that this can improve the practicability of TENG and accelerate the industrialization process of TENG.

References

[1] Tang Q, Yeh M, Liu G, Li S, Chen J, Bai Y, Feng L, Lai M, Ho K, Guo H and Hu C 2018 *Nano Energy.* **47** 74-80.
[2] Rubio-Romero J, Pardo-Ferreira M, Torrecilla-García J and Calero-Castro S 2020 *Safety Sci.* **129** 104830.
[3] Wang J, Shen J, Ye D, Yan X, Zhang Y, Yang W, Li X, Wang J, Zhang L and Pan L 2020 *Environ. Pollut.* **262** 114665.
[4] Yang Y, Zhu G, Zhang H L, Chen J, Zhong X D, Lin Z H, Su Y J, Bai P, Wen X N and Wang Z L 2013 *ACS Nano.* **7** 9461-68.
[5] Chen S W, Gao C Z, Tang W, Zhu H R, Han Y, Jiang Q W, Li T, Cao X and Wang Z L 2015 *Nano Energy.* **14** 217-225.
[6] Chen J, Tang W, Han K, Xu L, Chen B D, Jiang T and Wang Z L 2019 *Adv. Mater. Technol.* **4** 3-9.
[7] Li S M, Wang S H, Zi Y L and Wang Z L 2015 *ACS Nano.* **9** 7479-87.
[8] Fan F, Tian Z and Wang Z L 2012 *Nano Energy.* **1** 328-334.
[9] Wen Z, Chen J, Yeh M-H, Guo H Y, Li Z L, Fan X, Zhang T J, Zhu L P and Wang Z L 2015 *Nano Energy.* **16** 38-46.
[10] Zhong J, Zhong Q, Fan F, Zhang Y, Wang S, Hu B, Wang Z L and Zhou J 2013 *Nano Energy.* **2** 491-497.

Development and Application of Energy-Saving System for Academic Buildings

Jiaxing Lian, Shaofen Huang, Shengguang Lin, Guangping Qiu*

Zhujiang College, South China Agricultural University, Guangzhou, Guangdong, 510900, China

* Corresponding author: S01284540@acad.tri-c.edu

Abstract. Through the active investigation of load conditions and energy consumption of most academic buildings, combined with the characteristics of routine use of academic buildings, we designed and developed an energy-saving system for academic buildings. This project uses a microcontroller as the control chip to control the energy saving of the lighting system and temperature and humidity control system in the classrooms and hallways of the teaching building. Under the premise of meeting the daily needs of teaching and normal operation, the wasteful status of unoccupied lights and long-lasting lights is reduced. Experiments show that the design runs stably and causes a beneficial energy-saving effect, realizing energy-saving control in other stages of the teaching building.

1. Introduction

Since the 20th century, we have experienced five energy crises, which have resulted in a significant increase in the price of energy sources. The concept of energy conservation has been officially recognized in the world, and the concept of "building energy-saving and renovation" has been promoted. At present, the energy consumption of buildings accounts for one-third of the total consumption. As a substantial part of society, the energy consumption of schools cannot be ignored. At this time, reducing carbon emissions have become the consensus of various countries. The European Union studies and formulates the "European Climate Law". The EU plans to achieve the political goal of carbon neutrality and provides specific assessment measures and implementation suggestions. At the end of 2020, the Government of Japan issued a road map for carbon neutrality. The United States announced: the 2035 U. S. Power system to achieve a "zero carbon" national independent contribution plan. In 2021, China also put forward, two goals of "carbon peak" and "carbon neutralization" in the national "two sessions" work report meeting [1]. The status of carbon emissions in China in recent years is described in Figure 1 [2].

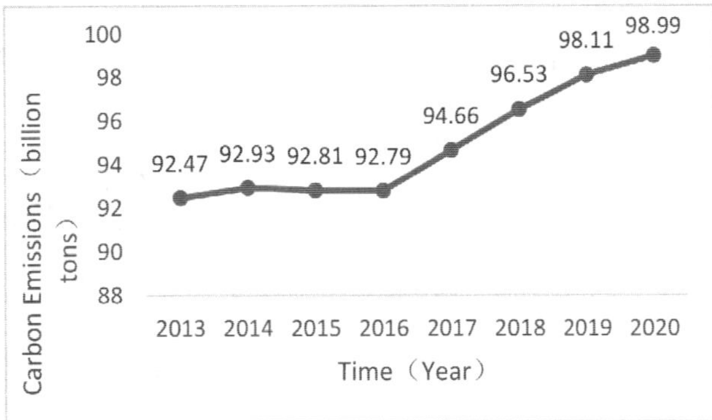

Figure 1. Total carbon emissions in China, 2013-2020

In response to the national policy, we began to investigate the way of energy-saving on the campus, and the school building is not only the incubation base of our scientific research, politics, and economy but also a core component in the social structure. Therefore, the energy consumption of academic buildings assumes non-negligible importance. Hence, this project takes the energy-saving of academic buildings as an entry point to examine intelligent energy-saving technology. The energy consumption of school buildings is shown in Figure 2 [3].

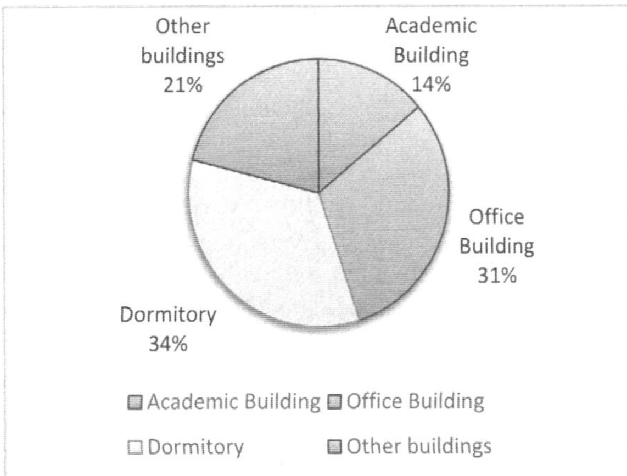

Figure 2. Energy consumption ratio of academic buildings

The number of schools in China has increased dramatically in recent years, and most schools have also carried out expansion projects, which leads to an increasing proportion of electricity consumption in academic buildings. However, some students gain a weak awareness of energy-saving and are not conscious of it. The phenomenon of not closing the power supply is widespread, resulting in a terrible waste of power resources. Therefore, the realization of energy-saving technologies in academic buildings will be of great help to the national energy-saving tasks. At this time, the power supply of most schools in China is manually controlled. The number of classrooms is massive, while the number of managers is limited. It makes it more difficult for managers to manage and causes immense waste of power resources, which is contrary to the concept of energy-saving and emission reduction advocated by the country.

Currently, energy-saving control technologies for buildings have been developed at home and abroad to some extent, and energy-saving has been achieved. For example, Yao J. et al. developed a lighting power-saving control system for university academic buildings, which can adjust the lights according to the luminous intensity and number of people in the classroom [4]. Liu F. et al. designed a microcontroller-based energy-saving regulation method for room temperature in public buildings, which can detect the indoor temperature according to the sensor and affect the indoor temperature in a certain range by controlling the solenoid valve [5]. Zhang Y. et al. designed a microcontroller-based energy-saving classroom control system, which can detect the luminous intensity and the number of people through sensors to manage the light resources in the classroom [6]. In addition, the Hermosillo Institute of Technology in Mexico developed a building automation control system in 2018, which uses Ethernet and ZigBee networks for communication, but requires the renovation of the teaching building by replacing lighting, ventilation, air conditioning, and other equipment to achieve this, which is costly [7]. Octavio et al. developed an IoT system for energy consumption to monitor energy consumption and energy management in buildings. It was tested for 6 months in their electromechanical laboratory building with good results, which proves that scholars develop energy-saving technologies from multiple perspectives [8]. In summary, the above studies only focused on the energy-saving control of a particular aspect of buildings, or the retrofitting costs are high. Although they exert some energy-saving effects, their effectiveness is not high.

This design aims to achieve energy-saving control while ensuring energy-saving efficiency. By using sensor technology and signal acquisition technology, a set of an energy-saving system of intelligent academic buildings based on the single chip microcomputer is designed. The design does not require major renovation of the buildings. It can perceive the simultaneous control of indoor temperature, humidity, and light resources according to the number of people in the classroom, light intensity, and temperature, so as to ensure comprehensive energy-saving control. Simultaneously, it can realize the centralized monitoring and threshold setting of the classroom environmental parameters, which has excellent convenience and operability.

2. General Design of the System

Many schools actively carry out energy-saving and consumption reduction actions, energy-saving measures obtain generally the following: the use of efficient energy-saving equipment; renovating the water circulation system; renovation of lighting facilities; strengthening the management of water and electricity; educating students, developing energy-saving habits, etc. These measures, and on the one hand, increase the cost of renovation investment, on the other hand, increase the number of managers and work intensity, in general, there are still some defects. Given these defects, we have developed an energy-saving control system that not only fulfills the daily needs but also saves energy and environmental protection. The most important thing is that there is no need for a large transformation and excessive costs. No structural changes to the building are required. This system can achieve the effect of energy-saving control by adding the use of energy-saving modules to the original building utility structure.

Students' awareness of energy saving is relatively weak, and the lighting system and air conditioning system of the teaching building are in working condition for a long time, which greatly wastes a lot of power resources. To address these two major problems, we designed the light control module and temperature and humidity control module through the single-chip microcomputer. This energy-saving control system is to achieve induction automatic control through sensors, without the need for long hours of manual intervention operation, to improve the effective management of lighting control and temperature and humidity control. The structure diagram of this energy-saving control system is shown in Figure 3.

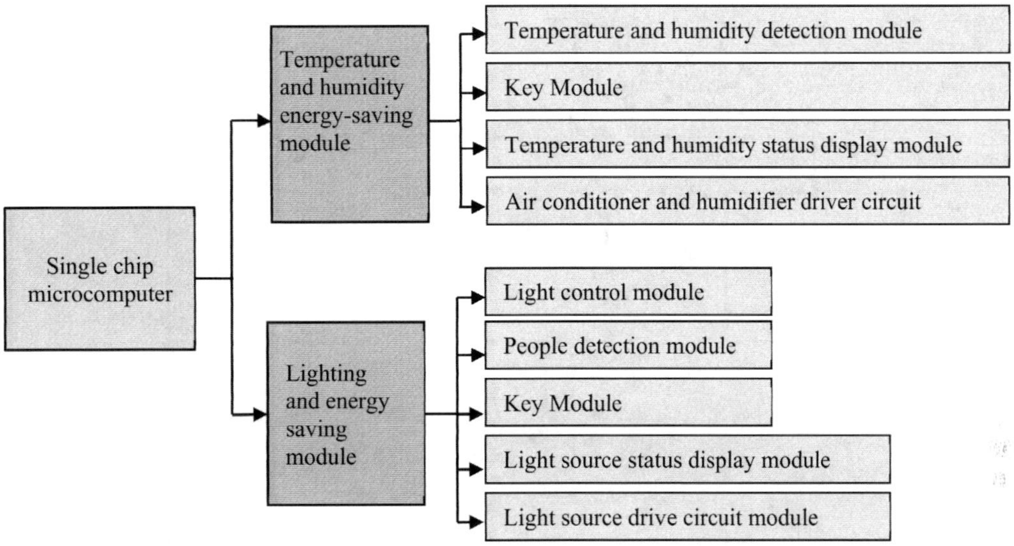

Figure 3. Structure of intelligent energy-saving control system of the teaching building

The microcontroller is inexpensive and has fast computing speed, low energy consumption, and super anti-interference capability. Therefore, we use the microcontroller as the main control chip, collect the signals through various sensors, and combine them with other hardware to realize signal transmission, data comparison, and manual monitoring to achieve intelligent mechanical economic scheduling operation of teaching building resources [9].

3. Hardware Design

3.1 Temperature and humidity control module

At present, many teaching buildings have installed air conditioners and electric fans in classrooms and offices. If students do not turn off the power of these devices in time after school, they need to wait for the administrator to turn them off one by one, which is a long waiting time and consumes a lot of power. The temperature and humidity control module developed in this system only requires the installation of temperature and humidity sensors in classrooms to control the use of air conditioners by setting the upper and lower temperature and humidity limits.

The temperature and humidity controller are typically composed of the sensor, the controller, and the heater. Its operating principle is as follows: the sensor detects the temperature and humidity information in the environment and transmits the information to the controller for analysis and processing. When the temperature and humidity of the environment reach or exceed the preset value, the relay contact in the controller is closed, the heater starts to work on the power supply, and the internal environment is heated or blasted. After a period of time, the temperature or humidity in the box is away from the set value, the relay contacts in the controller are disconnected, and heating or blowing stops [10]. The temperature and humidity control is primarily about collecting the parameters in the environment through the temperature and humidity module, which is displayed on the LCD screen. By comparing the threshold with the real-time parameters, the relevant modules are controlled to achieve the functions of cooling, heating, humidifying, and drying. The relevant threshold is set by the keypad module. The stability and accuracy of temperature and humidity detection are ensured by using the advantages of ultra-fast response and strong anti-interference capability of the DHT11 digital temperature and humidity sensor [11]. The control principle of the temperature and humidity control module is shown in Figure 4. By setting a suitable temperature and humidity threshold range, the single chip microcomputer controls the start and stop of the equipment according to this range, reduces the running time of the equipment, so as to reduce energy consumption and achieve the effect of energy saving under the premise of ensuring suitable environmental conditions. The control circuit designed according to the principle of temperature and humidity control is shown in Figure 5 below.

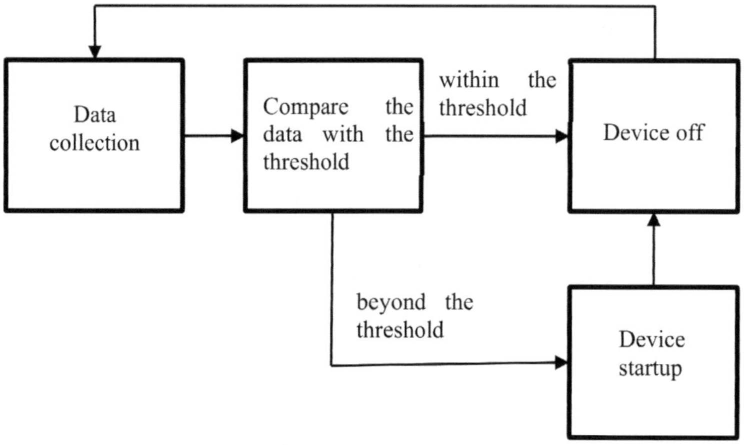

Figure 4. The control principle of the temperature and humidity control module

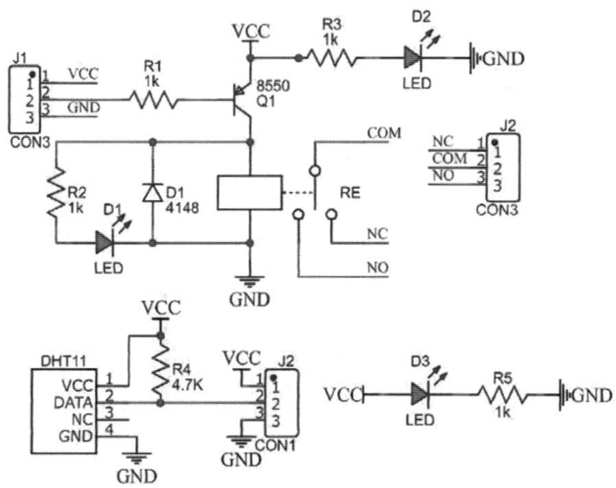

Figure 5. Temperature and humidity control module circuit diagram

3.2 Lighting control module

At present, most of the lighting systems in teaching buildings have problems such as complicated wiring and backward management. The building lights in the teaching buildings are in a long-on state at night. When students study in classrooms, the building lights can be properly turned off to save electricity resources. Some students have weak awareness of energy saving and turn on all classroom lighting even when only one person is in the classroom, which greatly causes a waste of resources. The light control module developed in this system controls the number of lights on and the light intensity through the number of people detection module and sets the status of the building lights on and off through the timer.

(1) The response threshold is set by using the photoresistor module to determine whether the outdoor luminous intensity is lower than the setting value. When it is higher than the setting value, it means that the brightness is sufficient. The output level is too low to prevent the classroom lights and the hallway lights from working and reduce unnecessary power loss. The output level is high when the luminous intensity is lower than the setting value. Then the classroom lights and the hallway lights will set the

working state according to the condition. The photoresistor is chosen as the measurement element mainly because it has the characteristics of strong and weak light measurement and a wide spectral response range [12]. The circuit diagram of the illumination control module is shown in Figure 6.

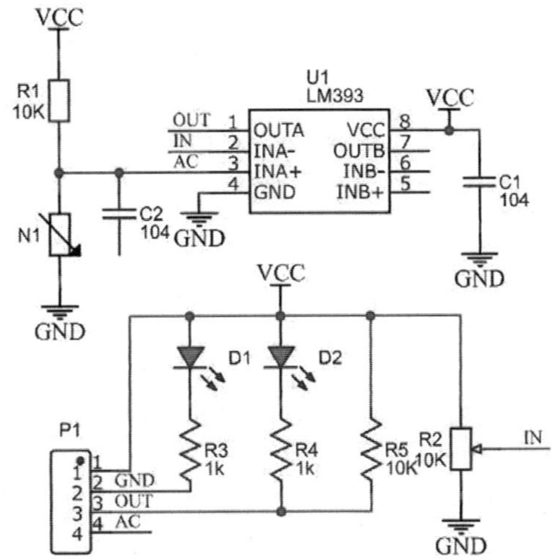

Figure 6. Light control module circuit diagram

(2) The number detection module is composed of two infrared obstacle avoidance chips. The infrared obstacle avoidance chip contains a transmitter and a receiver. Following the principle of light reflection, the infrared ray is continuously transmitted, and the ray will be received by the receiver when encountering obstacles. LM393 chip is used as a comparator in the system. When receiving the reflected infrared ray, the digital signal can be output after the comparator circuit is processed. The comparator is not affected by the non-polar capacitor and the high-frequency interference after filtering rectification, which effectively prevents the problem of high current and voltage. At this time, infrared obstacle avoidance chips are installed inside and outside the classroom door, and the number of people entering and leaving the classroom is detected by the program [13]. The partial circuit of the people detection module is indicated in Figure 7.

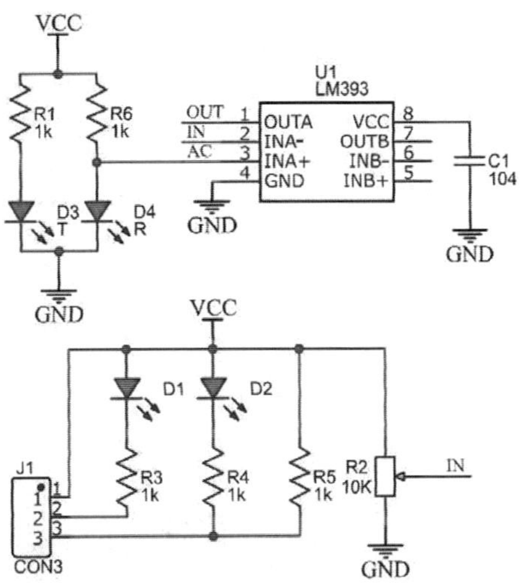

Figure 7. Circuit diagram of the headcount detection module

4. Software Design

The main task of the software design part is to complete the simulation experiments for temperature and humidity detection and the light detection circuits. This simulation experiment uses Proteus 8 to draw the simulation circuit diagram, and the Keil5 platform to write and compile the program files. The simulation runs normally on Intel Core i5-8500 CPU, 8GB RAM, and Windows 10 64-bit OS.

4.1 Temperature and humidity control

The temperature and humidity control process: the first step is to record the saved data from the EEPROM. And then the environmental temperature and humidity in real-time display on the LCD12864 are detected and the data is transferred to the microcontroller, in order to achieve an energy-saving effect by comparing the actual parameters of the environment with the threshold range, to achieve the indoor temperature and humidity control in the set appropriate conditions to reduce unnecessary power loss. The flow chart of temperature and humidity control is shown in Figure 8.

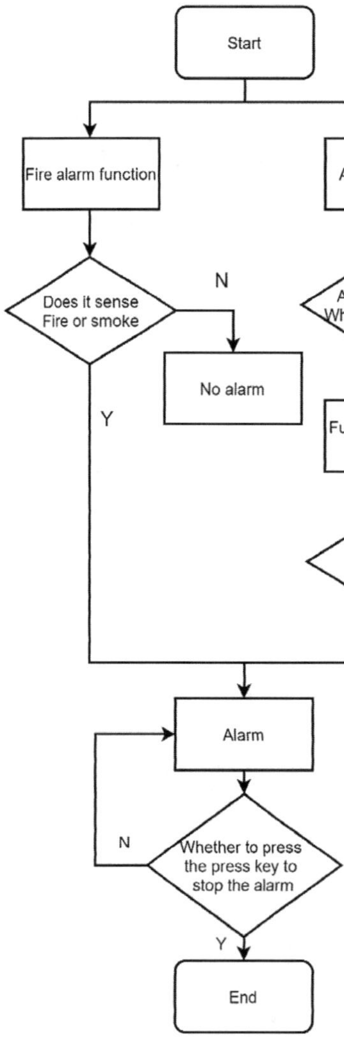

Figure 8. Temperature and humidity control flow chart

4.2 Lighting controls

Illumination control process: starting with the environmental illumination, detected by the photosensitive resistance, which is judged to confirm whether to reach the set threshold value. If the threshold is exceeded, the lights cannot be established to achieve the purpose of energy-saving. When the illumination does not reach the threshold value, the illumination module will evaluate the number of people, to determine the number of lights, and reduce the waste of resources. The flow chart of lighting control is shown in Figure 9.

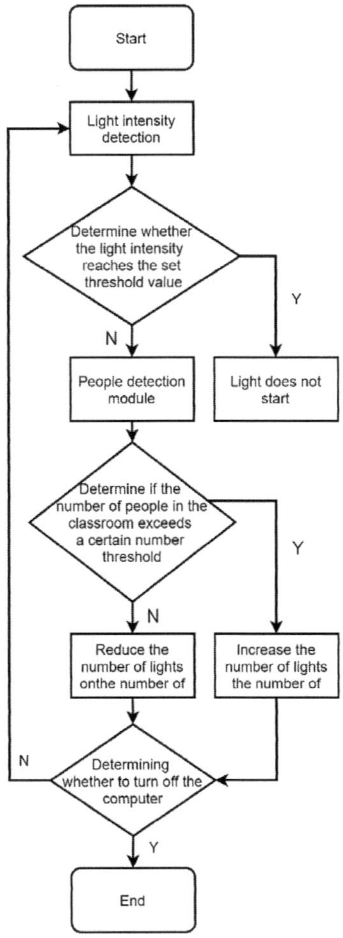

Figure 9. Flow chart of light control.

5. Results and Analysis

In order to verify the accuracy and operation test of the system, an environmental test analysis was carried out in five classrooms of a teaching building. This experiment combines the light control module and the number of people detection module for testing. The experiment is selected at night to ensure that the light level cannot reach the required threshold, at this time the number of people detection module for testing, employing the method of alternating in and out of the test, the sensitivity and accuracy of the calibration. In order to verify the accuracy of the equipment, a mercury thermometer and ordinary hygrometer are used to sample the temperature and humidity of the classroom environment, and then the measurement results of the system are compared. The temperature and humidity of the environment are adjusted according to the upper and lower limits of the system. Temperature and humidity test results are shown in Table 1 and Table 2.

Table 1. Temperature test results

rank	Mercury thermometer	Measuring by system	Temperature after operation
1	23 °C	22 °C	25 °C
2	29 °C	31 °C	27 °C
3	26 °C	26 °C	26 °C
4	22 °C	23 °C	25 °C
5	30 °C	32 °C	25 °C

Among them, the upper-temperature limit is 27 °C and the lower temperature limit is 24°C.

Table 2. Humidity test results

rank	Measurement of humidity	Measuring by system	Humidity after operation
1	45%	44%	58%
2	54%	57%	56%
3	67%	71%	70%
4	47%	46%	60%
5	41%	43%	59%

Where the humidity test is performed with a control threshold of 50%.

To verify that the retrofit cost of this energy-saving system is lower than other energy-saving options, several schools were investigated for retrofitting options. School 1 was through a major renovation of the entire school building's utilities and other facilities [16], and School 2 was through the replacement of lighting equipment, as well as the addition of sensor equipment [17]. The retrofit costs and energy savings of the two retrofit options with the proposed energy-saving system are shown in Table 3.

Table 3. Comparison of retrofitting options

Retrofit Plan	Retrofit Cost	Energy Efficiency
1. School 1 saves energy by reforming the circuit layout and replacing lighting equipment.	Renovation funds of 1.08 million yuan RMB, 68 classrooms, an average cost of 15,880 yuan RMB per classroom.	50% energy saving
2. School 2 by replacing lighting equipment, and adding sensor equipment for energy saving.	The renovation capital is 35,000 yuan RMB, 63 classrooms, and the average cost of each classroom is 550 yuan RMB.	20%-40% energy saving
3. The proposed method.	The reconstruction cost of each classroom is less than 50 yuan RMB.	25-50% energy saving

6. Conclusion

In the current situation, energy consumption continues to rise, coupled with the booming education industry, the increasing number of schools, and enrollment plans, resulting in a growing population density on campus. And energy-saving in academic buildings becoming more and more important. This design combines sensor technology, data visualization technology, and intelligent control technology to design an energy-saving model for academic buildings by analyzing the background of energy consumption in academic buildings and the demand for energy-saving. This energy-saving system can resolve some of the problems of wasted resources in the current school building without extensive renovation, improve the safety of the school building and the convenience of equipment operation, and greatly optimize the current school building facilities. In addition, this design can also be utilized in other civil buildings to improve the energy efficiency of the building, which has good portability and wide application.

Acknowledgments
This work was supported by the 2021 National Undergraduate Innovation and Entrepreneurship training program innovative training project (202112623004).

References

[1] Qin Rongjun. Foreign experience of carbon neutralization of carbon peak and its inspiration to the development of China's coal industry[J]. Coal Economic Research,2021,41(03):23-27. DOI:10.13202/j.cnki.cer.2021.03.005.
[2] Zhang Qilong, Liu Lu, Sun Binbin, Zhuo Wenyan, Jin Xiao, Lv Xin, Zhang Hongcheng, Wang Zhongyong. Analysis of the current situation of "carbon peaking and carbon neutral" and its implementation path in China[J]. China Building Materials Science and Technology,2022,31(01):83-86.
[3] Liu Jian. Research on energy-saving renovation model of existing academic buildings in colleges and universities[D]. Hebei University of Technology,2015.
[4] Yao Junhong. Research and development of lighting power saving control system for high school academic buildings[J]. Journal of Lighting Engineering,2011,22(04):84-86. DOI: 10.13223/j.cnki.ciej.2011.04.020.
[5] Liu F. Research on the design of a microcontroller-based method for energy-saving regulation of room temperature in public buildings[J]. Environmental Science and Management, 2020, 45(09):27-31.
[6] Zhang Yunli, Zhou Xiaoping, Zhu Shuangxia. Research on the design of microcontroller-based classroom energy-saving control system[J]. Manufacturing Automation,2012,34(04):103-105.
[7] Hoyo-Montaño J A, Valencia-Palomo G, Galaz-Bustamante R A, et al. Environmental impacts of energy-saving actions in an academic building[J]. Sustainability, 2019, 11(4): 989.
[8] Lasso-Lopez, Octavio et al. "Implementing an IoT Energy Monitoring System Using the Challenge-based Learning Model." 2020 IEEE Conference on Technologies for Sustainability (SusTech) (2020): 1-5.
[9] Wang Haining. Research on microcontroller-based temperature control systems [D]. Hefei.
[10] Zhou Le An. Temperature and humidity controller based on MSP430 microcontroller [J]. Sichuan Journal of Military Engineering, 2008, 029(002):113-114.
[11] Adiono, Trio et al. "A portable node of humidity and temperature sensor for indoor environment monitoring." 2018 3rd International Conference on Intelligent Green Building and Smart Grid (IGBSG) (2018): 1-5.
[12] Ye X, Su Y, Guo S, et al. Infrared Motion Guidance and Obstacle Avoidance of an ICPF Actuated Underwater Microrobot[C]// IEEE. IEEE, 2007:1851-1856.
[13] Zhang, Hao-Bin, et al. "A simple PVA/Cu(OAc)2 thermogel with an inherent near-infrared light response and its applications in smart windows and photoresistors." Journal of Materials Chemistry 8 (2020): 17800-17807.
[14] Mahzan N N, Enzai N I M, Zin N M, et al. Design of an Arduino-based home fire alarm system with GSM module[C]//Journal of Physics: Conference Series. IOP Publishing, 2018, 1019(1): 012079.
[15] Xie W L, Haoming H. Design and Application of Home Fire Alarm System[C]//Journal of Physics: Conference Series. IOP Publishing, 2021, 1871(1): 012125.
[16] Originality, 2019, University Comprehensive Energy Saving Renovation Program, https://max.book118.com/html/2019/0731/7043153023002044.shtm.
[17] Fang Ling, 2018, Research on Energy Saving Reconstruction Design of Existing Teaching Buildings on Campus - - Taking Guangdong Baiyun University as an Example, https://www.lunwen66.cn/article/f2179d21a05ffe1b697b88c7.html.

System Design of Temperature Measurement and Sampling Robot in Steel Plan

Yan Bian[1,a], **Dengli Yi**[2,b], **Yifan Cao**[3,c]

[1]Automation Research and Design Institute of Metallurgical Industry Beijing China

[2]Yingkou Institute of Technology Yingkou Liaoning China

[3]Linyi Collaborative innovation center of iron and steel industry Linyi Shandong China

[a]13501050309@163.com

[b]5050422@qq.com

[c]865812139@qq.com

[a*]Corresponding author: 13501050309@163.com

Abstract: In the traditional steel-making production process, the workers use a 20kg temperature measuring gun to insert into the ladle with more than 1500 degrees Celsius for temperature measurement and sampling. Each furnace must take the temperature at least 4 to 6 times, with sampling 3 to 5 times. The labor intensity is heavy. Due to the high temperature and dust at the furnace's mouth, the environment is bad, and there are unforeseen emergencies such as molten steel explosion, which has a large potential safety hazard. Therefore, using industrial robots' temperature measurement and sampling can reduce the labor intensity of workers and improve safety factors. This paper studies and designs the temperature measurement and sampling robot system in the steel plant, in order to avoid manual temperature sampling. Using industrial robots as the platform, using SolidWorks software to model and build the system. The path planning and simulation of the temperature measurement sampling robot is conducted through Robot Studio software, and then Adams software is used to conduct the dynamic simulation analysis, the results show the motion of the robot after visual simulation and obtain the dynamic parameters and verify the reasonable feasibility of the path.

1. Introduction

The temperature of molten steel in smelting determines the quality of the steel produced, etc. High temperature or too low will make the quality problems of the steel appear. Therefore, the monitoring of molten steel temperature plays a very important position in the process production. Steel plant temperature detection is usually done by the workers with a temperature measuring gun to the steel furnace sampling, workers need to work in a high-temperature environment for a long time, under difficult conditions and high-risk factors. Therefore, this paper designs the temperature measurement and sampling robot system, with the robot to replace manual work, improve efficiency, reduce risk, and realize the intellectualization of steel plants.

2. Overall system design

The temperature measuring and sampling robot system designed in this paper mainly includes industrial robots, temperature measuring guns, and fixtures, as shown in Figure 1. Relying on the fixture to connect the temperature measuring gun with the front end of the robot. Under the drive of the robot, the temperature measuring gun reaches inside the steel furnace to complete the temperature measurement process. The temperature is collected by the temperature measuring gun. In addition to the local display, the temperature value is also transmitted to the upper position computer monitoring interface through the control system, for storage, recording, alarm, etc.

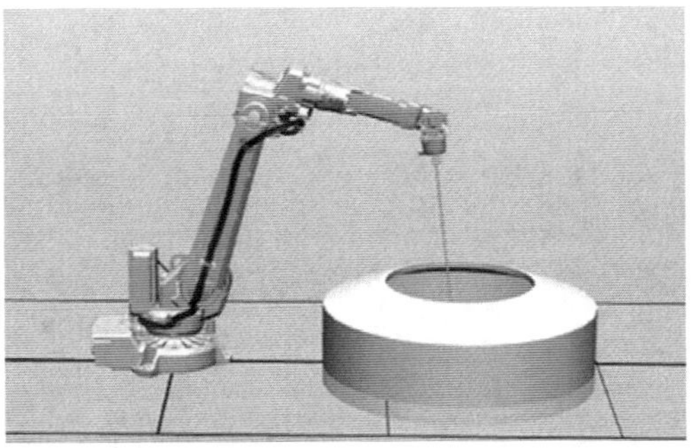

Figure 1. The overall layout of the system scheme

3. Hardware selection and design

3.1 Robot selection

This paper selects ABB's IRB-2600ID industrial robot. The main function is to drive the temperature measuring gun and move it to the designated position. The robot arm needs to have enough length to send the temperature measuring gun into the steel furnace, with an IRB-2600ID robot arrival distance of 1.85m, load weight of 15kg, and equipped with a flexible wiring harness. It covers a small area, suitable for material handling and loading and unloading operations.

3.2 Temperature measurement gun selection

The temperature is collected by the temperature gun and the KW602 temperature gun is selected. The temperature measuring gun consists of the gun base, screw, signal plug, compensation wire, test barrel, and copper head. The copper head is composed of a temperature-measuring even head, which is made of a positive and negative even wire welded on compensation wire, supported and protected by quartz. The outer part is equipped with a slag prevention cap, all assembled into the mud head and bonded into a whole with refractory filler, which can not be disassembled. The copper head is connected to the test barrel by threading to replace the copper head. The signal socket in the test barrel is inserted into the signal plug in the gun seat of the instrument to collect the temperature signal. Using SolidWorks software, the KW602 temperature measuring the gun model was established. The front end of the temperature measuring gun is connected with the copper head and the gun rod, as shown in Figure 2.

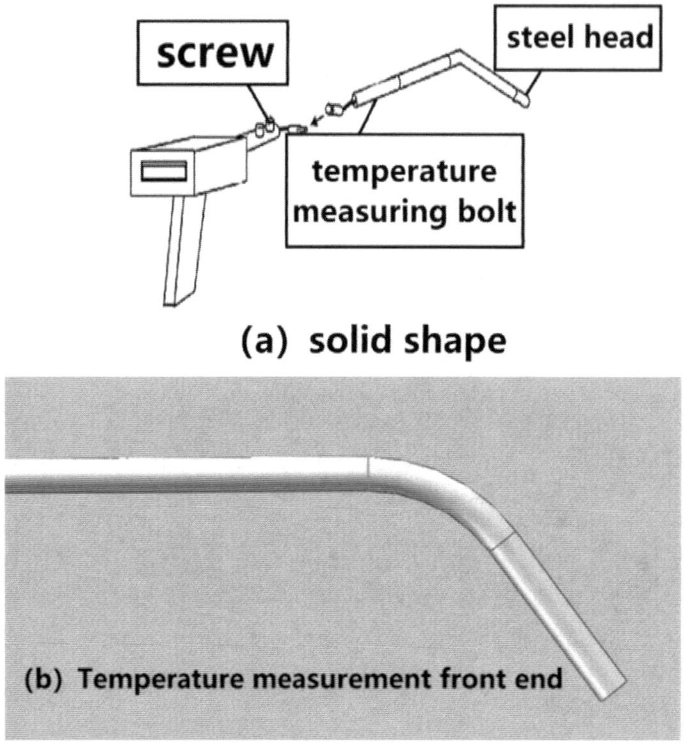

Figure 2. Outline structure diagram of temperature measuring gun

3.3 Design of Fixture

As the connection between the end of the robot and the temperature measuring gun, the clamp is used to ensure a close and reliable connection and convenient disassembly and assembly. The fixture is designed to be a hollow cuboid structure and equipped with a baffle, the clamp can put one end of the temperature measuring gun into and fix, and the baffle and the cube are connected by a thread. It not only solves the problem that the temperature measuring gun does not slide when the robot arm moves downward, but also solves the problem that the temperature measuring gun can be disassembled. The hollow-out device is designed to achieve lightweight under the condition of meeting the strength, thereby reducing the robot load and improving efficiency. Using the SolidWorks software, the fixture structure was established as shown in Figure 3.

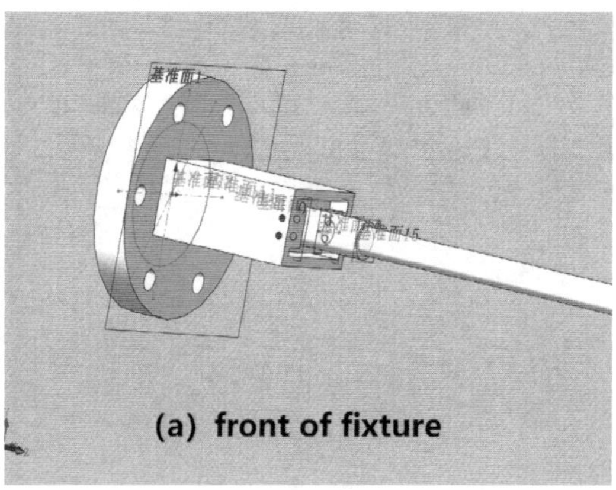

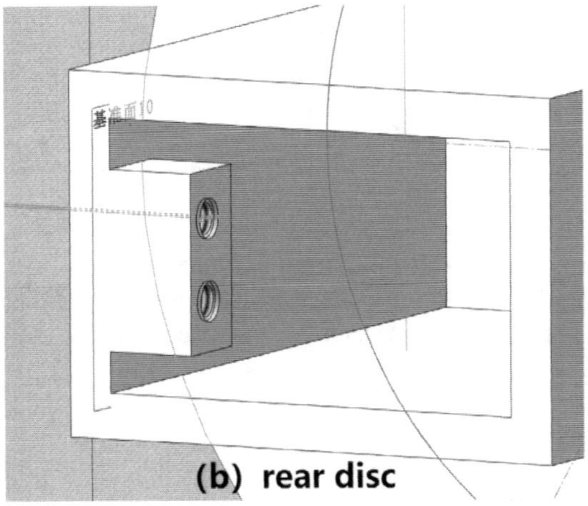

Figure 3. Fixture model

4. Robot Path Planning

Robot path simulation is designed using Robot Studio software. The model is imported into the Robot Studio, the geodetic coordinates are selected, and a workstation is established. Scaling is performed to match the connection of the robot model. Robot path planning and design can be conducted by showing the teaching device and programming, and the robot walking path can be preliminarily planned through the starting point, the key point, and the endpoint. The starting point is in FIG. 4. In order to protect the temperature gun, it is placed at the farthest end of the furnace to make the temperature gun the temperature impact and ensure the safety of the temperature gun to the maximum extent. B in FIG. 4 is a key point in the movement of the mechanical arm during temperature measurement. The mechanical arm drives the temperature measuring gun above the steel furnace mouth as the working safety point. After recording the rear position, the robot controls the temperature measuring gun into the steel furnace with an entry distance of about 20mm. After recording the position, as shown in C in Figure 4. After a temperature measurement and sampling work process, the robot is set to go back to the starting position, which represents the end of a temperature measurement process.

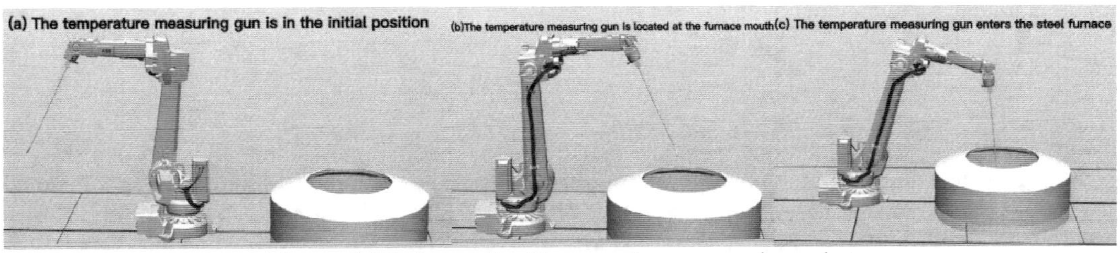

Figure 4. Motion path of temperature measuring robot

After the path simulation, the corner curve of the signal data file of Robot Studio software is shown in Figure 5.

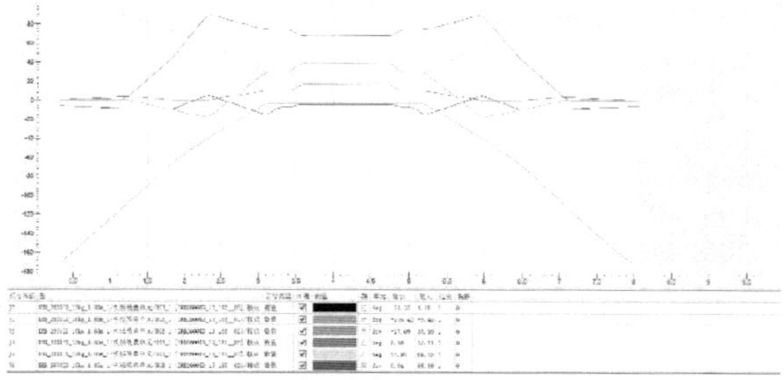

Figure 5. Movement path of the temperature measuring robot

5. Adams Dynamics simulations

The temperature measuring robot was moved according to the design path and simulated using Adams software. The built SolidWorks temperature measuring robot system model was imported into the Adams software to mark the parts. It is required to set the material in the tool list, add the rotation pair, and write the motion function for the simulation analysis. The displacement and velocity results of each axis were obtained after the simulation analysis, as shown in Figures 6 and 7. This shows that the axis of the robot is smoothly continuous, without mutations and fluctuations, and the whole is stable.

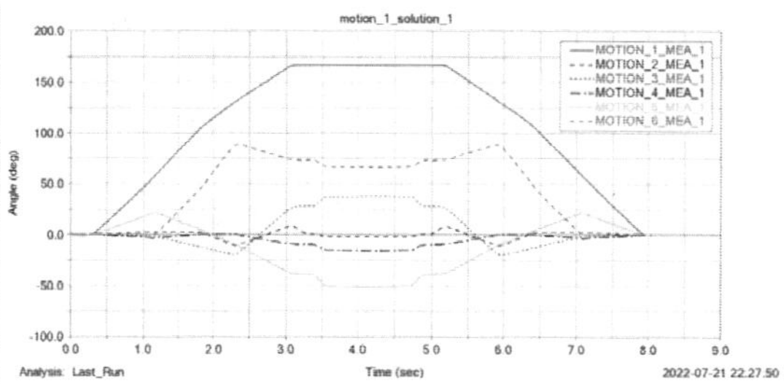

Figure 6. Displacement result diagram

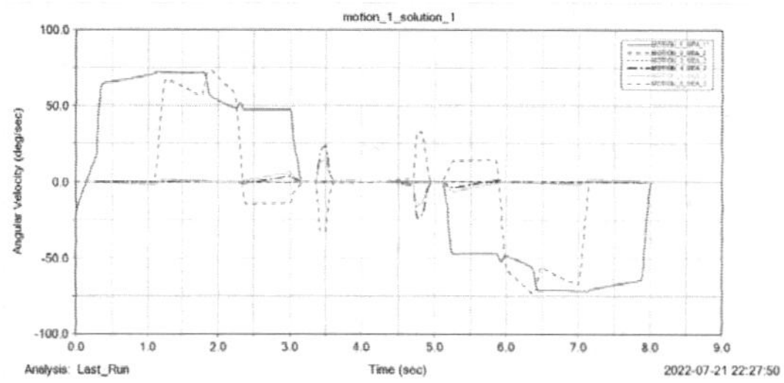

Figure 7. Speed result diagram

6. Conclusion

In this paper, the steel plant temperature measurement and sampling robot system is designed, using Solidworks/Robot Studio/Adams joint modeling and simulation technology, and using SolidWorks software to model and assemble in Robot Studio software. Through Adams's dynamic simulation analysis, the robot can run smoothly along this path and without interference. It provides a theoretical basis for the subsequent robot trajectory optimization and motion control.

Reference

[1] Yan W 2021 Research and development and application of intelligent temperature measurement and sampling multi-functional robot system. China Metal Bulletin vol 3 p2.
[2] Dong J 2019 Research and Application of Robot in Temperature Sampling. Electrotechnical

Technology vol 12 p2.
[3] Hong L. 2018 Structural design and motion analysis of a reconfigurable fixture robot. Internal Combustion Engine and Accessories vol 20 p2.
[4] Wang C, Zhang W, Jiang J, Ye X, Jiang L 2020 Joint kinematic analysis and control simulation of industrial welding robots based on MATLAB / Adams. Chinese Journal of Engineering Machinery vol 06 pp504-509.
[5] Xu W, Li J, Chen W 2022 Design and Research of the Inspection Robot System for Mine Use Coal Engineering vol 3 p5.
[6] Li Z 2021 Simulation Design of Robot Fixture Dynamic Replacement System Based on Robot Studio Software. Electromechanical Engineering Technology vol 12 pp 122-126.
[7] He Z, Yan B 2021 Modeling and rigid-body dynamics analysis of Kuka KR16 R1610-2 industrial robots. Inner Mongolia Science Technology & Economy vol 23 pp108-109.
[8] Jia L, Zhu Y 2020 Dynamic Simulation of Industrial Robot Based on SolidWorks and ADAMS Packaging Engineering vol 03 pp 207-210.
[9] Guo J, Zhu Z, Chen M 2018 Structural Design and Kinematics Modeling of Welding Robot. Welding Journal vol 39 pp 7.
[10] Hu J, Zhang Y, Wang F 2016 A kinematic Simulation Analysis and Practical Application of a Handling Robot. Machine Tool & Hydraulic vol 11 p5.

A Novel High Holding Voltage ESD device based on SCR and SiGe

Jing Liu, Huiting Liu, Yuansong Fan, Yan Zhao, Chun Liu

Department of Electronic Engineering, Xi'an University of Technology, Xi'an 710048, China

jingliu@xaut.edu.cn

Abstract. As a kind of efficient electrostatic protection device, SCR has been extensively used in on-chip electrostatic discharge protection for its strong current discharge ability and favorable internal radiating heat characteristic [1-7]. But there are low holding voltage and easy latch up happening two restrictions. A novel high-holding voltage ESD device based on SCR and SiGe is proposed in this paper. By introducing Ge composition in the P + region in the N-well of ordinary SCR, the holding voltage of the device is enhanced for that the current gain of the PNP parasitic transistor is reduced then the conductivity modulation is inhibited. The simulation results based on TCAD indicate that when Ge percentage composition is 30%, the holding voltage of new SCR with SiGe P+ region promotes four times from 2.06V to 10.32V. Therefore, there is no enlarging of the layout area of the novel structure when promoting the holding voltage and enhancing the anti-latch ability of SCR.

1. Introduction

With the development of semiconductor processing technology and the scaling down of transistors, integrated circuit circuits have become more and more susceptible to electrostatic discharge events [1-3]. The silicon-controlled Rectifiers (SCRs) are known as efficient Electrostatic Discharge (ESD) protection applications due to their high ESD robustness, low leakage current, and strong current discharging capability. However, NPN and PNP parasitic bipolar transistors in SCR establish positive feedback on each other, which causes conductivity modulation aroused by intense injection of electron (hole) current that can drive SCR into a deep snapback region with very low holding voltage. If the holding voltage of SCR is below the working voltage of the protected circuits, ESD protection devices will keep working even though the electrostatic discharge events are down. This can lead to latch-up risk and damage the performance of protection circuits [8-9].

There are three ways to improve holding voltage by restraining the positive feedback effect into SCR. First, the conductivity modulation can be restrained in the device by adding a section of a new structure to reduce the current gain. Second, current flows will be cut if particular un-snapback current discharge ways are infixed. Third, the current gain of parasitic transistors decreased, whose emitter junction injection efficiency was distributed by changing layout shapes and area ratio between anode and cathode [10-13].

A high holding voltage ESD device based on SCR and silicon-germanium (SiGe) is reported using the first method above. By adding a germanium component in the P+ region of the N-well, the current gain of the parasitic PNP transistor and positive feedback between transistors in this device are reduced. Thus, the anti-latch performance of this device is optimized. Studying current flow modes among devices with different germanium distribution forms by TCAD, the function of the Ge component to protection ability in the device is indicated.

2. Structure and working process of the device

A conventional Silicon Controlled Rectifier (SCR1) and its working process are shown in Figure 1. When intensive enough ESD current disturbance came, abundant electrons and holes were excited owing to the avalanche multiplication of reversed PN junction. Electron current flowed to anode across N-well and hole current on the contrary. The voltage drop caused by the above current can trigger the PNP and NPN parasitic transistor. These two transistors provided base currents for each other, enhancing conductivity modulation in the device. Thus, conventional SCRs have deep hysteresis with low holding voltage during working.

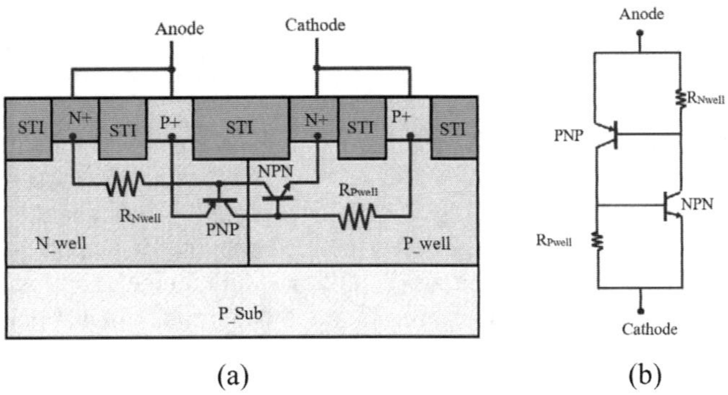

Figure 1. SCR device structure and equivalent circuit diagram:
(a) SCR device structure; (b) SCR device equivalent circuit diagram.

SCR holding voltage can be improved by decreasing the emitter junction injection efficiency of the PNP parasitic transistor, according to the principle that holding voltage decided to free carrier absorption density injected into the emitter. In addition to introducing the Ge component to the P+ region, there are no differences between the new structure and conventional SCR, shown in Figure 2. Energy bands of novel and convention devices are shown in Figure 3. By introducing the Ge component to the P+ region to drop band conduction level, the electron barrier from N-well to the P+ region is decreased, and electron current increases. Thus, positive feedback between parasitic transistors is weakened by decreasing the emitter junction injection efficiency of the PNP parasitic transistor. The new device's anti-latching ability is enhanced.

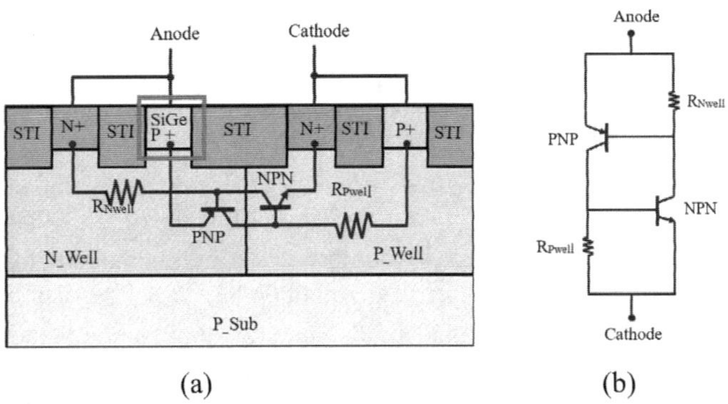

Figure 2. Structure and equivalent circuit diagram of new SiGe P+ SCR device:
(a) new SCR device structure; (b) new SCR equivalent circuit diagram.

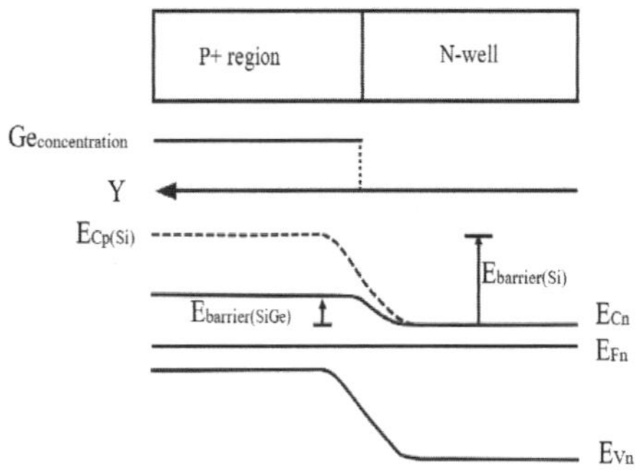

Figure 3. Energy band diagram in conventional SCR and novel SCR.

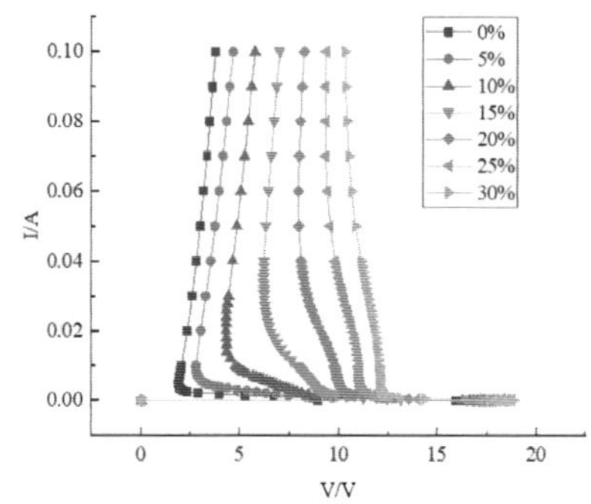

Figure 4. I-V characteristic curves of conventional and new SiGe P+ SCR.

Ge percentage composition remains constant from the surface to the junction. Considering the thickness of the SiGe epitaxial layer constrained by Ge percentage composition, Ge percentage composition is separately set to 5%, 10%, 15%, 20%, 25%, and 30% during simulation.

Curves about the I-V characteristics of the new structure under the Ge percentage composition above are separately drawn in Figure 4, and the parameters of holding voltage and trigger voltage are shown in Table 1. The conclusion is drawn by analyzing Figure 4 and Table 1 that trigger voltage is immune to Ge percentage composition. But as Ge percentage composition increases, holding voltage and current rise substantially, and output resistance decrease significantly. Compared with conventional SCR, holding voltage rises from 2.06V to 10.32V, four times that when the Ge percentage composition of the P+ region was set to 30%. Thus, the conclusion is drawn that the structure's holding voltage is optimized by efficiently adding Ge composition in the P+ region on the condition of the remaining device layer area.

Table 1. Key parameters of conventional and new SiGe P+ SCR.

Component percentage	0%	5%	10%	15%	20%	25%	30%
trigger voltage/V	18.22	18.35	18.42	18.53	18.66	18.75	18.84
holding voltage/V	2.06	2.80	4.31	6.23	7.97	9.32	10.32

3. Optimization Theory of New Device

Current density distribution when working with new and conventional devices is simulated for deeply studying the impact of the SiGe P+ region on the degree of conductivity modulation.

Figure 5 shows the current discharge ways of the two structures above during the holding state, which all contain way②, way③, and way④. Their components are as follows. Way② is made up of P+ regions in N-well, N-well, N+ regions in P-well, and P-well; Way③ is made up of N+ regions in N-well, N-well, and N+ regions in P-well; Way③ is made up of N+ regions in N-well, P-sub, P-well and P+ regions in P-well. During the holding stage, the magnification of the PNP parasitic transistor constituting way② in the new device lessens compared with conventional SCR. As the positive feedback of the parasitic transistor lessens, current distribution becomes irregular, and current is mainly discharged by way④.

As the current pulse increases, the device turns into the current discharging stage. Current flows are the same during the holding stage, but the current discharging ranges in the device are expanded. The positive feedback of PNP and NPN parasitic transistors in the new device gradually strengthens, but the current discharge performance of the new device is inferior to a conventional device.

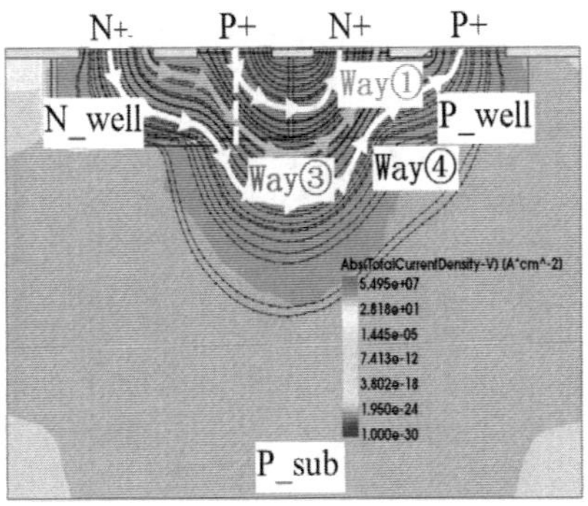

(a)

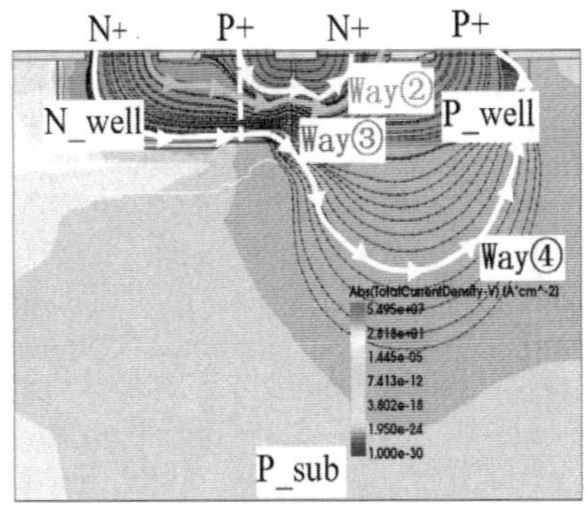

(b)

Figure 5. Current density distribution of the conventional and new SCR when holding: (a) current density distribution of conventional SCR when holding; (b) current density distribution of new SCR when holding.

The energy band of the new device cut along the dotted line in Figure 5 is shown in Figure 6. Electron current density distribution of two devices cut along the dotted line as Figure 5 during the holding stage is shown in Figure 7. By analyzing Figure 6 and Figure 7, the conclusion is drawn that Ge percentage composition in the P+ region reduces a potential barrier of the electron. Thus, the electron current injection emitter is enhanced, and the hole current diffusing base is decreased.

So compared with conventional, new device holding voltage is optimized. Because by decreasing the emitter injection efficiency of the PNP parasitic transistor, current amplification is weakened, then conductivity modulation during the hysteresis stage is reduced. In conclusion, the anti-latch performance of the new device is superior since the current gain of the PNP parasitic transistor can decrease by taking advantage of the SiGe P+ region.

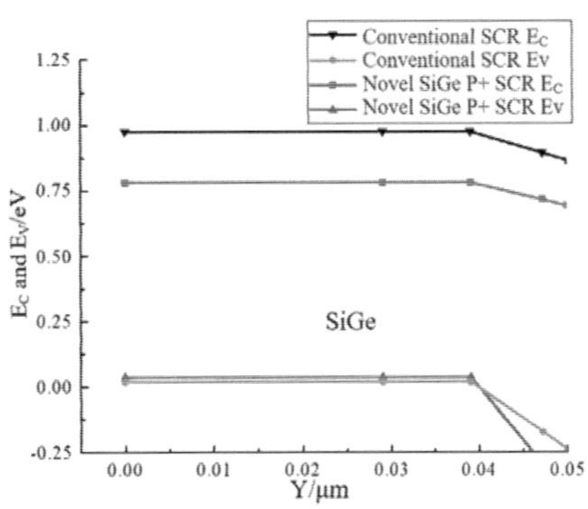

Figure 6. Energy band diagram of new SCR devices.

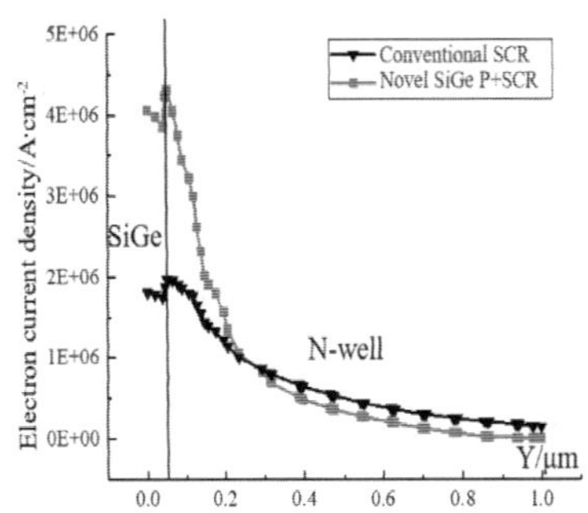

Figure 7. Electron density distribution of curves of conventional and new SCR when holding.

4. Optimal Design of Ge Composition Distribution

For researching the impact of the SiGe P+ region on the device, the holding voltage changes of the new device under different Ge composition distributions are compared to find optimum distributions. Trigonal Ge composition distributions which peak value separately at the surface of the P+ region and the PN junction between the P+ region and N-well, are discussed. These two kinds of trigonal and uniform Ge composition distributions are separately named Xup, Xdown, and Xaverage.

Holding voltages and trigger voltages of new structures under Xup, Xaverage, and Xdown distributions about different Ge composition peak values are shown in Table 2. One conclusion can draw by analyzing Figure 4 and Table 2 that holding voltage increases with the Ge composition percentage of the P+ region in the N-well. Ge composition percentage of uniform distribution and peak of trigonal distribution are set to 15% and 30% respectively. Figure 8 indicates curves about I-V characteristics under different Ge composition distributions. Exploring the impact of Ge composition distributions on the optimization effect of holding voltage, the conclusion can be drawn that uniform Ge composition distribution is the most effective, then is the Xup distribution, and the Xdown is last. The device working process under different Ge composition distributions will be discussed from two aspects of the internal field and barrier potential of the conduction band.

Figure 9 shows the conduction bands of conventional SCR and the conduction band of new SCR under different Ge composition distributions. The conduction bands of new SCR under uniform Ge composition distribution steadily reduce. For Xup distribution, Ge composition at the P+ surface is the highest while that at the PN junction is the lowest, so the conduction band reduction is the largest at the P+ surface but is the smallest at the PN junction. The electronic barrier of the PN junction under the Xdown distribution is lower than under Xdown. By this time, the conduction band increases linearly from the PN junction to the surface of the P+ region. According to the above analysis, the optimization of holding voltage under the Xdown distribution is better than under the Xup distribution in theory. But the fruits of simulation show that the phenomenon is the contrary. Because gradually changing Ge composition will take an extra electric field into the device. This extra field affects the performance of the device either.

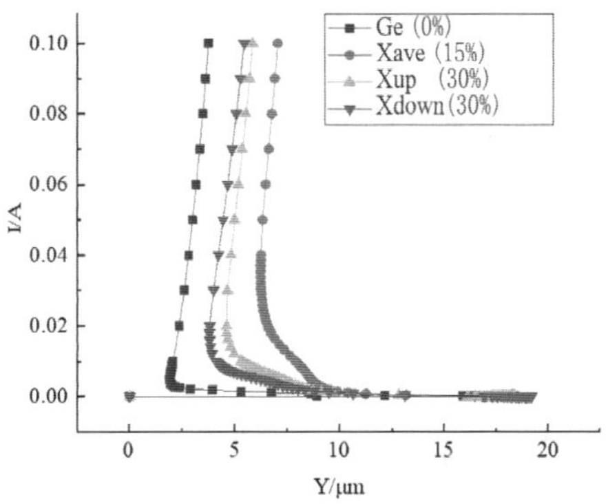

Figure 8. I-V characteristic curves of energy conventional SCR and new SCR under different Ge distribution forms.

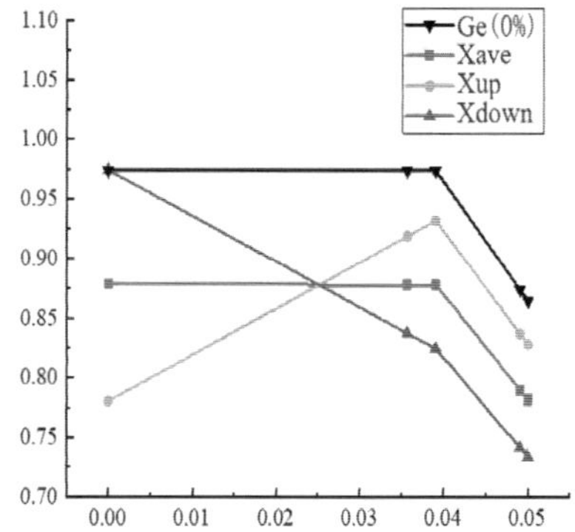

Figure 9. Conduction band levels in different Ge distribution forms.

Table 2. Key parameters under different Ge distribution forms from conventional SCR

Ge composition distribution		0%	5%	10%	15%	20%	25%	30%
Xaverage	Trigger Voltage/V	18.22	18.35	18.42	18.53	18.66	18.75	18.84
	Holding Voltage/V	2.06	2.80	4.31	6.23	7.97	9.31	10.32
Xup	Trigger Voltage/V	18.22	18.25	18.29	18.32	18.35	18.39	18.43
	Holding Voltage/V	2.06	2.36	2.76	3.27	3.73	4.15	4.60
Xdown	Trigger Voltage/V	18.22	18.32	18.43	18.57	18.75	18.97	19.23
	Holding Voltage/V	2.06	2.29	2.55	2.88	3.24	3.57	3.82

5. Conclusion

One novel ESD device with low holding voltage based on SCR and SiGe is analyzed. By using SiGe material as a P+ region, the current gain of PNP parasitic transistors is decreased, and the conductivity modulation caused by high injection is inhibited. Lifting holding voltage is thereby achieved. The results of researching device performances under three kinds of distribution of Ge composition indicate that the uniform distribution of Ge composition has the optimum effects. The holding voltage promotes four times from 2.06V to 10.32V when using uniform distribution Ge composition whose content is 30%. Therefore, there is no enlarging of the layout area of the novel device when promoting the holding voltage and enhancing the anti-latch ability of SCR.

References

[1] L. Yang, C. Yang, Y. Tu, et al. Field-circuit co-simulation method for electrostatic discharge investigation in electronic products[J]. IEEE Access, 2021,09:1-10.

[2] Q.Z. Ji, Z.L. Gao, N. Feng, et al. Complex integrated circuits ESD damage diagnosis technology and application [C], 2020 21st International Conference on Electronic Packaging Technology (ICEPT). IEEE, 2020:1-2.

[3] L. Zhu, H. Liang, X. Gu, et al. Design of a cascade-MOS-triggered SCR with high holding-voltage for high-voltage ESD protection[J], Solid-State Electronics, 2020, 171:1-4.

[4] Y. Wang, J. He, Y. Hu, et al. Study of silicon-controlled rectifier devices with different dimensions for ESD protection[C], 2020 IEEE 15th International Conference on Solid-State & Integrated Circuit Technology (ICSICT). IEEE, 2020:1-3.

[5] J. W. Jung, B. S. Lee, Y. N. Choi, J. W. Han and Y. S. Koo, Design of ESD protection with SCR-based structures for latch-up immunity[C], 2013 IEEE International Conference of IEEE Region 10 (TENCON), 2013:1.

[6] F. Ma, Y. Han, B. Song, S. Dong and M. Miao, Comparison of different MOS-triggered SCR structures for on-chip ESD protection[C], 2011 3rd International Conference on Computer Research and Development (ICCRD), 2011:331-335.

[7] C. Huang, F. Chiu, Q. Chen, M. Lai, and J. Tseng, An SCR-incorporated BJT device for robust ESD protection with high latch-up immunity in high-voltage technology[J]. IEEE Transactions on Device and Materials Reliability, 2012, 12(01):113-123.

[8] Z. Tong, H. L. Li and Z.W. Liu, A novel latch immune thyristor for on-chip electrostatic protection [J]. Microelectronics, 2018, 48(04):99-103.

[9] S.N. Ying, H.L. Li and Z.W. Li, A new type of SCR with high maintenance voltage [J]. Microelectronics, 2018, 48(06):811-814.

[10] Z. Qi, Research on the new mode of ESD protection device structure optimization in anti-latch-up[D]. Chengdu: University of Electronic Science and Technology of China (UESTC),2020:5 .

[11] Z. W. Liu, J. J. Liou, J. Vinson. Novel silicon-controlled rectifier (SCR) for high-voltage electrostatic discharge (ESD) Applications[J]. IEEE Electron Device Letters, 2008, 29(7):753-755.

[12] Y.C. Huang, M.D. Ker. A latch-up-immune and robust SCR device for ESD protection in 0.25-μm 5-V CMOS process[J]. IEEE Electron Device Letters, 2013, 34(5) 67: 4-676.

[13] Z. Qi, M. Qiao, X Zhou, et al. Investigation of a latch-up immune silicon-controlled rectifier for robust ESD application[J]. International Symposium on Power Semiconductor Devices and IC's (ISPSD), IEEE, 2017:183-186.

N-buffer design and optimization in 4500V class IGBT

Ruifen Nie*[ab], Rui Jin [ab], Yaohua Wang [ab], Li Li [ab], Mingchao Gao [ab], Yueyang Liu [ab], Feng He [ab], Jiang Liu [ab], Cui Li [ab], Wanru Sun [ab], Baohua Tian [ab]

[a]State Key Laboratory of Advanced Power Transmission Technology

[b]Beijing Institute Of Smart Energy

* Corresponding author: nieruifen@126.com

Abstract: In this paper, the effects of doping structure of multi-proton injection (MPI) buffer layer on static and short circuit characteristics of IGBT were studied, and the relationship between donor concentration of the buffer layer and injection dose and annealing temperature was studied. The MPI buffer layer suitable for 4500V IGBT was optimized by comparative experiments. This MPI buffer layer can solve the problem of high-temperature activation of the traditional IGBT buffer layer and lower thermal budget and can be fabricated on large wafers for better parameter consistency.

1. INTRODUCTION

To produce devices with the lowest possible static on-state and dynamic turn-off losses, while maintaining a sufficiently high blocking voltage capability, IGBT designers have proposed different N+ field stopping concepts over the years to meet the need for thin N drift layers. FS-IGBT, based on deep diffusion wafer technology, has demonstrated this capability, but the fabrication of long stop layers (FS) with this technology presents special challenges. Conventional donors, such as phosphorus, require high diffusion temperatures, usually greater than 1000 ° C, and diffusion can only be carried out before other processes begin in order to avoid wafer bending or adverse effects on other diffusion processes [1] [2] [3].

The high voltage and high power module package has high requirements on the consistency of chip parameters. Therefore, it is a better choice to process 4500V IGBT chips processed on wafers with diameters over 200mm. Because wafers larger than 200mm are more likely to bend at high temperatures, it is necessary to maintain a high wafer thickness (> 750μm) during the machining process. In order to achieve a low saturation pressure drop, the back of the wafer is usually thinned after the front process. The FS-IGBT process is characterized by making the FS layer on the back side first, which is not suitable for the IGBT process of the large wafer that requires later thinning. The hydrogen injection buffer layer has become a research hotspot in recent years because of its low activation temperature, it can be made after the IGBT chip frontal process is finished and thin, without affecting the doping region already formed. Table 1 lists the process sequence of FS-IGBT and IGBT using hydrogen injection buffer layer technology. It can be seen that the buffer layer can be fabricated after wafer thinning, which can achieve thinner processing results on large wafers.

Continuous hydrogen injection and low-temperature annealing (< 450ºC) of a variety of energies can produce multiple proton injection (MPI) buffers[4]. Stable hydrogen-related donor formation can be maximized by optimizing proton injection dose, annealing temperature, and time. This design provides a high degree of flexibility to shape the field barrier layer according to the desired application, resulting in devices with appropriate static and switching characteristics [5]. In this paper, a variety of

MPI buffer layers prepared by low-temperature annealing with continuous injection of medium and high-energy hydrogen were designed. The relationship between the number of MPI buffer peaks and static parameters and device robustness was studied by simulation. The relationship between donor activation concentration and injection dose and annealing temperature was also studied. Finally, the optimal MPI buffer layer design scheme is selected for 4500V IGBT through the above simulation experiments.

Table1. The summary process sequence of FS-IGBT and IGBT with hydrogen injection buffer layer

Process sequence	FS-IGBT	IGBT with hydrogen injection buffer layer
1	Buffer	Ring
2	Ring	Field Oxide
3	Field Oxide	Gate Oxide
4	Gate Oxide	Gate
5	Gate	P-body
6	P-body	N Source
7	N Source	P+ Ohmic Contact
8	P+ Ohmic Contact	Insulating Layer
9	Insulating Layer	Contact
10	Contact	Metal
11	Metal	passivation layer
12	passivation layer	Grind
13	-	Buffer
14	Back process	Back process

2. H BUFFER EXPERIMENT AND ANALYSIS

In order to study the influence of MPI buffer peak number and different combination modes on device performance, the setting comparison experiment is shown in Table. 1. Four buffer layer injection methods with different energy and dose were designed, and different injection combinations were selected for 5 groups of experiments. The injection conditions selected by different experimental groups were marked with * in Table. 2. In the process of hydrogen injection, due to the limitation of the capacity of the injection equipment, high energy, and high dose often cannot exist at the same time. Therefore, the high injection dose matches the low injection energy, and the low injection dose matches the high injection energy. Moreover, the injection energy of the general equipment is less than 2MeV, so the maximum injection depth is limited.

Table. 2 Experimental grouping I

Process	Split	001	002	003	004	005
H implat1	H/1.45MeV/1E13/Tilt7	*				*
H implat2	H/1.1MeV/1.3E13/Tilt7	*	*			*
H implat3	H/850KeV/2.5E13/Tilt7	*	*	*		*
H implat4	H/450KeV/2.5E14/Tilt7	*	*	*	*	

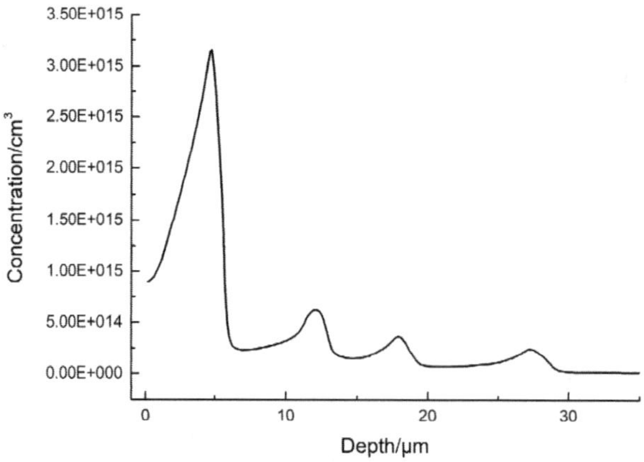

(a) 001

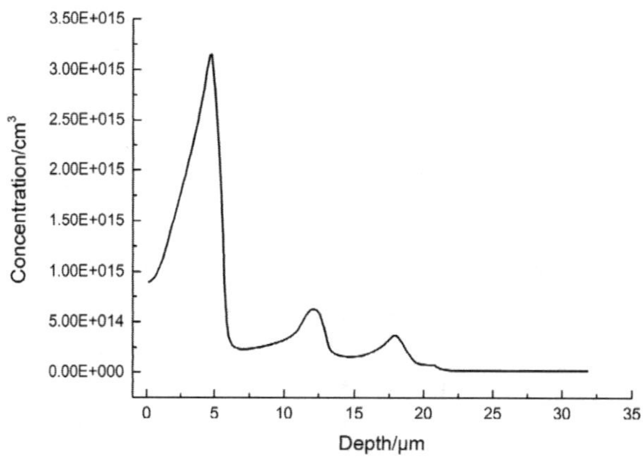

(b) 002

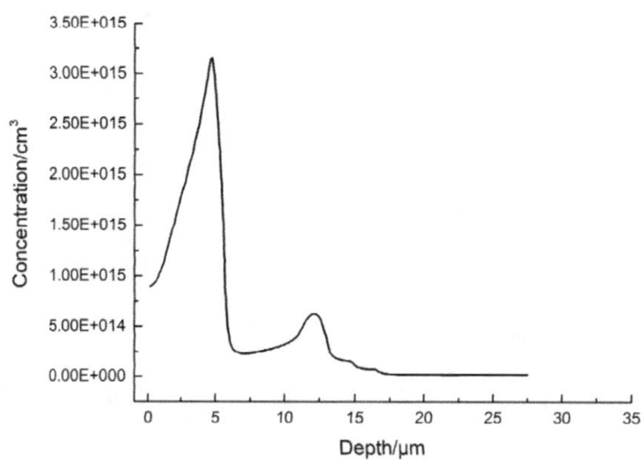

(c) 003

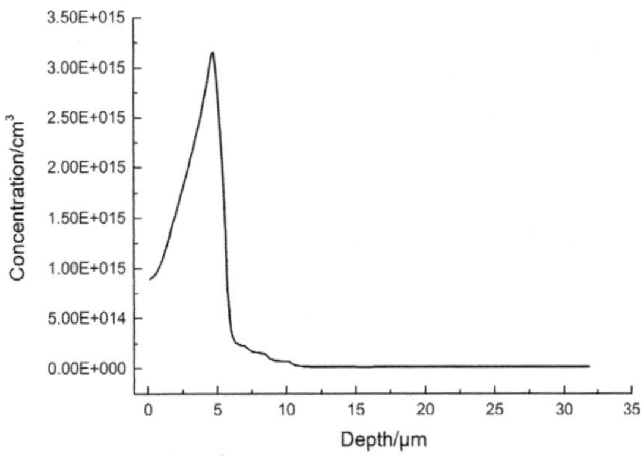

(d) 004

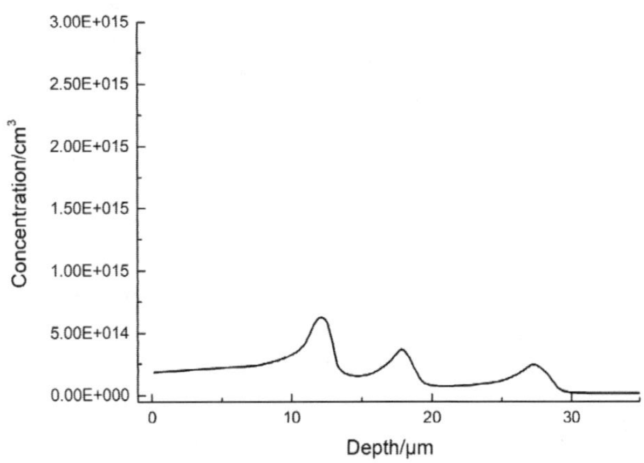

(e) 005

Figure. 1 SRP results of groups 001 to 005

The activation temperature and time of all MPI buffer layers were 400°C and 30 min, respectively. The test results of the Spreading resistance profile (SRP) of the buffer layer made from group 001 to group 005 are shown in Figure. 1. MPI buffers with different injection times showed different numbers of donor concentration peaks in SRP results. The shallowest peak has the highest donor concentration for the injection and activation results of H implanT4 because the injection dose is the largest and the depth is small and easy to activate. Similarly, the peak with the deepest depth corresponds to the injection and activation results of H implant1. Since the injection dose is the smallest and the depth is large, it is not easy to activate, so the donor concentration is the smallest. The static parameter simulation results of IGBT devices with different MPI buffer layers are shown in Table. 3. These IGBT chips have the same design and process except for the buffer layer structure.

Needless to say, the buffer layer on the back has no effect on the threshold of the IGBT device, so all IGBTs have almost the same threshold. In the blocking state, the buffer layer on the back of the IGBT chip can realize larger blocking voltage and smaller leakage current in the thin drift layer by blocking the electric field and can affect the saturation voltage drop, on-state and turn-off loss of the device by filtering the holes from the collector in the open state.

Table. 3 Static simulation results of different groups I

Item	001	002	003	004	005	Unit
Threshold voltage	6.59	6.59	6.59	6.59	6.58	V
Saturation voltage (At 50A)	2.52	2.52	2.52	2.51	2.48	V
Breakdown voltage	5306	5309	5303	5218	2385	V
Leakage current (At 4500V)	56.7	63.8	61.8	103	-	μA

According to the simulation results of 001, 002, and 003, the donor concentration peak of the minimum injection depth is the same, the overall peak number is different, and the breakdown voltage difference is very small. When the three deep donor concentration peaks are removed and only the minimum injection depth peak is retained, the breakdown voltage of 004 decreases slightly and the leakage current increases obviously. 005 has three deep peaks, after removing the peak of the minimum injection depth, the breakdown voltage decreases obviously. According to the static simulation results, the donor concentration peak of the minimum injection depth peak has the greatest contribution to the breakdown voltage, and the leakage current can be reduced by increasing the number of donor concentration peaks.

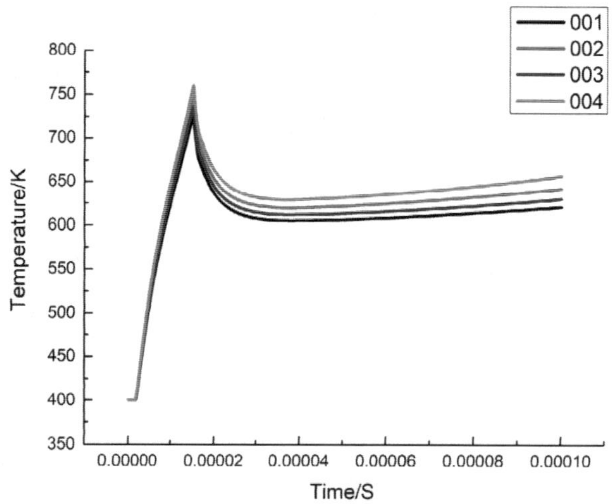

Figure. 2 Simulation results of the maximum temperature of devices with different buffer layer(001 002 003 004) structures in short-circuit state

Figure. 2 shows the simulation results of the maximum temperature of the IGBT devices with 001 002 003 and 004 buffer layers in a short circuit state. The simulation conditions are as follows: Vge=15V, Vce=3400V, pulse length 15 microseconds. Devices with buffer layer 005 cannot simulate short circuit conditions due to a breakdown voltage below 3400V. All the IGBT device temperatures reached a maximum value at the end of the 15 microseconds. Compared with other buffer layer devices, with the 001 buffer layer structure of the IGBT device at the end of the short circuit temperature minimum, this is because the buffer layer can filter the holes from the back, and the buffer layer with more peaks can filter more, so the current rises slowly and the temperature rises relatively low in short circuit [6] [7]. Therefore, devices with more buffer layer peaks have better robustness. However, the number of buffer layers cannot be increased indefinitely. Firstly, the injection depth is limited due to the limited capacity of injection equipment. Secondly, too deep a buffer layer will lead to the narrowing of the drift zone and static parameter change [8].

3. OPTIMAL DESIGN OF H BUFFER LAYER OF 4500V IGBT DEVICE

Through the above experiments, it can be concluded that for 4500V IGBT devices, devices with four peaks of MPI buffer layer have good static characteristics and short-circuit characteristics. In order to obtain a more optimized MPI buffer layer design of 4500V IGBT, four groups of MPI buffer layers formed by four injection processes were fabricated on the back of the chip respectively. The process and structure design of the experimental chip was identical except for the buffer layer. In order to obtain a more optimized MPI buffer layer design of 4500V IGBT, four groups of MPI buffer layers formed by four injection processes were fabricated on the back of the same chip respectively, as shown in Table 4, and marked * as the injection conditions selected in the experiment. All the experiments were dose-differentiated in the shallowest (H IMPLat1) and deepest (H Implat4) hydrogen injection, respectively. The two injections in the middle are the same, and the dose is differentiated in the shallowest (H IMPLat1) and deepest (H Implat4) hydrogen injection respectively, and activation is differentiated at 420°C for 60min and 400°C for 60 min respectively. The purpose of the experiment is to study the relationship between donor concentration of buffer layer and injection dose and annealing temperature and select the optimal scheme.

Table. 4 Experimental grouping II

Process	Split	001	006	007	008
H implat1	H/1.45MeV/1E13/Tilt7			*	*
	H/1.45MeV/1.3E13/Tilt7	*	*		
H implat2	H/1.1MeV/1.3E13/Tilt7	*	*	*	*
H implat3	H/850KeV/2.5E13/Tilt7	*	*	*	*
H implat4	H/450KeV/2.5E14/Tilt7			*	*
	H/450KeV/4E14/Tilt7	*	*		
H buffer active	400°C/60Min	*			*
	420°C/60Min		*	*	

Figure. 3 is the SRP measurement result of 001 006 007 and 008 buffer layers. The shallowest and deepest injection doses of the 006 buffer layer are higher than that of the 007 buffer layer, and the activation conditions are the same. However, the SRP results show that the 006 buffer layer is only slightly higher than the 007 buffer layer in the first peak, and there is no significant difference in the other three deeper peaks.

All injection conditions of the 001 buffer layer and 007 buffer layer are identical, but 001 is activated at 400°C for 60min while 007 is activated at 420°C for 60min. The SRP results show that all peak concentrations of the 001 buffer layer are significantly higher than 007, indicating that the activation rate of 400°C is higher than that of 420°C. The same conclusion can be drawn from the comparison of 008 and 006. It should be noted that the shallowest and deepest injection doses of 008 were relatively small, but the activation concentration was the highest in the SRP test results.

The above experiments show that it is limited to increase donor concentration solely by increasing the injection dose of hydrogen, and it is more important to select an appropriate activation temperature. If the injection dose and activation temperature are properly matched, a small injection dose may also generate a higher donor concentration.

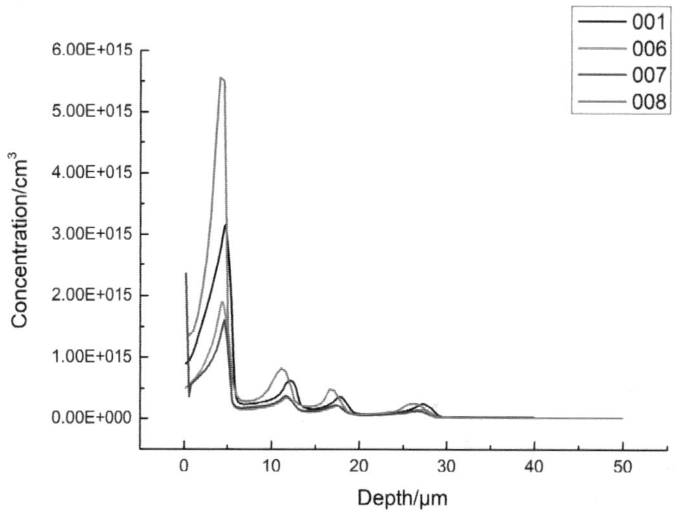

Figure. 3 SRP results of groups 001and 006 to 008

Parameters such as leakage current at higher operating temperatures and the ruggedness under short circuit conditions are heavily influenced by the bipolar transistor gain, which is largely dependent on the shape of the buffer design and the anode strength [9] [10].

The static parameter simulation results of IGBT devices matching 001 006 007 and 008 MPI buffer layers are shown in Table. 5. These IGBT chips have the same design and process except for the buffer layer structure. The IGBT device with a 007 buffer layer has the lowest breakdown voltage, while the IGBT device with a 008 buffer layer has the highest breakdown voltage, which proves once again that the shallowest peak height mainly determines the breakdown voltage, and the higher the shallowest peak height, the higher the breakdown voltage. The IGBT device with different buffer layer structures has little difference in on-state voltage, so the IGBT device with a 008 buffer layer structure is the best choice from the static parameter consideration.

Table. 5 Static simulation results of different groups II

Item	001	006	007	008	Unit
Threshold voltage	6.59	6.59	6.59	6.59	V
Saturation voltage（At 50A）	2.52	2.5	2.5	2.54	V
Breakdown voltage	5306	5053	4926	5391	V
Leakage current（At 4500V）	56.7	124	135	38.7	μA

Figure. 4 shows the simulation results of the maximum temperature of IGBT with 001, 006, 007, and 008 buffer layers on the back in a short circuit state. The simulation conditions are also as follows: Vge=15V, Vce=3400V, and the pulse length is 15 microseconds. It can be seen that the device with a 008 buffer layer structure has the lowest temperature rise and the best firmness after the short circuit. According to the static and short circuit simulation results, the buffer layer structure formed by 008 injection and activation conditions is most suitable for 4500V IGBT production.

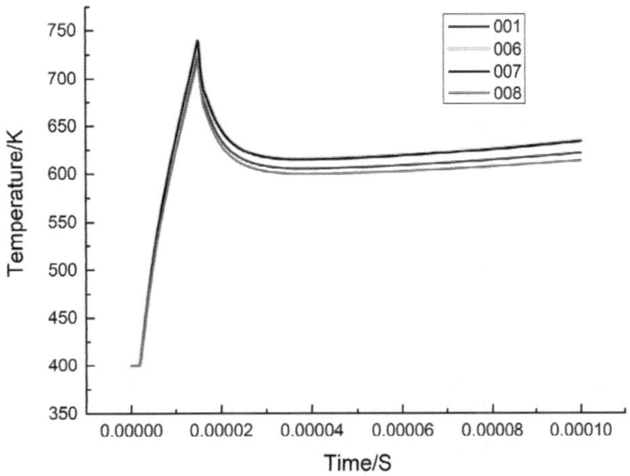

Figure. 4 Simulation results of the maximum temperature of devices with different buffer layer(001 006 007 008) structures in short-circuit state

4. CONCLUSION

In this paper, buffer layers with different numbers of donor concentration peaks (SRP test results) are designed and fabricated to study the influence of the number of donor concentration peaks on the static and short circuit characteristics of IGBT devices. The shallowest peak has a decisive effect on the breakdown voltage of IGBT devices. The short circuit characteristics of IGBT can be optimized by increasing the number of buffer layer peaks, but the number of buffer layer peaks cannot be increased indefinitely. Firstly, the injection energy of injection equipment is limited, and secondly, too deep a buffer layer will make n-type drift too small, resulting in the degradation of other electrical parameters. The activation concentration of the buffer layer has no linear relationship with injection dose and activation temperature, and a reasonable combination of the two can obtain a higher activation concentration. In this paper, static and short circuit simulation of four IGBTs with different MPI buffer layers is carried out to find the optimal buffer layer process scheme for 4500V IGBT.

ACKNOWLEDGEMENTS

This work was financially supported by the State Grid Co., Ltd. Science and Technology Project: Research on Key Process Technology of 4500V IGBT Chips with High Current Density and High Junction Temperature (NO. 5500-202058473A-0-0-00).

REFERENCES

[1] J. Vobecký, M. Rahimo, A. Kopta, S. Linder "Exploring the Silicon Design Limits ofThin Wafer IGBT Technology: The Controlled Punch Through (CPT) IGBT," Proceedings of the 20th International Symposium on Power Semiconductor Devices & IC'sMay 18-22, 2008 Oralando, FL

[2] T. Laska et al "Short circuit properties of Trench/Field Stop IGBTs design aspects for superior robustness", Proc. ISPSD'2003, p. 152.

[3] M. Rahimo et al "Extending the boundary limits of high voltage IGBTs and diodes to above 8kV", Proc. ISPSD'2002, p. 41.

[4] F. . -J. Niedernostheide et al., "Tailoring of field-stop layers in power devices by hydrogen-related donor formation," 2016 28th International Symposium on Power Semiconductor Devices and ICs (ISPSD), 2016, pp. 351-354, DOI: 10.1109/ISPSD.2016.7520850.

[5] H K. Nakamura et al "Advanced wide cell pitch IGBT having Light PunchThrough (LPT) structures", Proc. ISPSD'2002, p. 277. [2] E. Buitrago et al., "A Critical View of IGBT Buffer Designs for 200 °C Operation," 2019 31st International Symposium on Power Semiconductor Devices and ICs (ISPSD), 2019, pp. 47-50, DOI: 10.1109/ISPSD.2019.8757638.
[6] K. Suzuki, K. Nishi, M. Kaneda and A. Furukawa, "N-buffer design optimization for short circuit SOA ruggedness in 1200V class IGBT," 2018 IEEE 30th International Symposium on Power Semiconductor Devices and ICs (ISPSD), 2018, pp. 128-131, DOI: 10.1109/ISPSD.2018.8393619.
[7] H. P. Felsl, F. . -J. Niedemostheide and H. . -J. Schulze, "IGBT field-stop design for good short circuit ruggedness and a better trade-off with respect to static and dynamic switching characteristics," 2017 29th International Symposium on Power Semiconductor Devices and IC's (ISPSD), 2017, pp. 143-146, DOI: 10.23919/ISPSD.2017.7988931.
[8] A. Nakagawa et al "MOSFET-mode Ultra Thin Wafer PTIGBTs for Soft Switching Application – Theory and Experiments", Proc. ISPSD'2004, p.103.
[9] M. Rahimo et al "Novel Enhanced–Planar IGBT Technology Rated up to6.5kV for lower Losses and Higher SOA Capability", Proc. ISPSD'2006,p. 33.
[10] M. Rahimo et al " Switching-Self-Clamping-Mode "SSCM", a breakthrough in SOA performance for high voltage IGBTs and Diodes ", Proc.ISPSD'2004, p. 437.

Supercapacitor Control System Based on Fuzzy PID and Active Equalization Technology

ZiChen Guo, ChangMing Zhang*

School of Electrical and Information Engineering, Zhengzhou University, Zhengzhou, China

*Corresponding author: iecmzhang@zzu.edu.cn

Abstract. In this paper, a supercapacitor control system is designed to improve the motion performance of small unmanned vehicles. The supercapacitor is used as the auxiliary power supply for the chassis of the unmanned vehicle. When working at low power, the lithium battery supplies power to the chassis and charges the supercapacitor. When climbing or accelerating, the supercapacitor is used to assist the power supply to improve the body's movement ability. The current and voltage values are collected by high-precision sensors and the system charging control is completed by fuzzy PID. It has a good effect on the test, which significantly improves the motion performance of the unmanned vehicle.

1. Introduction

Supercapacitors, also known as electric double-layer capacitors, have large capacitance, high power density, and high energy storage efficiency [1]. Therefore, supercapacitors have a wide range of applications in energy storage systems. In paper[2], a photovoltaic energy storage system based on a supercapacitor is designed, the charging and discharging model of a supercapacitor is established from the perspective of energy control and signal control, and a coordinated control method is proposed to ensure higher energy utilization. while having better robustness. Electric vehicles have developed rapidly in recent years, and the capacity of lithium batteries, which are the main energy storage devices of electric vehicles, will decrease rapidly under the impact of high-power loads and affect the performance of the energy storage system of electric vehicles. The supercapacitor has a high cycle life and can discharge a large current in a short period of time. These characteristics enable it to improve sports performance when the electric vehicle starts and climbs, and reduce the impact of high-power work on the lithium battery. In paper[3], the supercapacitor electric vehicle is modelled and simulated, and compared with the electric vehicle without a supercapacitor, it is concluded that the supercapacitor can help improve acceleration performance and reduce power consumption. In paper[4], an energy-saving vehicle based on supercapacitors is designed, which can only be powered by supercapacitors and can be quickly charged and coasted for long distances.

This paper designs a supercapacitor-lithium battery power system for small unmanned vehicles. The STM32F103 is used as the main controller, and the fuzzy PID algorithm is used for the charging control of the supercapacitor. The supercapacitor replaces the lithium battery to supply power for the unmanned vehicle during high-power operation, which improves the motion performance of the unmanned vehicle.

2. System Design

2.1. System implementation process
This paper adopts the lithium battery-supercapacitor hybrid power supply mode. When working at low power, it is powered by a lithium battery. When the unmanned vehicle is climbing or accelerating, it is switched to the supercapacitor for a short-term high-power power supply, which improves the motion performance of the unmanned vehicle or small robot. The system block diagram is shown in Figure 1. When the chassis power of the unmanned vehicle is less than the power value set by the program, the lithium battery supplies power to the chassis and the supercapacitor. On the contrary, when the chassis power of the unmanned vehicle is greater than the power value set by the program, it is powered by the supercapacitor.

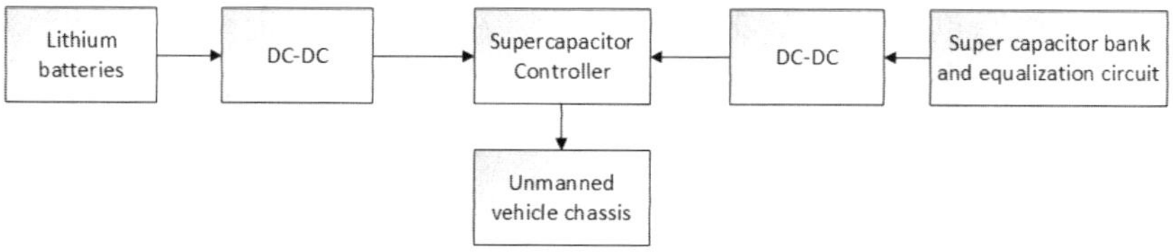

Figure 1. System chart

2.2. Modeling and Calculation of Supercapacitors
The model of supercapacitors is different from that of ordinary capacitors. The simplified model of a supercapacitor is shown in Figure 2, where R_{es} is the equivalent series impedance. Energy loss in the supercapacitor will occur during charging and discharging, and the higher the current, the greater the loss. R_{ep} is the equivalent impedance in parallel. When the leakage current flows through R_{ep}, a certain static loss will be generated. In this paper, ten supercapacitors of type CXHP2R7506R are used in series, with a rated voltage of 2.7 V, a capacitance of 50 F, $R_{es} = 24\text{m}\Omega$, and a maximum peak current of 30.68 A.

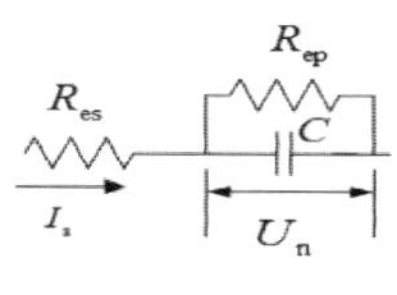

Figure 2. Supercapacitor Equivalent Circuit

The following is the analysis and derivation of the equivalent circuit, Assuming that the charging current is I_s, the rated voltage is U_n, W is the energy stored after reaching the rated voltage and the charging time is t.

$$I_s = C \times \frac{du}{dt} = C \times \frac{U_n - I_s \times R_{es}}{t} \tag{1}$$

$$W = \frac{1}{2} \times C \times (U_n - I_s \times R_{es})^2 \tag{2}$$

For the discharge process, the load equivalent resistance is generally much smaller than the parallel equivalent resistance, so the discharge model can be ignored R_{ep}. The voltage across the capacitor is:

$$U(t) = U_C(t) - I(t) \times R_{es} \tag{3}$$

Assuming that the working current is 1 I_s during discharge, V_1, V_2 are respectively the highest voltage and the lowest voltage that can work normally. Disregarding other energy losses, the discharge duration is:

$$T = \frac{C\left(V_1^2 - V_2^2\right)}{\left(V_1 + V_2\right) I_1} \tag{4}$$

2.3. The design of the discharge path selection part

Sections should be numbered with a dot following the number and then separated by a single space: The discharge path selection part is used to select the supercapacitor or lithium battery for power supply. Figure 3 shows the design of the discharge path selection section in the supercapacitor controller.

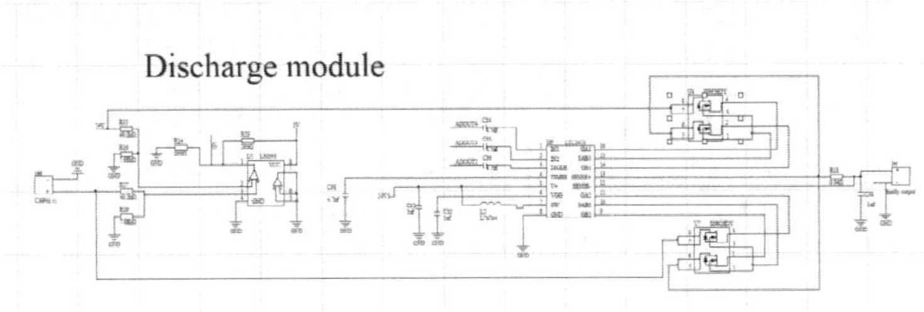

Figure 3. Discharge path selection part design

The lithium battery and the supercapacitor are connected to the supercapacitor controller at the same time, and the controller controls the opening or closing of one of them. This article uses two back-to-back MOSFETs to turn on and off. Compared with using only one MOSFET, the method of connecting the source and source of two MOSFETs can solve the influence of parasitic diodes and effectively prevent current inflow. The LTC1473 is used to receive the MCU control signal and control the four MOSFET.

2.4. Supercapacitor Equalization Circuit Design

Supercapacitors often need to be used in series or in parallel in actual use. Due to differences in production and manufacturing, the supercapacitor group will have voltage inconsistencies during the charging and discharging process, which affects the service life of the supercapacitor [5]. Therefore, supercapacitors usually need to be equipped with corresponding equalization circuits. The equalization circuit can be divided into active equalization and passive equalization. In paper[6], a sensor-free active equalization circuit is introduced, which is composed of supercapacitors, inductors, and balance capacitors, and the equalization voltage is turned off alternately by switches. This article refers to the above and designs the circuit shown in Figure 4.

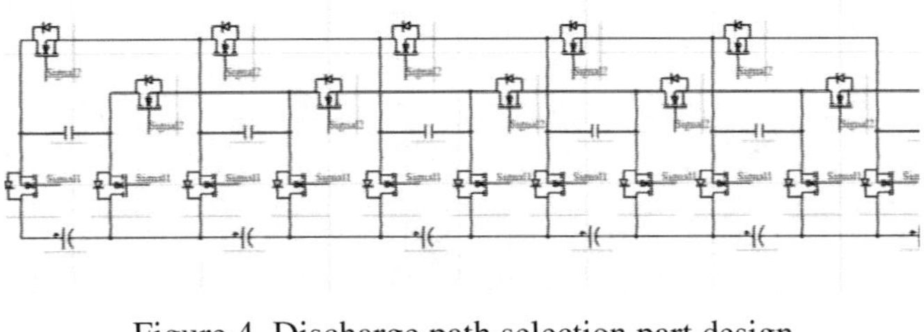

Figure 4. Discharge path selection part design

3. Charging and discharging mode and PID algorithm

3.1. Charging and discharging method selection
Common charging methods for supercapacitors include constant current charging, constant voltage charging, constant power charging, and pulse charging [7]. Among them, the charging speed of constant current charging is faster, but the temperature will rise in the later stage, and the charging efficiency will become lower. Constant voltage charging may cause excessive charging current in the early stage of charging and damage the supercapacitor, but overcharging can be avoided in the later stage. In this paper, two-stage charging is adopted, that is, constant current charging is used at the beginning of charging, and then constant voltage charging is used. In addition, to increase the charging speed without affecting the motion of the unmanned vehicle, the value during constant current charging or constant voltage charging is adjusted according to the power required by the unmanned vehicle. Taking the constant current charging stage as an example, assuming when the power is higher than $P_{\max}$, it is powered by a supercapacitor, and the current power required by the unmanned vehicle is $P_1(P_1 < P_{\max})$. At this time, the charging power of the supercapacitor should satisfy $P_1 + P_2 = I_s \times C(t) + P_1 < P_{\max}$. The voltage and current values and the power required by the current unmanned vehicle are collected in real time by the supercapacitor controller, and the parameter values are dynamically adjusted. The discharge mode depends on the chassis requirements and only limits protection for excessive current or voltage.

3.2. figDesign of Control Algorithms
There are many research results on supercapacitor charging control, among which the PID control algorithm is more commonly used. In paper[8], a PID control algorithm is used to wirelessly charge the supercapacitor, and the process of adjusting the parameters is described. In paper [9], a fuzzy PID-based supercapacitor control system was designed, and it performed well in the test. In this paper, the fuzzy PID algorithm is used to control the charging current and voltage. The specific plan is to collect the voltage and current value of the supercapacitor through a high-precision sensor and obtain the power of the unmanned vehicle through a power meter. Taking the constant current charging process as an example, the constant current value is.:

$$I_c = \frac{P_{\max} - P_1}{U_C} \tag{5}$$

This value is achieved by controlling the DC-DC module with constant current and constant voltage functions through the DAC on the STM32. Figure 5 shows the algorithm structure. The current loop and the voltage loop are independently controlled and adjusted by the feedback values of voltage, current, and power.

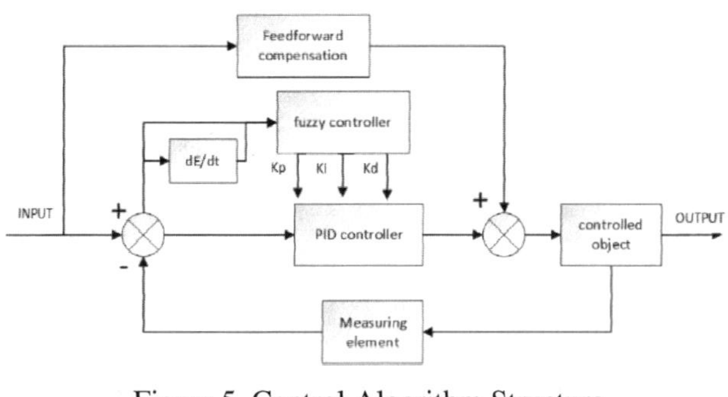

Figure 5. Control Algorithm Structure

4. Simulation and testing

4.1. Simulation of Equalization Circuits
Figure 6 shows the Simulink model and simulation results of the equalization circuit. The capacitance value is set to 50 F, and the switching frequency is 10 kHz.

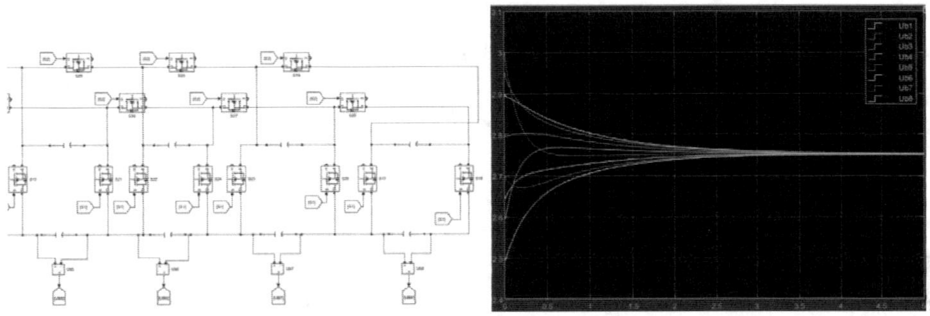

Figure 6. Simulation process and results of the equalization circuit

4.2. Charging test
Install the supercapacitor in the unmanned vehicle for the charging and discharging tests. After the test, the voltage of the supercapacitor is about 23.65 V when fully charged. When the voltage is lower than 15.6 V, the chassis power supply cannot be maintained. It takes about four minutes to discharge to 16 V until the supercapacitor is fully charged. The capacitor bank voltage during charging is shown in Figure 7.

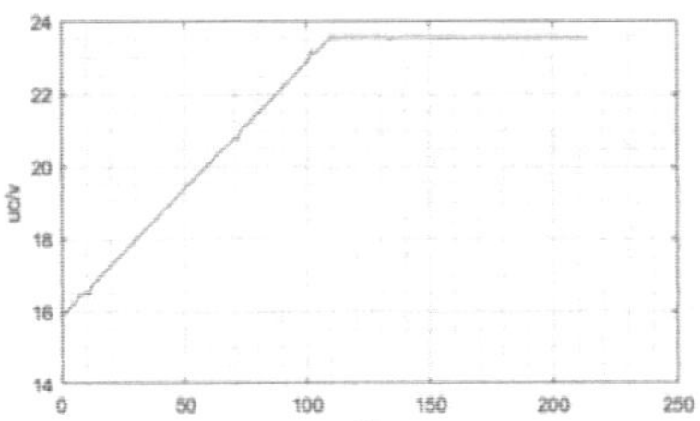

Figure 7. Voltage changes during charging

4.3. Practical effect test
As shown in Figure 8. The unmanned vehicle has a full load of about 10 kg, a slope of 15, and a ramp length of about 3 m. The pre-filled supercapacitor module is installed in the unmanned vehicle, and the unmanned vehicle is started at a distance of 12 m from the slope, and it takes about 7.6 s to reach the top of the slope. Remove the supercapacitor module and test it under the same conditions. The unmanned vehicle completes the movement for about 9.7 s. The use of supercapacitors significantly improves the motion performance of unmanned vehicles.

Figure 8. Test of unmanned vehicle movement effect

5. Conclusion

This paper designs a supercapacitor control system for small unmanned vehicles, and introduces the main circuit design and charging control scheme of the system. Finally, the supercapacitor was installed in the unmanned vehicle for experiments. The test results showed that the use of the supercapacitor module can significantly improve the motion performance of unmanned vehicles. In the next step, we will conduct more experiments to further verify the use effect of the system, and improve the charging circuit and charging algorithm in order to achieve better results.

References

[1] F. S. Liu, J. Peng, L. Li, X. Gong and H. Lu, *An MPC based energy management strategy for battery-supercapacitor combined energy storage system of HEV,* 2016 35th Chinese Control Conference (CCC), (Chengdu, China), pp. 8727-8731.

[2] T. F. Li and H. Yu, *Coordination control for supercapacitor charge and discharge in PV energy storage system based on EL and backstepping*, 2018 Chinese Control and Decision Conference (CCDC), (Shenyang, China), pp. 1299-1304.

[3] N. Jinrui, W. Zhifu and R. Qinglian, *Simulation and Analysis of Performance of a Pure Electric Vehicle with a Super-capacitor*, 2006 IEEE Vehicle Power and Propulsion Conference, Windsor, UK, pp. 1-6.

[4] F. Li, *Design of energy-saving vehicle based on supercapacitor*, 2021 IEEE 5th Advanced Information Technology, Electronic and Automation Control Conference (IAEAC), (Chongqing, China), pp. 1735-1739.

[5] Z. Gao, F. Jinang, H. Gao and X. Liao, *A new zero-voltage switching equalization system for series super-capacitor string*, 2015 34th Chinese Control Conference (CCC), (Hangzhou, China), pp. 7838-7842.

[6] P. Yu, F. Ge and K. Li, *A Sensor-less Active Voltage Equalization Method for Series-Connected Supercapacitors*, 2019 Chinese Control and Decision Conference (CCDC), (Nanchang, China), pp. 3196-3201.

[7] P. Dhabe, M. K. Agrawal and M. Rode, *Controlled Charging of Ultra Capacitor by PWM Technique for Electric Vehicles*, 2018 International Conference on Smart Electric Drives and Power System (ICSEDPS), (Nagpur, India), pp. 13-17.

[8] S. C. Shu, R. Juanatas and J. Niguidula, *Design of PID Algorithm in Super-Capacitors Wireless Charging System*, 2021 Innovations in Power and Advanced Computing Technologies (i-PACT), (Kuala Lumpur, Malaysia), pp. 1-6.

[9] X. Chen, S. Liu, G. -f. Li and H. -y. Zhang, *Study of the fuzzy PID controlled super-capacitor auxiliary power system of the electric vehicle,* 2010 International Conference on Computer Design and Applications, (Qinhuangdao, China), pp. V3-114-V3-118.

Exploration and Analysis of the Application of "1+1" UPS Redundant Power Supply in a Large Science and Technology Park in Beijing

Dongpeng Zhou, Hongjun Yang, Di Qu

Power Administration Department of CAST

Zhoudp2022@163.com

Abstract. Based on the actual test of the "1+1" UPS redundancy system, the causes of some abnormal data are analysed. Subsequently for the purpose of avoiding these abnormalities and improving power supply reliability, the key points are described. The park should pay attention to when to apply the "1+1" UPS redundancy system.

1. Introduction

A large science and technology park in Beijing (hereinafter referred to as the "park") uses a large number of stand-alone UPSs in the power supply and distribution system. These UPSs better ensure the high-reliability power demand of the park's important equipment. However, with the increasing development of the park, a large number of new loads have been used. The park faces unpredictable security risks of single UPS. So the park plans to use a parallel power supply for some important equipment and select two UPSs for parallel connection tests in the early stage. This paper mainly tests the performance of the two parallel UPSs in the pilot operation. Combined with the actual test and the current situation of the power supply system in the park, we find out the points that need to be paid attention to when to choose a "1+1" UPS supply in the park, so as to ensure the safety and reliability of the parallel UPS[1].

2. Test content and analysis

2.1 Schematic diagram of parallel UPS connection

The so-called parallel UPS is that the two UPSs operate at the same frequency, phase and output voltage amplitude. The outputs are directly connected, and power is supplied to the load through the output power distribution cabinet. When the system is running normally, the two UPSs each bear 50% of the load (current sharing operation). The connection diagram is shown in Figure 1.

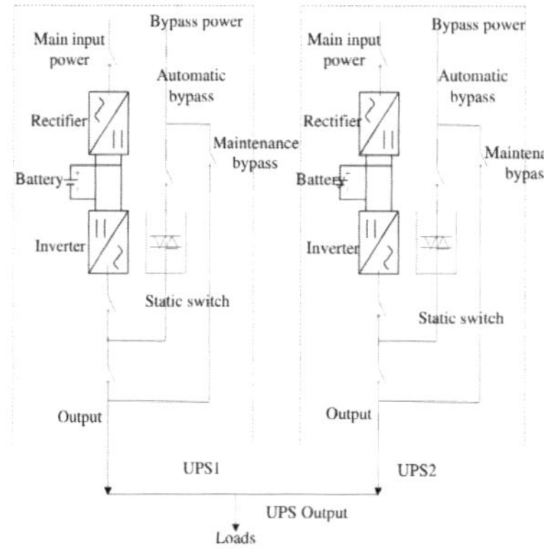

Figure 1. "1+1" parallel redundant UPS connection

2.2 Test instruction
The capacity of the two tested parallel UPSs is 40 kVA, and the selected test equipment is "Rizhi 3197 Power Quality Analyzer". The test contents include a static test (output neutral voltage, output waveform) and a dynamic test (simulated mains fault, simulated UPS fault test).

In Figure 1, A is the lower end of the mutual switch of the parallel UPS output cabinet, which represents the test connection point.

2.3 Static test analysis
The so-called static test refers to the test when the parallel UPS equipment enters the normal system state.

2.3.1 Zero-ground voltage test of the parallel output terminal
The excessive zero-ground voltage of the UPS output terminal will affect the accuracy of some equipment such as precision instrument equipment[2]. The zero-ground voltage of the two 40 kVA parallel UPS output terminals is shown in Figure 2. The testing load is "false load".

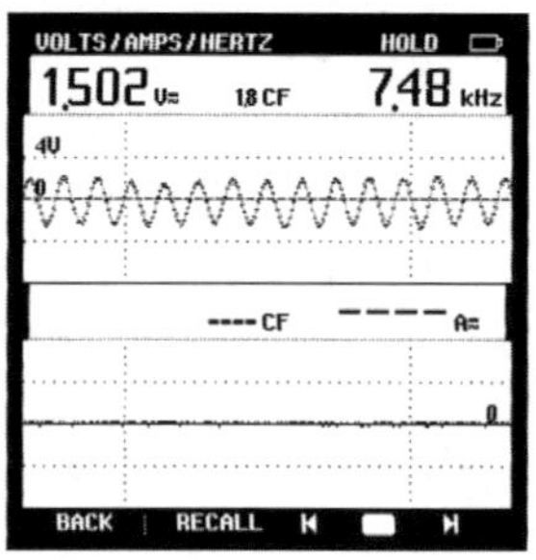

Figure 2. Output zero-ground voltage of the two 40 kVA parallel UPSs

According to Figure 2, the zero-ground voltage at the output end of the two 40 kVA parallel UPSs is 1.502 V, which is greater than the UPS zero-ground voltage "1 V" of the IT industry recommendation. This test shows that the output zero-ground voltage of the two parallel UPSs do not meet the specification requirements.

2.3.2 The output voltage waveform test of Parallel UPS
The "false load" is the same as the above "zero-ground voltage" testing load. The output voltage waveform of the two 40 kVA parallel UPSs is shown in Figure 3.

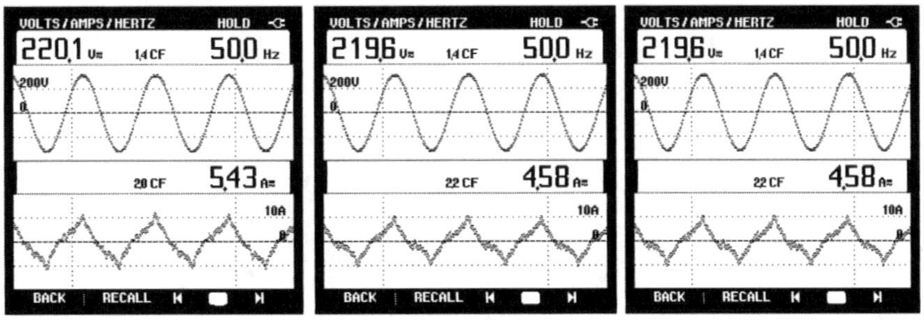

Phase A Phase B Phase C

Figure 3. Output voltage waveform of the two 40 kVA parallel UPSs

According to Figure 3, the effective output voltages of the two 40 kVA parallel UPSs can be obtained, that is, A-phase 220.1 V, B-phase 219.6 V, and C-phase 219.3 V, which are smaller compared with the normal voltage of 220 V. In addition, a waveform diagram can be seen from Figure 3 intuitively in which the output voltage waveform and characteristics of the two 40 kVA parallel UPSs are normal.

2.4 Dynamic test analysis

The so-called dynamic test refers to the test that the parallel UPSs are in an abnormal state. In this paper, it is assumed that the two tested UPSs are UPS1 and UPS2, respectively.

2.4.1 UPS1 faults simulation test

The UPS1 failure is simulated. The output voltage and the current waveforms of UPS2 under all load processes are tested. The test results are shown in Figure 4.

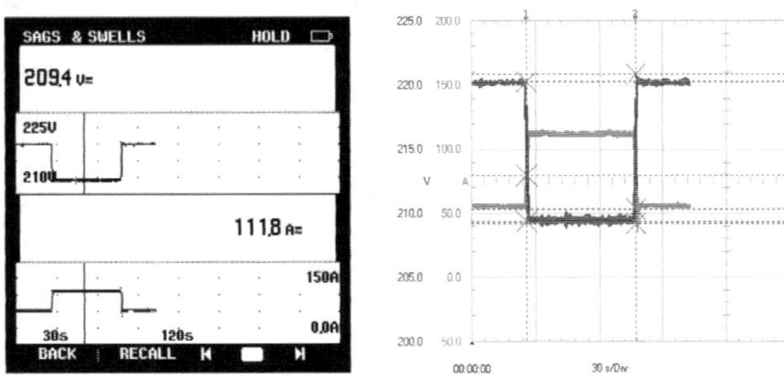

Macrograph Specific graphs

Figure 4. Output voltage and current waveforms of the UPS2 (supply all the loads)

According to Figure 4, when the two UPSs run at about 26 seconds, UPS1 fails, and UPS2 supplies all the loads at this time. But the output voltage is slightly reduced, between about 209.2 V and 209.3 V. The output current of UPS2 doubles due to the instantaneous load, and its value rises from about 55 A to 110 A instantaneously. This state continues for nearly 55 seconds until the UPS1 returns to a normal state, especially when the output voltage and the current of UPS return to the original state. According to the test, when UPS1 fails, UPS2 successfully supplies all the loads, and the output characteristic of UPS2 is good.

2.4.2 Input power faults simulation test

The test output voltage and the current waveforms are shown in Figure 5 when the input power is at fault.

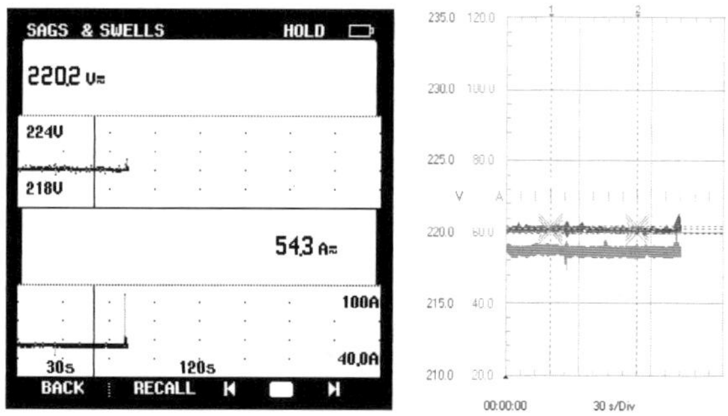

Macrograph Specific graphs

Figure 5. Output voltage and current waveforms of the two 40 kVA parallel UPS

According to the two graphs in Figure 5, when the two UPSs run together at about 18 s, a test fault is simulated. At this time, the battery supplies all the loads; the output voltage is maintained at about 220 V; the output current remains unchanged, and this state lasted for about 36 s until the input power is restored. It can be seen from the test that when the input power is at fault, the battery successfully supplies all the loads.

3. Test evaluation and analysis

Through the above test, the overall performance of the two parallel UPSs is good, and the power supply reliability is better than that of a single UPS. In an emergency, the parallel UPS can supply continuous power supply for the important equipment in the computer room, and meet uninterrupted power demand[3]. However, two problems are also exposed during the test: first, the output zero-ground voltage is too high; second, the output voltage of a single UPS is lower than that of the parallel UPS. The specific reasons are analyzed as follows:

Due to the limited space of the test UPS installation area and the need to reduce workload, the "false load" is still placed far away from the UPS. If the distance is too long, the zero-line impedance will increase[4]. If the cross-sectional area of the zero-line is too small, the zero-line impedance will further increase. Then the zero-line current will generate a large potential difference, which is the zero-ground voltage of the UPS output cabinet (Due to the limitation of space, this article will not repeat them) [5]. Factors such as "false load" imbalance, eventually lead to the large zero-ground voltage of the parallel UPS during the test.

4. Issues that need attention

Through the above test and analysis, if the park chooses the "1+1" UPS system to supply power to important equipment in the future, it must try to avoid the problems below. The power supply cable of the UPS is too long, and the cross-section of the cable (especially the cross-section of the zero-line cable) is too small[6]. The cable should be selected as large as possible, especially since the two connected UPSs belong to different types. Another problem that needs to be explained is that the park's internal specification "power supply load capacity should exceed 50% of the maximum power load, and power failure is not allowed during the test". Therefore, when the "1+1" UPS redundant power supply mode is specifically applied, the UPS rated capacity should not be less than 1.5 P/cos & (kVA) (P is the load power; cos & is the power factor).

5. Conclusions

Through the actual test of the "1+1" redundant UPS in a park, it is found that the zero-ground voltage is too large and that the output voltage of single UPS is too small in the application. The author analyzes the detailed reasons with the aid of the actual test, and finally puts forward the factors that should be considered when applying a parallel UPS, so as to avoid the recurrence of such problems and achieve the reliability of the power supply.

References

[1] Zhou Dongpeng etc. Exploration and analysis of methods to improve the reliability of UPS operation [J]. *Power Industry*, 2012. 09.
[2] Fu Jiajun etc. Equivalent input disturbance-based H∞ repetition control of UPS inverter [J]. *Journal of Electronic Measurement and Instrumentation*, 2022. 06.
[3] Wang Fugang. Strategy Analysis of Improving UPS Power Supply Reliability[J]. *Telecom Power Technology*, 2020. 08.
[4] Zhai Yujie. Reliability Analysis of UPS Redundant Parallel Power System[J]. *Telecommunications Technology*, 2007.05.
[5] Lin Jia. Design and Implementation of UPS Monitoring System Based on Internet of Things Technology [J]. *Automation Application*, 2012. 03.
[6] UPS product user training manual [G]. Shenzhen: Emerson Network Power Co. Ltd.

A Compact Reflectionless Low-Pass Filter Based on TSV Technology

Fengjuan Wang*[a], Rui Wang[a], Ruiqi Li[a], Ningmei Yu[a], Yuan Yang[a]

[a]School of Automation and Information Engineering, Xi'an University of Technology, Xi'an, 710048, China

*Corresponding author: wfjxiao4@163.com

Abstract: This paper presents a compact reflectionless low-pass filter based on through-silicon via (TSV). Its overall structure is composed of TSV inductors and redistribution layer(RDL) capacitor. The high-frequency structure simulator (HFSS) validates the proposed reflection filter. The results show that the filter operates at a frequency of 0-6.6GHz, with The return loss of input and output is less than -14.6dB in all test frequency bands(0-30GHz). In addition, the size of the reflectionless filter is only $0.19\times0.23\times0.0139\ mm^3$ ($0.014\times 0.017\times0.01\ \lambda g^3$).

1. INTRODUCTION

As an important device in the front end of the RF module, the filter adjusts the signal in the spectrum according to the different working frequency bands, most of the traditional filters are reflective filters. Their working feature is to filter the stopband signal by reflection. Therefore, the stopband signal will be reflected back to the signal source, This will adversely affect circuits that are very sensitive to reflected signals. Different from the traditional filter, the reflectionless filter absorbs the stopband signal and greatly reduces the signal level reflected back to the source, which is reflected in the filter index that its reflection coefficient is at a lower level whether in the passband frequency or stopband frequency(S11<-10dB).This feature provides more protection for the source [1] and has a significant impact on the performance improvement of the frequency mixing system. So it is important to study the RLF.

In recent years, there are various ways to study RLF. A various passband RLF based on lumped element design is presented based on Odd-Even Mode Analytical Method in [2], and realized by transmission line in a few years [3]. Besides,In [4], the proposed reconfigurable reflectionless filter based on duplexer architecture can realize reflectionless filtering of various passbands, even multi bands. In addition to input RLF, two-port RLF was further explored[5-6], which can achieve reflectionless behaviors at their input port and output port. However, the current RLF is large in size, so TSV technology is introduced to miniaturize the device without affecting its performance.

This paper presents a compact reflectionless low-pass filter based on TSV technology, which consumes energy of unnecessary frequency components in the signal by adding a lossy element resistance. It can also be used in gigahertz applications (Wi-Fi 6 GHz band). Section 2 presents a third-order RLF structure and its design steps, Section 3 analyzes the simulation results and Section 4 presents the final conclusions.

2. DESIGN OF REFLECTIONLESS FILTER BASED ON TSV

2.1 The circuit topology of low-pass reflectionless filter

The circuit topology of theRLF is shown in Figure 1(d). The design steps are as follows.

In a two ports network, the passband and stopband of the even-mode equivalent circuit and the whole circuit are opposite[3]. That is to say, the frequency reflected in the even-mode circuit is transmitted in the final two-port network. Therefore, if we want to design a low-pass filter, we must ensure that the even-mode equivalent circuit is a high-pass structure, reflecting low frequency instead of high frequency. Since the input impedances of even-mode and odd-mode circuits are reciprocal to each other, the odd-mode part is obtained by replacing inductors with capacitors, series elements with shunt elements, and resistors with their complementary conductance, as shown in Figure 1(a).

The dual network obtained can't be immediately seen as the two halves of the symmetrical two-port network. Therefore, it is necessary to simplify the even-mode and odd-mode parts by adding open circuit elements and short circuit elements and exchanging the positions of series elements to make them symmetrical (the concepts of symmetry and duality are the fundamental principles upon which all RLF topologies presented in this paper ultimately derive).

Firstly, in Figure 1(b), the even-mode side: exchange the position of the last capacitor and the series terminal resistance. Odd-mode side: change the grounding connection of the inductor and terminal resistance from absolute grounding to virtual grounding.

Then, in Figure 1(c), even-mode side: add an inductor between the input port and the symmetry plane (which is an open circuit). Similarly, an open circuit connection is added between the resistor and the symmetry plane. Because the symmetry plane here is an open circuit, these elements have no effect on the circuit. Odd-mode side: add a capacitor between the symmetry plane and the ground. Because both ends of the capacitor are short-circuited, the capacitor has no effect on the circuit.

Finally, Figure 1(d) shows the final structure, where g_k is the value of the prototype element that needs to satisfy the duality condition , and R is the 50 Ω impedance.

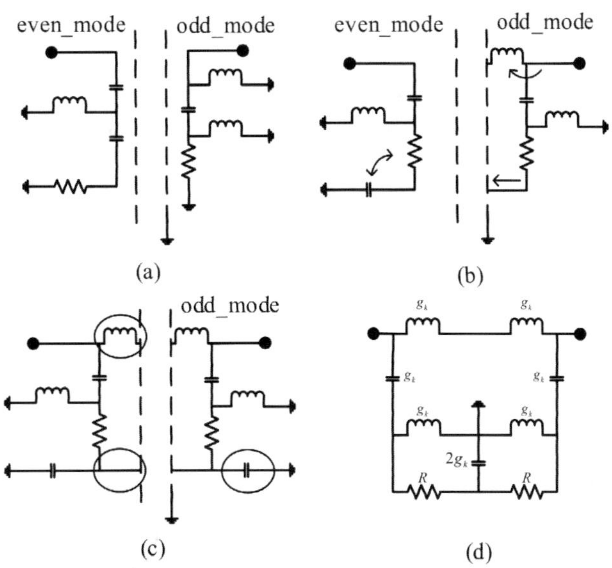

Figure 1. Derivation of a low-pass reflectionless filter

When we determine that the unit normalized value g_k is equal to 1, we design a low-pass filter with a transmission zero of 10GHz. Where the inductor-capacitor value is as follows(1)(2). The simulation results in ADS (Advanced Design System) are shown in Figure 2.

$$L = \frac{g_k Z}{\omega} = \frac{1 \times 50\Omega}{2\pi \times 10\text{GHz}} = 0.796\text{nH} \tag{1}$$

$$C = \frac{g_k Y}{\omega} = \frac{1 \times (50\Omega)^{-1}}{2\pi \times 10\text{GHz}} = 0.318\text{nH} \tag{2}$$

From Figure 2, it can be seen that the transmission zero point is at 10GHz, the 3dB bandwidth is 0-6.6GHz, the stopband suppress is -14.5dB, and the return loss of both ports is less than -60dB.

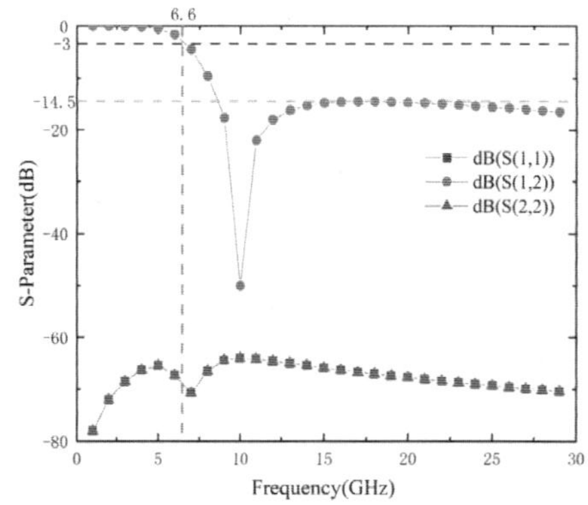

Figure 2. Simulation results of ADS

2.2 Entire 3D structure of the reflectionless low-pass filter

The inductance and capacitance of RLF designed in this paper are realized by TSV technology. The principle is due to the coupling effect between TSV copper columns. In the overall structure, the RDL part of capacitance and inductance is wrapped by silicon dioxide, and the copper column part of TSV is wrapped by silicon.

The TSV inductor consists of two layers of RDL and a middle TSV. The inner column of TSV is copper connected to RDL at top and bottom and the outer column is silica, as shown in Figure 3 (a). On the basis of our previous research on the TSV filter. Electromagnetic simulation software Q3D is used to extract the three-dimensional structure of the required inductor and capacitor, and the key parameters of the inductor are displayed in Table 1.

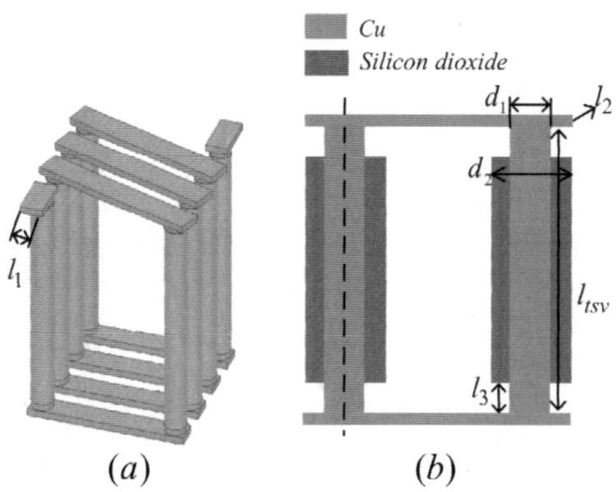

Figure 3. (a) Three-dimensional Overall structure of TSV inductor. (b) plane structure of TSV inductor.

Table 1. Structural parameters of TSV inductor

	Structure parameter	Symbol	Value (μm)
TSV	Inner diameter	d_1	9.6
	Outer diameter	d_2	10
	Height of TSV	l_{tsv}	100
	Interval Height	l_3	3
RDL	Width of RDL	l_1	10
	Thickness of RDL	l_2	3

The capacitor is composed of two layers of finger shaped or comb shaped RDL. Because of the coupling between copper, it can produce larger capacitance in smaller spaces, with Relatively compact structure and better quality factor. The TSV coupling capacitor has large size and large parasitic effect at the same capacitance value. Therefore, the capacitor structure selected in this paper is a double-layer interdigital capacitor. The three-dimensional structure of RLF is shown in Figure 4. Its overall size is only 0.19×0.23×0.139 mm^3 (0.014× 0.017×0.01 λg^3).

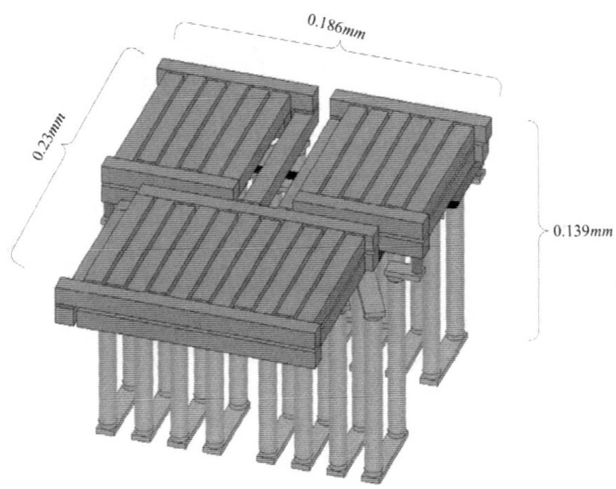

Figure 4. 3D view of the overall filter

3. RESULTS AND ANALYSIS

Figure 5 shows the finite element method (FEM) simulation results of the reflectionless filter model in HFSS (High-Frequency Structure Simulator) software.

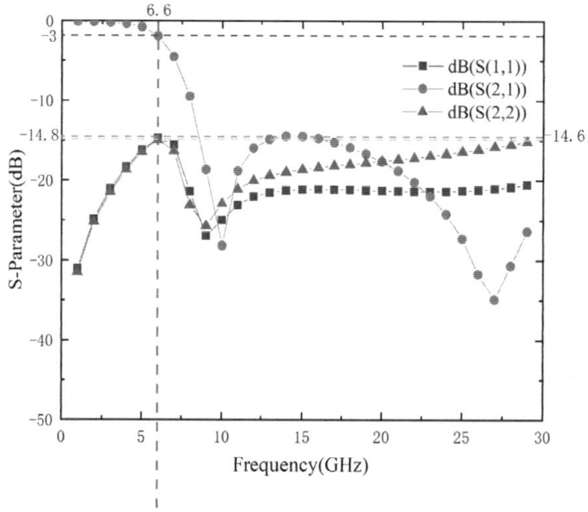

Figure 5. Simulation results of HFSS.

As can be seen from the figure, the filter operates at a frequency of 0-6.6GHz, can be used in Wi-Fi 6 GHz band(In the United States, FCC Commissioner and FCC Chairman Ajit Pai voted unanimously in October 2018 to release more Wi-Fi bands up to 1.2 GHz around and near 6 GHz.), the stopband suppress is -14.5dB. In all test frequency bands 0-30GHz, the input return loss less than -14.6dB besides output return loss less than -14.8dB, realizing the reflectionless of dual ports.

Finally, Table 2 tabulates a comparison with some previous works, through comparison, this work has a good return loss, which is lower than -14.6dB in the full frequency band at both ports, and the size is only 0.19×0.23×0.139mm^3 with high integration degree.

Table 2. comparison with some previous works

Design	Filter type	-3dB bandwidth (GHz)	Bandwidth of S11	Reflectionless port	Size (mm × mm)	Design technology
[1] work_1	LPF	0-0.11	DC-0.3<-15	all-port	40×40	lumped
[1] work_2	BPF	0.6-1.2	DC-0.3<-13	all-port	40×40	lumped
[5] work_1	BPF	1.17-1.7	0.76-2.01<-10dB	all-port	42×59.5	microstrip
[6] work_1	BPF	1.29-2.78	0-4GHz<-14.3dB	all-port	27×27	microstrip
[7] work_1	BPF	1.52 - 2.55	0.01-5.5<-13.52dB	input	100×70	microstrip
[7] work_2	BPF	1.14-2.87	0.01-5.5<-13.98dB	input	130×60	microstrip
[7] work_3	BPF	1.2-2.8	0.01-5.5<-14.55dB	input	130×140	microstrip
[8]	LPF	DC-1.31	DC-11.86<-11.21dB	input	54.6×43.5	microstrip
This work	LPF	DC-6.6	0-30GHz<-14.6dB	all-port	0.19×0.23×0.139	TSV

4. CONCLUSION

This paper presents a miniaturized, compact and high performance reflectionless filter. Compared with the state-of-the-artwork, the size of the filter is greatly reduced, the miniaturization of the device is realized, the two ports achieve better return loss within the entire test band, and better out of band suppression. It has high application value in Wi-Fi 6 GHz band.

REFERENCES

[1] Jin, X., et al., Four-Port Symmetrical Reflectionless Lumped Filter/Diplexer: Prototype and Design Table. IEEE Transactions on Microwave Theory and Techniques, 2021. 69(4): p. 2211-2219.
[2] Morgan, M.A. and T.A. Boyd, Theoretical and Experimental Study of a New Class of Reflectionless Filter. IEEE Transactions on Microwave Theory and Techniques, 2011. 59(5): p. 1214-1221.
[3] Morgan, M.A. and T.A. Boyd, Reflectionless Filter Structures. IEEE Transactions on Microwave Theory and Techniques, 2015. 63(4): p. 1263-1271.
[4] Psychogiou, D. and R. Gomez-Garcia, Reflectionless Adaptive RF Filters: Bandpass, Bandstop, and Cascade Designs. IEEE Transactions on Microwave Theory and Techniques, 2017. 65(11): p. 4593-4605.
[5] Fan, M., et al., Frequency-Tunable Constant-Absolute-Bandwidth Single-/Dual-Passband Filters and Diplexers With All-Port-Reflectionless Behavior. IEEE Transactions on Microwave Theory and Techniques, 2021. 69(2): p. 1365-1377.
[6] Xu, K., et al., Quasi-Reflectionless Filters Using Simple Coupled Line and T-Shaped Microstrip Structures. IEEE Journal of Radio Frequency Identification, 2022. 6: p. 54-63.
[7] Yang, L., et al., Multilayered Input-Reflectionless Quasi-Elliptic-Type Wideband Bandpass Filtering Devices on Diplexer-Based Structures. IEEE Transactions on Microwave Theory and Techniques, 2022. 70(1): p. 122-138.
[8] Yang, L., et al., Input-Reflectionless Low-Pass Filter on Multilayered Diplexer-Based Topology. IEEE Microwave and Wireless Components Letters, 2020. 30(10): p. 945-948.

A 3-D Integrated Common Mode Filter Based on Through Silicon vias Technology

Fengjuan Wang[1], Haoxiang Huang[1], Xiangkun Yin[*2], Ningmei Yu[1], Yuan Yang[1]

[1]School of Automation and Information Engineering, Xi'an University of Technology, Xi'an, 710048, China

[2]School of Microelectronics, Xidian University, Xi'an 710071, China

[*]Corresponding author: yinxkcn@163.com

Abstract: Common mode filters (CMF) are widely used to ensure signal integrity. However, due to the large area, traditional CMFs cannot meet the eager demand for the miniaturization of modern communication systems. In this paper, a three-dimensional integrated CMF based on through silicon vias (TSV) technology is proposed. The adoption of TSV technology greatly reduces the area of CMF. The structure is modeled and simulated by HFSS software. The stopband of the common mode signal is 7.9 GHz ~ 18.8 GHz. In the operating frequency band, the effect of group delay jitter on frequency is less than 10 ps. Moreover, the plane area of the CMF is only 0.195 mm^2.

1. INTRODUCTION

In a modern high-speed communication system, differential mode signal is gradually applied to various signal transmission systems because of its good anti-interference ability. However, due to various non-ideal factors, the common mode (CM) noise will be generated in the process of differential signal transmission.

In the past few years, various circuits and devices for common mode suppression have been reported, including CM choke using ferrite core, defected ground structure (DGS) and mushroom-shaped electromagnetic band-gap structure (EBG) or transmission line resonators based on print circuit board (PCB) [1] [2], and low temperature co-fired ceramic (LTCC)[3], etc. These designs can achieve a good level of suppression on a wide common mode stopband. Nonetheless, they cannot be well integrated with modern chip processes due to their large flat area. Therefore, it is necessary to design a miniaturized CMF with a good common mode suppression level. As a mature technology, through silicon vias (TSV) has been broadly acceptable to the layout of 3-D integrated circuits, for a vertical interconnection between chips, which greatly reduces the plane area of the chip. In this paper, a novel 3D-integrated CMF is proposed. It is designed with integrated passive devices (IPD) based on TSV technology, which achieves a small area and maintains a good level of common mode suppression.

This paper is organized as follows. In Section 2, a fifth-order circuit topology based on a K-type low pass filter and a model of CMF is proposed. In Section 3, the characteristics of simulation are analyzed and Section 4 is the concluding part.

2. FILTER DESIGN

2.1 Circuit topology of CMF

As shown in Figure 1, we established the circuit topology of the proposed CMF. Two low-pass filters (LPF) connect the ground terminals and the actual ground through an inductor. Because of its good response to the transmission signal and simple structure, this design adopts LPF to stimulate the signal transmission channels. Due to the symmetric structure required by CM suppression, the fifth-order K-type LPFs are used to simulate the transmission channel of the signal. The inductance and capacitance values are shown in Figure 1.

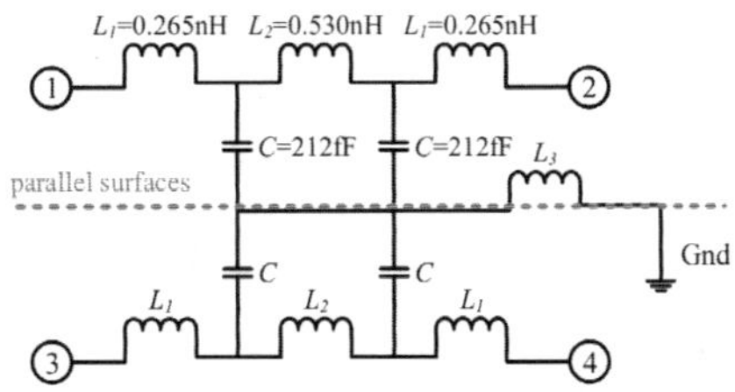

Figure 1. Ideal circuit topology of CMF

This circuit has a symmetric structure. When Port-1 and Port-3 or Port-2 and Port-4 are excited by the DM signals, L_3 on the parallel surface is considered as a short circuit because of the perfect electric conductor (PEC) boundary. At this point, the DM equivalent circuit network can be regarded as an LPF. In the passband, the DM signal can pass through. At the same time, when the ports are excited by CM signals, L_3 is equivalent to twice its inductance because of the perfect magnetic conductor (PMC) boundary. In this case, the CM equivalent circuit network is regarded as a band-stop filter [4]. As shown in Figure 2, the CM signal is suppressed in the stop band. The range of $S_{CC21} < -10dB$ is maximized by changing the inductance value of L_3. The value of L_3 as 0.428 nH is determined.

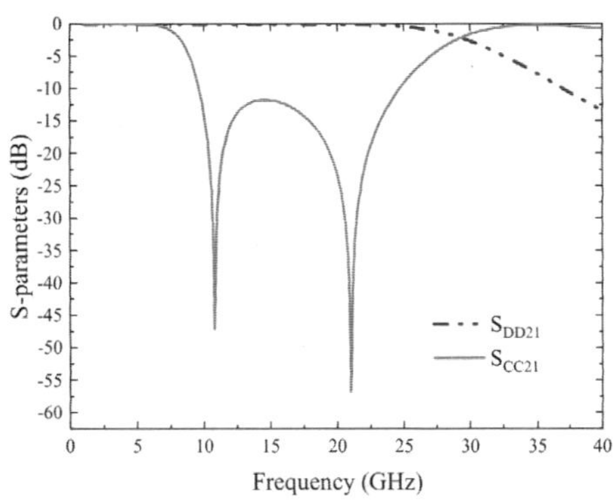

Figure 2. Simulation results of the ideal circuit

2.2 Model of filter

Considering the connection between the inductors and the capacitors, RDL inductors and TSV capacitors are adopted in this design. The design is laid out using a two-layer structure. The upper layers are RDLs buried in silicon dioxide, and the lower layers are TSVs wrapped in a silicon substrate, as shown in Figure 3.

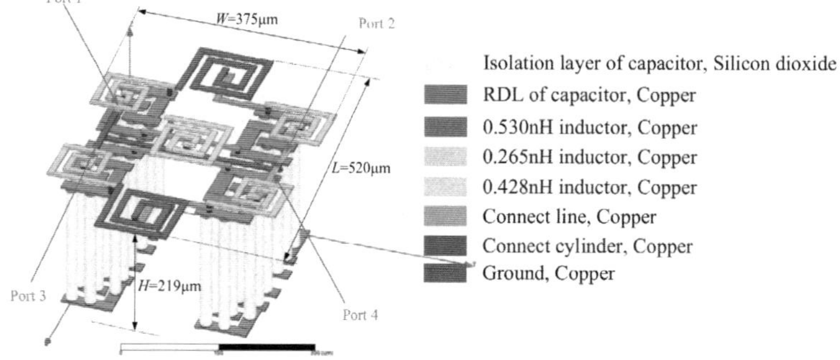

Figure 3. Model of CMF

2.3 Analysis of equivalent circuit

Table 1. Values of non-ideal effect

Symbol	Parameter	Value
R_{s1}	Parasitic resistance of L_1	244 mΩ
R_{s2}	Parasitic resistance of L_2	397 mΩ
R_{s3}	Parasitic resistance of L_3	348 mΩ
L_s	Parasitic inductance of connecting line	69.81 pH
R_{ss}	Parasitic resistance of L_s	66.76 mΩ
L_{m_left}	Mutual-inductance between L_1 and L_2 (case 1)	6.9 pH
L_{m_right}	Mutual-inductance between L_1 and L_2 (case 2)	15.1 pH

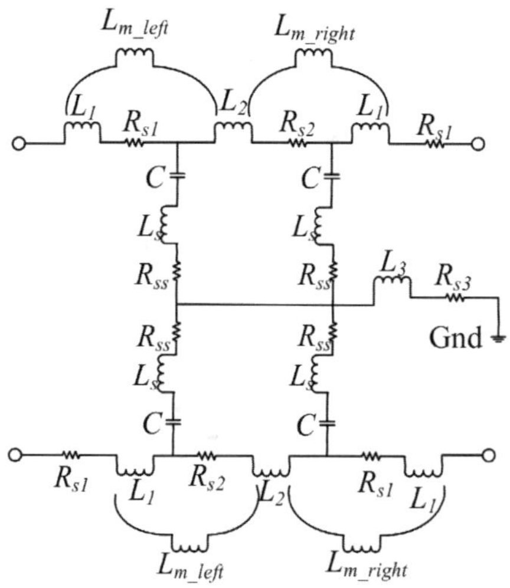

Figure 4. Equivalent circuit of CMF

Due to the non-ideal effect, the simulation results of the actual model are different from those of the ideal circuit. The parasitic inductance of the connecting lines is an important factor. In addition, the parasitic resistance of the inductors and the mutual inductance between the inductors also affect the filtering characteristics of CMF. The parasitic parameters can be obtained by HFSS software, as shown in Table 1. And Figure 4 shows the equivalent circuit we established.

3. RESULTS AND ANALYSIS

Figure 5 (a) shows the finite element method (FEM) simulation results of the CMF model in HFSS software, whereas (b) shows the comparison between the FEM model and the equivalent circuit. It can be seen that the band range of common mode suppression ($S_{CC21} < -10dB$) is 7.90 GHz~18.84 GHz. In the range of common mode suppression, DM signals maintain good integrity. FBW can be calculated to be 81.9%. At the same time, the simulation results of the FEM model agree well with those of the equivalent circuit. The results show that the established equivalent circuit can reflect the characteristics of the actual model.

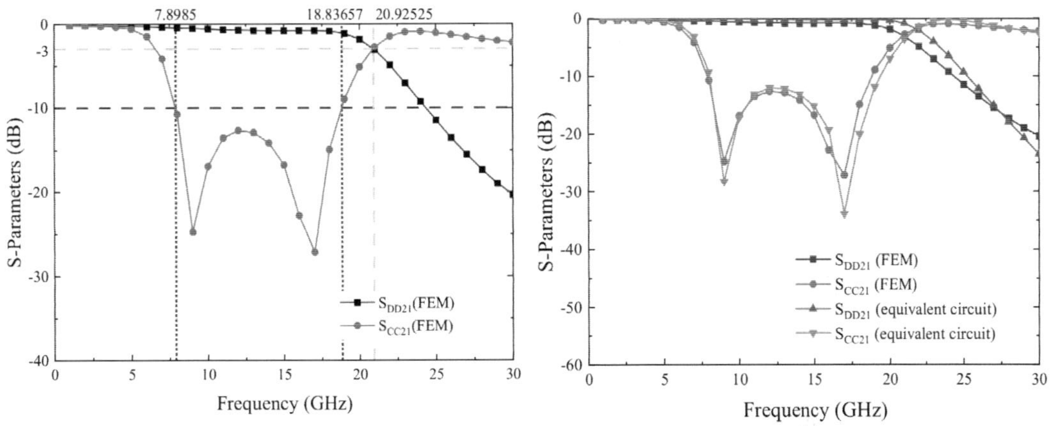

(a) (b)

Figure 5. Simulation results of S-parameters
(a) Simulation results of FEM model (b) Caparison between FEM model and equivalent circuit

The group delay of differential mode is about 25 picoseconds, as shown in Figure 6. And it changes little with the frequency in the CM stopband (within 10 ps). Figure 7 shows the eye diagrams of the FEM model and ideal circuit model. The test environment of eye diagram analysis is the signal of a pseudo-random sequence. The bit rate of the signal is 5 Gbps. The rise and fall time is 50 ps. The high level and low level are 1 V and -1 V, respectively. It can be seen that the shape of the eye is complete and consistent with the simulation results of the ideal circuit, which illustrates the CMF would not reduce the transmission quality of the digital signal.

This design is compared with other advanced CMFs, as exhibited in Table 2. It can be noticed that the plane area of this design is smaller while ensuring a wider CM stopband and good DM signal integrity.

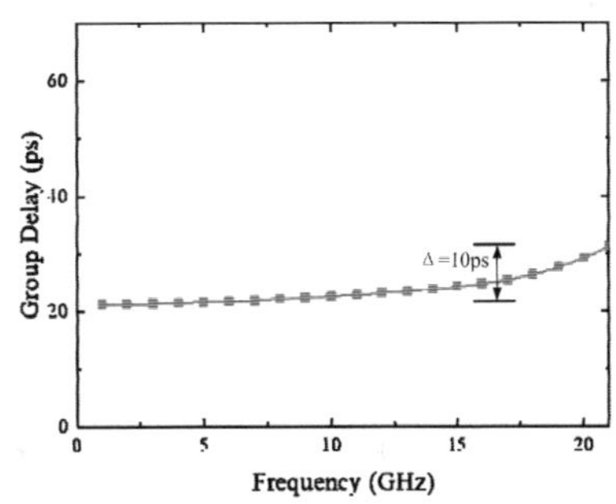

Figure 6. Simulation result of group delay

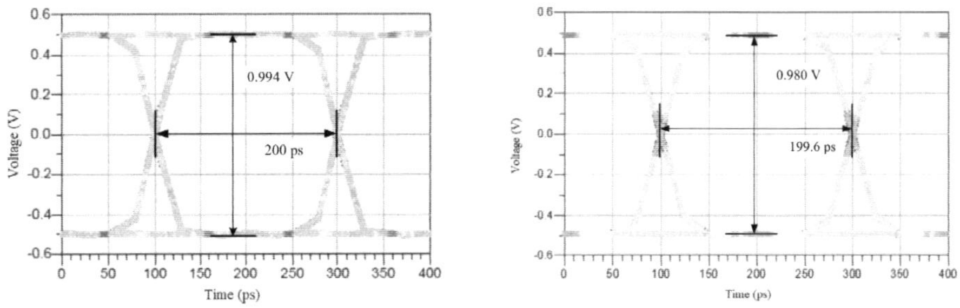

(a) (b)

Figure 7. Simulation results of eye diagram analysis
(a) Eye diagram of ideal circuit model. (b) Eye diagram of FEM model

Table 2. Comparison of this design with other advanced designs

Design	CM stopband (GHz)	DM passband (GHz)	Area (mm^2)
[1]	3.76~16.09	0~20	100
[2]	5.3~9.74 (Defined as S_{CC21}<-20dB)	0~10	158.4
[3]	2.8~10.5	0~10	27
[4]	1.6~4.7	0~10	0.95
Here	7.9~18.8	0~21	0.195

4. CONCLUSION

In this paper, a 3-D CMF based on TSV is proposed. Fifth-order K-type LPF is adopted to realize an equivalent signal transmission channel. In the CM stopband, this CMF has both good CM signal suppression and DM signal passing characteristics, and the area is very small (0.195 mm^2). In addition, due to the good compatibility of TSV and CMOS technology, the CMF can be integrated into the chip to meet the needs of chip miniaturization. In terms of future work, a better circuit topology should be carried out to enhance the bandwidth.

REFERENCES

[1] D. -B. Lin, C. -K. Yu, C. -K. Lin and H. -P. Lin, "On the Simulation Analysis for a Wideband Common-mode Filter with Defected Corrugated Reference Plane Structure". *2021 IEEE International Conference on Consumer Electronics-Taiwan (ICCE-TW)*, 2021, pp. 1-2.

[2] J. Lim, S. Oh, *et al.* "Enhanced broadband common-mode filter based on periodic electromagnetic bandgap structures". *Microwave and Optical Technology Letters*, vol.60, no.12, pp. 2932-2937, Oct. 2018.

[3] Y. Shi, P. Zhou, W. Feng, Y. Guo and W. Tang, "Quad-Mode LTCC Surface Mounted Packaging Common-Mode Filter Based on the Asymmetric Short-Stub Loaded Resonator". *IEEE Transactions on Circuits and Systems II: Express Briefs*, vol. 67, no. 10, pp. 1944-1948, Oct. 2020.

[4] C. -Y. Hsiao, Y. -C. Huang and T. -L. Wu, "An Ultra-Compact Common-Mode Bandstop Filter with Modified-T Circuits in Integrated Passive Device (IPD) Process". *IEEE Transactions on Microwave Theory and Techniques*, vol. 63, no. 11, pp. 3624-3631, Nov. 2015.

Improved Lightweight YOLOv3 model for Target Detection Algorithm

JianSheng Hu[a], JunJie Ma*[a], Bin Xiao[a], Rui Zhang[a]

[a]XJ Electric Co., Ltd., Xuchang, Henan, China

* Corresponding author: 277808516@qq.com

Abstract: When detecting small objects in interior situations, the classic object detection algorithm performs poorly in terms of real-time detection task and high precision detection task. This paper suggests an optimized tiny-YOLOv3-Shufflenetv2 light-weight model based on indoor scenes. The scheme adopts the fusion light-weight network which combines ShuffleNetv2 and YOLOv3, it reduces the complexity of the model to meet the lightweight requirements while ensuring good detection results for deployment to mobile robots. Also in this paper, an indoor small target object dataset, indoor-2022, is created to improve and optimize the model for the data images. YOLOv3, YOLOv3-Shufflenetv2, and tiny-YOLOv3-Shufflenetv2 are trained and tested on the indoor-2022 small target dataset in the Pytorch framework. The experimental findings indicate that in the indoor-2022 dataset. Compared with the single YOLOv3 model for object detection tasks, the fusion improved model used in this article improves the recognition ability of small objects in indoor images, With a 10-fold reduction in model size and a 4-fold increase in detection speed, only results in 1.6% reduction in the mean accuracy (mAP), and the comparison experiments with the current stage of traditional target detection algorithms validate the proposed tiny-YOLOv3-Shufflenetv2 model is verified to be superior and feasible. The optimized model in this article reduces mannequin parameters and model size while additionally ensuring the accuracy and velocity of inspection, and meets the requirements for deployment on indoor mobile robots.

1. INTRODUCTION

Indoor mobile robots have become a research hotspot for researchers because they can assist humans in daily operations in more complex application scenarios. Therefore, it is essential to learn about the accuracy of indoor mobile robot target detection algorithms and lightweight network structure.

For purpose of meetting the high frame rate and real-time detection requirements of object detection task, it is essential to simplify the network and then embed it into mobile devices, thus resulting in the popular lightweight network in recent years. S. Liu et al. designed a face detector[1], which has the gain of excessive velocity and accuracy and marks the realistic software of deep getting to know in the discipline of goal detection. L. Yang et al. proposed the Hier R-CNN method[2] for human detection, they proposed Hier R-CNN algorithm and made new data based on COCO dataset for side-to-side training, which can detect every body part. C. Jiang et al. proposed a fast and high-performance object proposal algorithm[3], which is very effective in recognizing targets and is robust to different shapes of targets and achieved the championship. K. -H. Shih et al. proposed a model compression method and compressed ZF-Net[4], and the results show that this compression method can compress the network parameters by more than 70% and save more than 60% of the computation. X. Zhang et al. propose a detection method of defect insulator based on YOLO and SPP-Net[5], and the detection accuracy of the two models can reach to 89%. W. Fang et al. proposed Tinier-YOLO[6], a further compression model based on Tiny-YOLOV3, which introduced dense connections and refined design to enhance feature propagation and make certain the most go with the flow of information in the network. Redmon et al. proposed YOLOv2[7], which added a BN layer, introduced a multi-scale training method and added a k-means clustering algorithm to further make the detection speed, and removed the pooling and fully connected layers, so the network can be further deepened faster.

Due to the complex structure of CNN and a large amount of parameters, when implemented on mobile devices, the data must be processed in the cloud, and then the results are transmitted back to the mobile device, but this cannot meet the real-time requirements. The first lightweight network is SqueezeNet[8] proposed by Forrest et al. The network structure reduces the AlexNet parameters by a factor of 50 without reducing the network accuracy and proposes the Fire Module. The Google team proposed Xception[9], whose main idea is Depth Separable Convolution (DSC), which splits a standard convolution operation into Depthwise Convolution and Pointwise Convolution. Face++ team Xiangyu Zhang proposed ShuffleNet[10], which uses Group convolution and channel-by-channel convolution for ResNet. The experimental consequences exhibit that ShuffleNet has higher overall performance and pace than MobileNet.

It has been shown that in order to embed algorithms into mobile devices or FPGAs, future network structures will tend to be more lightweight, therefore, the study of lightweight target detection network algorithms will be an essential research route in the subject of target detection in the future.

2. TINY-YOLOV3-SHUFFLENETV2

2.1 Overall structure

The tiny-YOLOv3-ShuffleNet model is shown in Figure 1. The three detections in YOLOv3-ShuffleNetV2 are simplified to two, and the 32-fold downsampling is removed, keeping only the 8-fold downsampling and 16-fold downsampling. This is done to further simplify the network structure and make the model further reduced. The improved model contains 42 layers of standard convolutional layers and 12 layers of depth-separable convolution. This method allows the branch used to detect small targets to extract deep features, thus improving the accuracy of detecting small targets. y2 is output after six CBR units and one layer of convolution after Concat, which is used for small size target detection tasks.

The GIoU loss function is used in the prediction stage for estimating the hole between the expected body and the actual frame, and there are three channels of output after three convolution layers in the prediction. Taking the enter decision of 416×416, the output dimension is 13×13×225, 26×26×225. GIoU solves the trouble of no longer being in a position to evaluate two pix when they are no longer intersected, the place A denotes the actual box, B denotes the expected box, C denotes the vicinity of the smallest outer rectangle of the anticipated and actual boxes, and its definition is as follows:

$$\mathrm{GIoU} = \mathrm{IoU} - \frac{|C - (A \cup B)|}{C} \tag{1}$$

$$L_{GIoU} = 1 - \mathrm{GIoU} \tag{2}$$

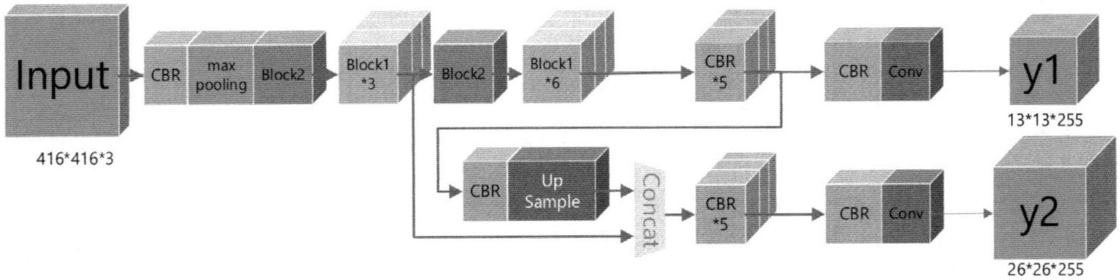

Figure 1. Tiny-YOLOv3-ShuffleNetv2 model

CBR means conv+BN+ReLU operation. Block1 cell is a structure in the ShuffleNetv2 model, as established in Figure 2, The Channel Split operation divides data into two parts, and one part goes through the CBR+Dwconv+BN+CBR. The contact operation differs from the Add operation in that the number of channels in the tensor expands after the contact operation, while the Add operation only adds the values directly, the number of channels in the tensor remains unchanged, and finally, the Channel Shuffle operation is performed; the block2 cell is the second structure of the ShuffleNetv2 model. The block2 unit is the two-structure of the ShuffleNetv2 model, as proven in Figure 3.

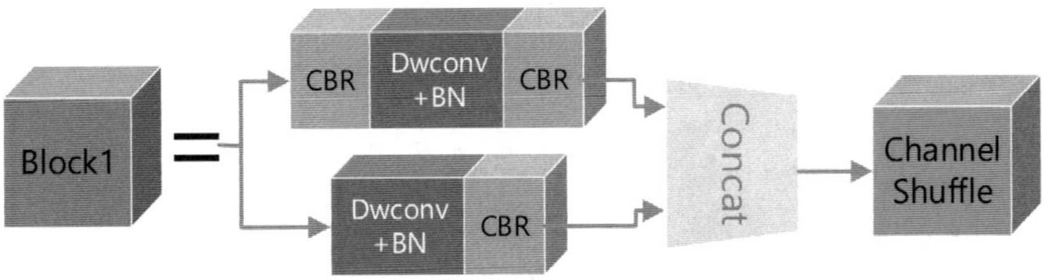

Figure 2. Block1 structure

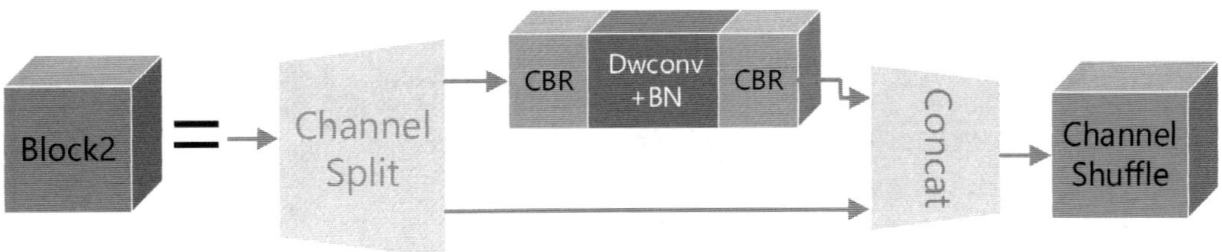

Figure 3. Block2 structure

2.2 ShuffleNetv2 block

The backbone network structure of the YOLOv3-ShuffleNetv2 model is ShuffleNetv2, which contains 13 block1 cells and 3 block2 cells. The first part consists of one block2 cell and three block1 cells for 8-fold downsampling, the second part consists of one block2 cell and seven block2 cells for 16-fold downsampling, and the third part consists of one block2 cell and three block1 cells for 32-fold sampling. The y1, y2, and y3 are the output networks of YOLOv3. The designer of the model designed the multi-scale detection network to decorate the accuracy of the algorithm for identifying small targets. Upsample and feature fusion methods were used, and finally, three scales of features were fused, 13x13, 26x26, and 52x52, respectively. The y1 detects large targets, y2 detects medium-sized targets, and y3 detects small targets. Since the target data sets in this experiment are all small and medium-sized targets, only the 8-fold downsampling and 16-fold downsampling structures are reserved for detecting small and medium-sized targets, and the model parameters are further reduced to simplify the model while ensuring that the detection accuracy is not degraded. The improved model contains a total of 42 standard convolutional layers and 12 depth-separable convolutional layers.

2.3 Depthwise separable convolution

The tiny-YOLOv3-ShuffleNetv2 model uses the channel-by-channel convolution in the depthwise convolution to simplify the model. The parameters of the depthwise convolution are much less. Suppose the measurement of the input feature map is $W_F \times W_F \times M$, W_F represents the same magnitude of the input feature map, M represents the number of image channels of the input feature, the dimension of the convolution kernel is $W_K \times W_K$, and N denotes the quantity of convolution kernels. Then the parameters of the ordinary convolution are calculated as:

$$P_{conv} = W_K \times W_K \times M \times N \times W_F \times W_F \tag{3}$$

As shown in Figure 4, depthwise convolution is a aggregate of depth convolution (DW) and point convolution (PW). Depth convolution performs a convolution which is a type of channel-by-channel, and then combined with a type of point-by-point convolution, and then outputs the feature map, so the parameters of depth convolution and point convolution are computed as:

$$P_{dw} = W_K \times W_K \times M \times W_F \times W_F \tag{4}$$

$$P_{pw} = M \times N \times W_F \times W_F \tag{5}$$

The parameters of the deeply separable convolution are computed as:

$$P_{Depthwise\ separable} = P_{dw} + P_{pw} = W_k \times W_k \times M \times W_F + M \times N \times W_F \times W_F \tag{6}$$

The parameters of depthwise convolution and ordinary convolution are proven in Equation (7). From the analysis, it can be considered that the more convolution kernels used in the network, the more 1/N converges to 0. Theoretically, the ordinary convolution computation is W_k^2 times more than the depthwise separable convolution.

$$\frac{P_{depthwise\ conv}}{P_{conv}} = \frac{1}{N} + \frac{1}{W_K^2} = \frac{1}{N} + \frac{1}{W_K^2} \tag{7}$$

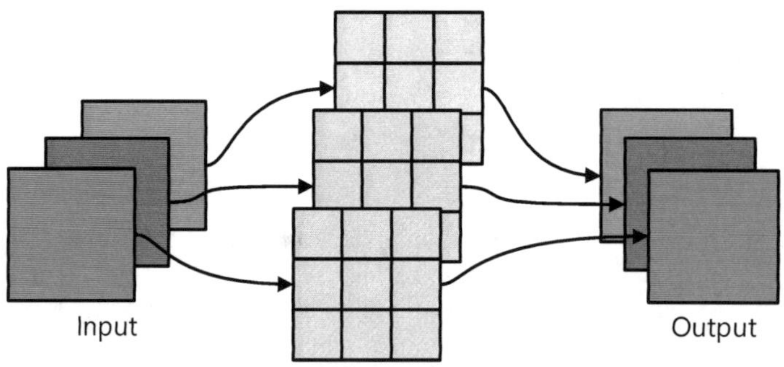

Figure 4. Deeply separable convolution

3. DATASET PREPARATION

In this study, the Indoor2022 dataset was created as the dataset for network training and testing. As proved in Figure 5, the Indoor2022 dataset includes a complete of 1000 photos with five sorts of targets, of which 900 are used for detection network training and 100 for validation. The objectives in the Indoor2022 dataset are all small and medium-sized, and the targets occupy no more than 1/100th of the image size.

Figure 5. Indoor2022 dataset

4. EXPERIMENTS

4.1 Experiment platform and protocol

The experiments were conducted in an Amax server environment with Ubuntu 16.04 with 8 Tesla K80 graphics cards, each with 12 G. The network models were built, trained, and tested with the Pytorch deep learning framework, using the CUDA 10.1 computing framework and cudnn to accelerate the deep learning computations.

This paper uses mAP (average accuracy), FPS (frames per second), and model size to measure the models' detection accuracy, speed, and computational complexity. Accuracy is used to assess the predicted correctness as in Equation (13). In this article, we compare the performance of imparity models on the Indoor2022 dataset by training YOLOv3, YOLOv3-ShuffleNetv2, and tiny-YOLOv3-ShuffleNetv2 models respectively, and compare the experimental results.

4.2 Experimental results

Figure 6 shows the training results of YOLOv3, YOLOv3-ShuffleNetv2, and tiny-YOLOv3-ShuffleNetv2 on the Indoor2022 dataset. It represents the variation of mAP of the three models with the wide variety of iterations, the horizontal coordinate is the wide variety of reiterations and the perpendicular coordinate is the mAP value. From the figure, it can be viewed that the mAP value step by step will increases with the increase of the range of iterations. YOLOv3 has the best result and the mAP can reach up to 0.8, YOLOv3-ShuffleNetv2 also has the highest mAP of 0.8, tiny-YOLOv3-ShuffleNetv2 has a slightly worse effect, but its mAP can still reach up to 0.78. From the experimental results, we can conclude that the YOLOv3 model has the best results on Indoor2022, the YOLOv3-ShuffleNetv2 model can still achieve almost the same accuracy as YOLOv3 with a lighter model, and the tiny-YOLOv3-ShuffleNetv2 has a slight loss of accuracy, but This is achieved by greatly reducing the model size.

The parabole of the training consequences of YOLOv3, YOLOv3-ShuffleNetv2, and tiny-YOLOv3-ShuffleNetv2 models on the Indoor2022 dataset, respectively, is proven in Table 1.

Table 1. Experimental product of the three models

Model	YOLOv3	YOLOv3-ShuffleNetv2	tiny-YOLOv3-ShuffleNetv2
Model size(MB)	235.2	82.14	20.23
mAP	0.8038	0.8022	0.7878
FPS	12	25	45

Tiny-YOLOv3-ShuffleNetv2 has the smallest range of parameters. replacing the geneal convolution with a deepwidth convolution in the ShuffleNetv2 structure can substantially decrease the quantity of parameters. The wide variety of parameters in the improved tiny-YOLOv3-ShuffleNetv2 model is reduced again, and the remaining model size is solely 1/10 of the YOLOv3 model, it is in line with the deployment requirements of indoor mobile robots. The tiny-YOLOv3-ShuffleNetv2 model has the largest FPS and tiny-YOLOv3-ShuffleNetv2 is the fastest.

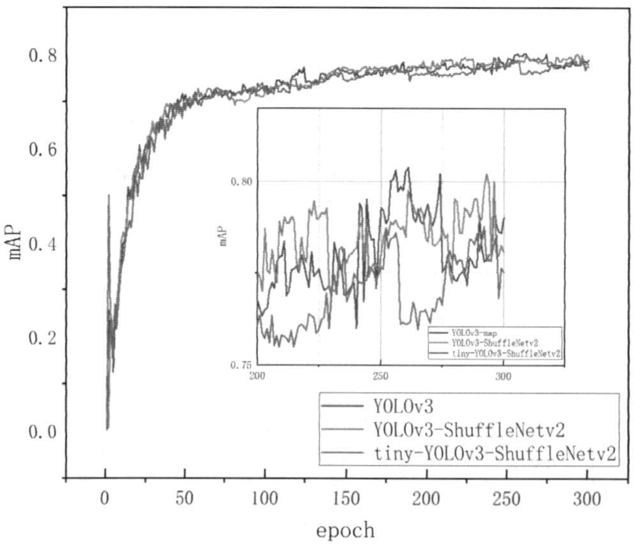

Figure 6. Comparison of mAP values of the three models

5. CONCLUSION

In this article, we suggest a light-weight target detection model primarilt based on the target detection model YOLOv3 and the light-weight model ShuffleNetv2 normally for small targets. The main experimental scheme is to train and test the indoor dataset on YOLOv3, YOLOv3-ShuffleNetv2, and tiny-YOLOv3-ShuffleNetv2 models respectively. The YOLOv3-Shufflenetv2 model reduces the training parameters by 2/3 based on YOLOv3. The tiny-YOLOv3-ShuffleNetv2 model reduces the size to 1/10 of the original size, allowing it to be deployed in mobile devices with an FPS of 45, satisfying the real-time requirement. The model replaces the YOLOv3 backbone network darknet-53 with ShuffleNetv2 and trims the detection department of the network for massive targets to gain a light-weight network model relevant to small targets, which both reduces the network parameters and builds up the accuracy of the detection tasks for small targets. From the above experimental data, it can be considered that the light-weight target detection network model designed in this article for small targets can effectively reduce the parameters and can be easily implemented in portable mobile equipment.

REFERENCES

[1] S. Liu, G. Tian, Y. Zhang and P. Duan, "Scene Recognition Mechanism for Service Robot Adapting Various Families: A CNN-Based Approach Using Multi-Type Cameras," in IEEE Transactions on Multimedia, vol. 24, pp. 2392-2406, 2022, doi: 10.1109/TMM.2021.3080076.

[2] L. Yang, Q. Song, Z. Wang, M. Hu and C. Liu, "Hier R-CNN: Instance-Level Human Parts Detection and A New Benchmark," in IEEE Transactions on Image Processing, vol. 30, pp. 39-54, 2021, doi: 10.1109/TIP.2020.3029901.

[3] Felzenszwalb P, McAllester D, Ramanan D. A discriminatively trained, multiscale, deformable part model[C]//2008 IEEE conference on computer vision and pattern recognition. Ieee, 2008: 1-8.

[4] K. -H. Shih, C. -T. Chiu, J. -A. Lin and Y. -Y. Bu, "Real-Time Object Detection With Reduced Region Proposal Network via Multi-Feature Concatenation," in IEEE Transactions on Neural Networks and Learning Systems, vol. 31, no. 6, pp. 2164-2173, June 2020, doi: 10.1109/TNNLS.2019.2929059.

[5] X. Zhang, Y. Zhang, M. Hu and X. Ju, "Insulator defect detection based on YOLO and SPP-Net," 2020 International Conference on Big Data & Artificial Intelligence & Software Engineering (ICBASE), 2020, pp. 403-407, doi: 10.1109/ICBASE51474.2020.00092.

[6] W. Fang, L. Wang and P. Ren, "Tinier-YOLO: A Real-Time Object Detection Method for Constrained Environments," in IEEE Access, vol. 8, pp. 1935-1944, 2020, doi: 10.1109/ACCESS.2019.2961959.

[7] Redmon J, Farhadi A. YOLO9000: better, faster, stronger[C]//Proceedings of the IEEE conference on computer vision and pattern recognition. 2017: 7263-7271.

[8] Iandola F N, Han S, Moskewicz M W, et al. SqueezeNet: AlexNet-level accuracy with 50x fewer parameters and< 0.5 MB model size[J]. arXiv preprint arXiv:1602.07360, 2016.

[9] Chollet F. Xception: Deep learning with depthwise separable convolutions[C]//Proceedings of the IEEE conference on computer vision and pattern recognition. 2017: 1251-1258.

[10] Zhang X, Zhou X, Lin M, et al. Shufflenet: An extremely efficient convolutional neural network for mobile devices[C]//Proceedings of the IEEE conference on computer vision and pattern recognition. 2018: 6848-6856.

Theoretical Study on Gas Sensing of SO_2 on TH-Si_xC_y

Rukun Li, Ying Yang*, Guihong Luo, Mengyang Zhang, Tingting Zhang

School of Automation and Information Engineering, Xi'an University of Technology, Xi'an, Shaanxi, 710048, P. R. China

* Corresponding author: yangy@xaut.edu.cn

ABSTRACT: The adsorption process of SO_2 on TH-Si_xC_y is studied using first principles and the adsorption behavior is evaluated by parameters such as adsorption energy, band structure, and DOS. By analyzing the adsorption properties of SO_2, it is found that the chemical adsorption of SO_2 on TH-SiC makes TH-SiC change from semiconductor to metal. Thereby changing the electrical conductivity of TH-SiC and enabling sensing becomes feasible. However, SO_2 and TH-SiC_2 are formed with physical adsorption, and the band structure of TH-SiC_2 is not changed, so it can effectively resist the pollution of SO_2.

1. INTRODUCTION

Two-dimensional materials have a great specific surface area, easy to control electrical properties and other characteristics, and have broad development prospects in gas sensing devices [1]. SO_2 is a gas pollutant. In today's serious air pollution problem, it is of great practical significance to explore an effective method for monitoring SO_2. The adsorption of gas molecules on 2D materials such as silicon carbide [1] [2], black phosphorus[3], molybdenum disulfide [4], and graphene [5] can significantly affect their band structures, resulting in changing the conductivity to realize the gas sensing function which has been reported.

As a representative, by introducting the silicon's doping can improve the gas sensing properties of graphene effectivly. In addition, silicene is a highly sensitive gas sensor [6] which is considered great potential as a gas sensor [1]. A recently proposed TH-Si_xC_y is a new type of 2D semiconductor material with broad application prospects. It has large surface areas and great changes in electrical conductivity before and after gas adsorption, so they may be promising for effective gas detection and gas sensing potential candidates.Tetragonal (T) rings and hexagonal (H) rings are combined to form the TH-Si_xC_y, where TH-SiC is an indirect bandgap semiconductor of 1.80 eV and TH-SiC_2 is a 0.02 eV direct band gap semiconductor [7]. Due to the great area of specific surface and the very sensitive surface properties, it is of great significance to study the gas-sensing properties of TH-Si_xC_y adsorbed SO_2.

Here, the change of molecular band structure of SO_2 adsorbed on TH-Si_xC_y is studied by density functional calculation and simulations.

2. COMPUTATIONAL METHODS

In this paper, the PAW projected plane-wave pseudopotential of the VASP package [8] [9] is used for geometric optimization calculation. The HSE06 [10] hybrid functional method, which predicts semiconductor band gap more accurately, is used to calculate TH-SiC, TH-SiC_2 and the bands of TH-SiC and TH-SiC_2 after adsorption of SO_2. In this paper, the interval of the vacuum layer is set as 20Å. The truncation energy of the wave function is 500eV, and the k point grid of $8\times6\times1$ is adopted in the relaxation process. The criterion of optimal convergence is that the force of each atom is no more than 0.02eV/Å. Van der Waals force bonding is adopted in the calculation of the SO_2 adsorption model, and the distance between SO_2 and TH-Si_xC_y is 3Å, ensuring physical adsorption between gas and TH-Si_xC_y. Spin polarization is not considered in the calculation.

Adsorption energy is an important index to measure adsorption performance and is defined as:

$$E_{ad} = E_{molecule} + E_{TH-si_xc_y} - E_{complex}$$

Where $E_{molecule}$, $E_{TH-si_xc_y}$, and$E_{complex}$ are the energy of the gas molecule, TH-Si_xC_y, and adsorption system.

3. RESULTS AND DISCUSSION

The structure of pristine TH-SiC optimized crystal in different directions is shown in Fig. 1(a), with lattice constants a = 5.53 Å and b = 7.63 Å and d = 1.60 Å. Fig. 1(c) is the optimized calculation of pristine TH-SiC_2 crystal structure, with lattice constants a =5.52 Å and b =7.20 Å and d = 1.34 Å. C and Si atoms are in brown and blue, respectively. Both C and Si atoms can be hybridized with sp^3 or sp^2. TH-SiC consists of 6 C atoms and 6 Si atoms. TH-SiC_2 contains 8 C and 4 Si atoms. An sp^3 Si atom is sandwiched between two sp^2 C atoms.

TH-SiC with an indirect band gap of 1.8 eV [Fig. 1(b)], point X of the Brillouin zone is the valence band maximum (VBM), while point Y is the conduction band minimum (CBM). The band is not symmetric about the extreme point. The band structure of TH-SiC_2 with 0.02 eV direct band gap is shown in Fig. 1(d), where CBM and VBM are both located at the Γ point.

Table 1 The adsorption energy (E_{ad}), adsorption height (h), and covalent bonds formed by SO_2 with TH-Si_xC_y.

Complex	E_{ad}(eV)	h(Å)	Bonding
SO_2+TH-SiC	1.6	1.73	Si-O
SO_2+TH-SiC_2	0.06	3.91	—

The adsorption properties of TH-SiC and TH-SiC_2 for SO_2 are studied. SO_2 is initially placed at a height of 3 Å above the TH-Si_xC_y, and the most stable adsorption conformation is obtained after optimization [Fig. 2]. In addition, the adsorption parameters are shown in Table 1. The boundary between physisorption and chemisorption is the adsorption energy of 0.4 eV, which is also an important indicator to judge whether the material can achieve gas sensing. As shown in Fig. 2, SO_2 is chemically adsorbed on TH-SiC, and two O atoms of SO_2 form Si-O chemical bonds with Si3 and Si5 atoms on TH-SiC, respectively. The physisorption energy of SO_2 on TH-SiC_2 is only 0.06 eV, the adsorption height is large, and the influence on the configuration of TH-SiC_2 is small. In addition, since changes in band gap can cause significant changes in electron conduction, the sensitivity of TH-Si_xC_y to adsorbed gas is determined by the bandgap, which enables the sensing of adsorbed gas molecules to be realized.

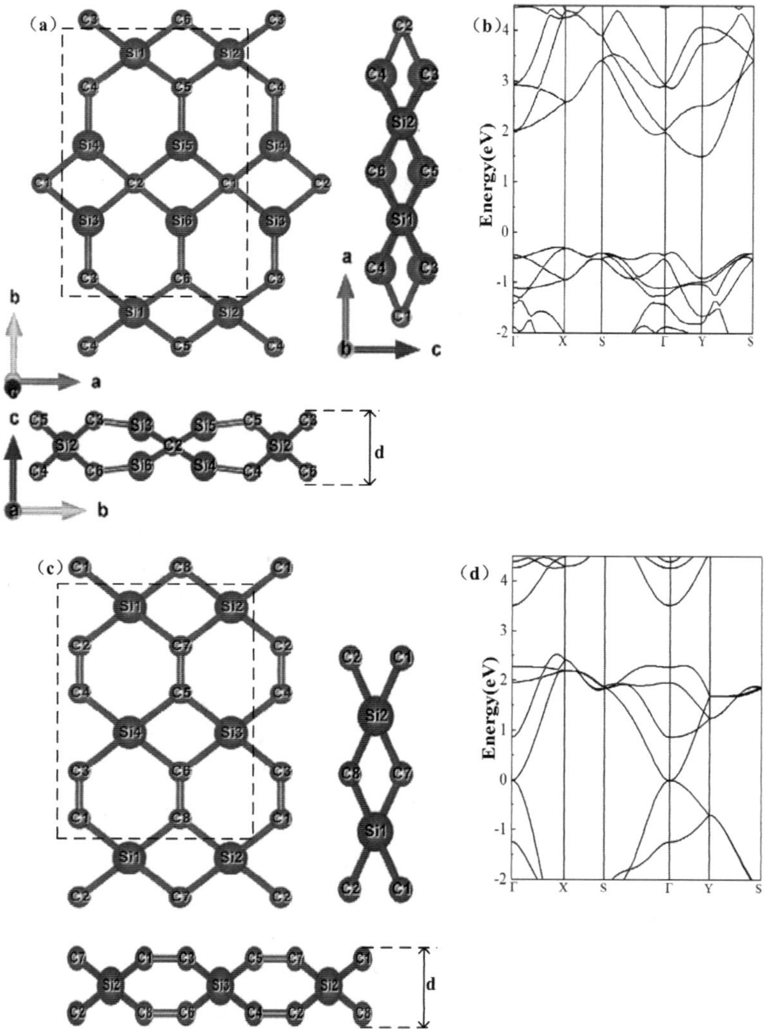

Fig. 1. (a) 2D TH-SiC structures with different orientations. (b) Band structure of TH-SiC. (c) 2D $TH\text{-}SiC_2$ structure from different orientations. (d) Band structure of $TH\text{-}SiC_2$.

The energy band structure and density of state (DOS) of TH-SiC adsorbed by SO_2 are shown in Fig. 3. It is found that TH-SiC changes from semiconductor structure to metal structure after chemical adsorption of TH-SiC with SO_2. One band passes through the Fermi level and covers an energy region ranging from -0.1 to 0.3 eV [Fig. 3(a)], causing the material to change from semiconducting to metallic. Further studies of the band properties of transitions found that the energy bands near the Fermi level are the contributions of C3 and C5 atoms of TH-SiC [Fig. 3(b) (c)], and the contributions of C3 and C5 are consistent. This is because the two pairs of sp^2 silicon-carbon bond atoms C3-Si3 and C5-Si5 of TH-SiC break after the adsorption of SO_2 by TH-SiC. O atoms form covalent bonds with Si3 and Si5, leaving dangling bonds on C3 and C5, due to the generation of unpaired electrons, resulting in an energy band crossing Fermi level. TH-SiC exhibits metallic characteristics. The adsorption of SO_2 makes the conductivity of TH-SiC much greater than that of semiconductor characteristics of TH-SiC, so TH-SiC is a good material for SO_2 sensing.

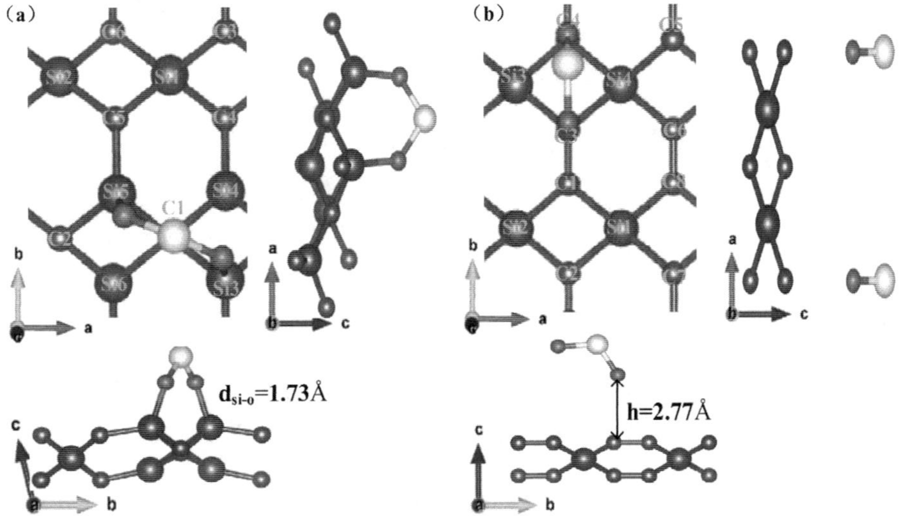

Fig. 2. Adsorption structures of SO_2 on (a) TH-SiC and (b) TH-SiC_2. The distance between the chemisorbed SO_2 and TH-SiC is shown as d, the height of the physisorbed SO_2 and TH-SiC_2 is shown as h. The spheres in yellow and red represent sulfur and oxygen atoms, respectively.

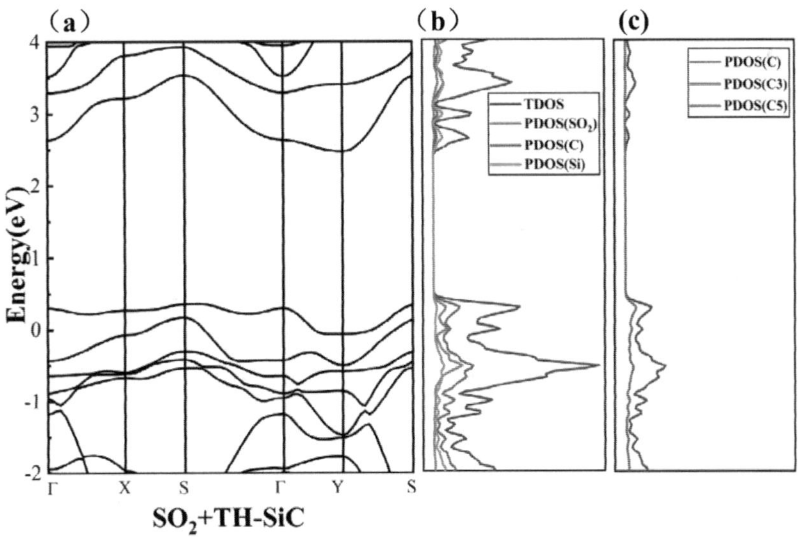

Fig. 3. (a) Band structures of SO_2 adsorbed TH-SiC, Fermi level is adjusted to 0 eV. (b) The DOS of SO_2 adsorbed on TH-SiC. The black line denotes the TDOS of the adsorbed complex, the red line denotes the PDOS of the adsorbed SO_2, the blue line denotes the PDOS of C in TH-SiC, and the green line denotes the PDOS of Si in TH-SiC. (c) The DOS of C in TH-SiC, the black line denotes the TDOS of the C in TH-SiC, and the red line and the green which denote the PDOS of C3 and C5 in TH-SiC line are highly overlapping.

Band structure and DOS of physical adsorption of TH-SiC_2 by SO_2 are shown in Fig. 4. It is found that the bandgap of TH-SiC_2 is only enlarged by 0.02eV after adsorption of SO_2, and the CBM and VBM have no change, that is to say, the adsorption of SO_2 does not have a significant effect on the conductivity of TH-SiC_2. The total density of state (TDOS) of TH-SiC_2 adsorbed by SO_2 and partial density of state (PDOS) of SO_2 are shown in Fig. 4(b). The results show that SO_2 does not affect the CBM and VBM of TH-SiC_2, clearly indicating that the band gap of TH-SiC_2 cannot be opened by SO_2 adsorption. In other words, TH-SiC_2 material can resist SO_2 gas pollution.

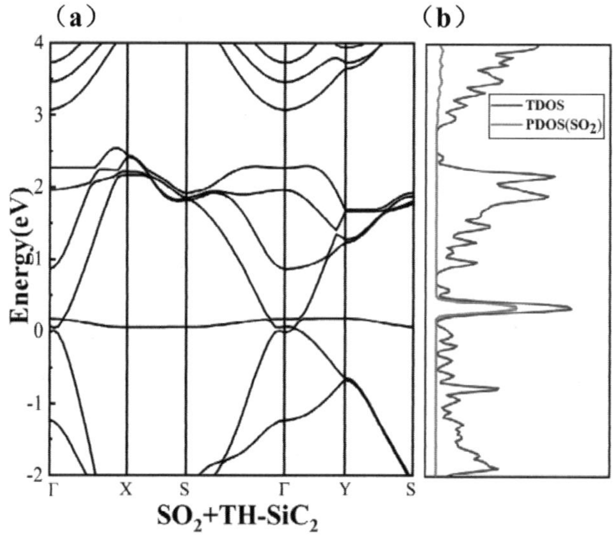

Fig. 4. (a) Band structures of SO_2 adsorbed TH-SiC_2. (b) The DOS of SO_2 adsorbed on TH-SiC_2.

4. CONCLUSION

In this paper, the adsorption performance of SO_2 on TH-Si_xC_y is studied by DFT. Results show that SO_2 can form stable chemisorption on TH-SiC. Due to the covalent bond between the O atom of SO_2 and the Si atom of TH-SiC, the two pairs of sp^2 silicon-carbon bond atoms C-Si of TH-SiC is broken. Unpaired electrons are generated on C, leaving empty bonds, and forming an energy band beyond the Fermi level. TH-SiC exhibits metallic characteristics, thus adjusting the conductance and making the sensing feasible. In addition, SO_2 is physically adsorbed with TH-SiC_2, and the SO_2 has no effect on the CBM and VBM of TH-SiC_2. The band gap of TH-SiC_2 does not change, which clearly shows that TH-SiC_2 material can resist the pollution of SO_2.

ACKNOWLEDGMENT

This work acknowledged funding from the National Natural Science Foundation of China (Grant No.51207128 and 62174134) and the Shaanxi Innovation Capability Support Project (Grant No. 2021TD-25).

REFERENCES

[1] Dong H, Wang L, Zhou L, et al. Theoretical investigations on novel SiC_5 siligraphene as gas sensor for air pollutants[J]. Carbon, 2017, 113: 114-121.
[2] Davydov S Y, Posrednik O V. On the adsorptive properties of silicon carbide: gas molecule adsorption[J]. Fullerenes Nanotubes and Carbon Nanostructures, 2019:1-3.
[3] Abbas A N, Liu B, Chen L, et al. Black phosphorus gas sensors[J]. ACS Nano, 2015, 9: 5618~5624.
[4] F.K. Perkins, A.L. Friedman, E. Cobas, P. Campbell, G. Jernigan, B.T. Jonker, Chemical vapor sensing with monolayer MoS_2, Nano Lett. 13 (2) (2013) 668e673.
[5] Y. Dan, Y. Lu, N.J. Kybert, Z. Luo, A.C. Johnson, Intrinsic response of graphene vapor sensors, Nano Lett. 9 (4) (2009) 1472e1475.
[6] J.-w Feng, Y.-j Liu, H.-x Wang, J-x Zhao, Q.-h Cai, X.-z Wang, Gas adsorption on silicene: a theoretical study, Comput. Mater. Sci. 87 (0) (2014) 218e226.

[7] Wei Q, Yang Y, Yang G, et al. New stable two-dimensional silicon carbide nanosheets[J]. Journal of Alloys and Compounds, 2021, 868: 159201.
[8] G. Kresse, J. Furthmüller, Efficient iterative schemes for ab initio total-energy calculations using a plane-wave basis set, Phys. Rev. B 54 (1996) 11169–11186.
[9] W. Kohn, L.J. Sham, Self-consistent equations including exchange and correlation effects, Phys. Rev. 140 (1965) A1133–A1138.
[10] J. Heyd, G.E. Scuseria, M. Ernzerhof, Hybrid functionals based on a screened Coulomb potential, J. Chem. Phys. 118 (2003) 8207–8215.

Structural and photoelectric properties of Sn-doped ZnO thin film with the co-sputtering method

Yantao Liu[a]*, Wenxia Wang, Jianping Ma[b]*

Department of Electronic Engineering, Xi'an University of Technology, Xi'an 710048, China

[a]liuyt@mail.xaut.edu.cn, [b]majp@mail.xaut.edu.cn

Abstract. Sn-doped ZnO thin films have been prepared with magnetron co-sputtering. The results indicate that this method can improve the micro-structure of films and enhance its photoelectric performance. The structure and surface morphology results indicate that all the doped films show the preferred orientation (0 0 2), the roughness can decrease to 1.81 nm at 15 W with increasing DC power, the carrier concentration, resistivity decreases, and mobility can reach 7.99×10^{19} cm^{-3}, 1.27×10^{-4} $\Omega\cdot cm$ and 1047.5 cm^2/Vs, respectively, and the observed optical band gap can increase to increase to 3.85 eV.

1. Introduction

ZnO is suitable for short-wavelength photovoltaic and photovoltaic devices [1]. It has the advantages of high conductivity, high transparency, a simple preparation process, various raw materials, and relatively low cost [2]. Doping can artificially adjust the characteristics of the energy band edge, quasi-Fermi level, and work function. It can change the morphology and optical properties of semiconductors to enhance their electrical performance, which has great potential application value for optoelectronic devices and gas sensors, etc. Recently, Al, Ti, Ga, F, Co, Ti, Mn, Ag, Nb, and Sn-doped ZnO have been reported [3]. Among these elements, Sn is a suitable doping element because it has two available free electrons to change conductivity. Furthermore, Sn^{4+} ions can replace Zn^{2+} in the ZnO crystal structure[4,5].

In this paper, Sn-doped ZnO (SZO) films were deposited using co-sputtering. With changing the sputtering power on the tin dioxide target, the Sn content can be changed. The characteristic of microstructure, electricity and optics of the Sn-doped ZnO films were systematically analyzed.

2. Experimental

2.1. Method

The deposition of Sn-doped ZnO thin films have been completed using magnetron co-sputtering, the ZnO target uses RF power supply, and the tin dioxide target uses DC power supply. In first step, glass substrates were ultrasonically cleaned with acetone and ethyl alcohol, each for 10 min. In second step, SZO thin films were deposited using ZnO ceramic target and SnO_2 target. RF magnetron sputtering power was set to a continuous 100 W. Meanwhile, DC power was set in the range between 0 W and 20 W, the flow rate of oxygen and argon gas were 10 sccm and 60 sccm, the pressure of vacuum chamber was 1.0 Pa. Then, films samples were heat treated in O_2 atmosphere for 1 hour.

2.2. Measurements

The structure of SZO films were tested by an UltimaIV X-ray diffractometer, and the surface morphologies were measured with a dimension icon microscope. The atomic percent of SZO films was measured with Inspect F50 spectrometer. Electricity of films was obtained by CH-100 Hall system. Optical transmission measurements were performed by Lamda950 spectrophotometer.

3. Results and Discussions

3.1. Microstructure of SZO films

The XRD patterns of SZO thin films with different DC sputtering power was shown in Figure 1 . Only the (0 0 2) diffraction peak was detected as the phase of ZnO, this result indicates that Sn ions did not fundamentally change the crystal structure. All the films showed strong (0 0 2) peaks in the c-axis direction. With the increase of Sn doping DC power, the diffraction intensity of the (0 0 2) peak position increases, which indicates that Sn doping ameliorates the crystallinity and c-axis orientation of SZO film, the strongest peak at the DC sputtering power of 15 W, and it begins to weaken rapidly after 15 W, this may be caused by excessive doping concentration, in this case, the incorporation of Sn atoms forms segregation at the grain boundaries, and resulting in hindering the growth of ZnO crystal structure [6].

From Figure 2, the FWHM and grain size of SZO films can be obtained with Scherrer formula at (0 0 2) peak [7]. The FWHM increases with the increase of the Sn-doped DC sputtering power after Sn atoms are doped in the ZnO lattice. The calculated grain size of the undoped ZnO film is 14.91 nm, the grain size is reduced to 11.39 nm when DC power is 15 W, the crystal lattice shrinks, and the grain size becomes smaller after substitution because of the smaller radius of Sn atoms than Zn atoms, as DC power continued to increase, grain size increases, this is because excessive doping would segregate at the grain boundaries to form defects, resulting in larger grain size.

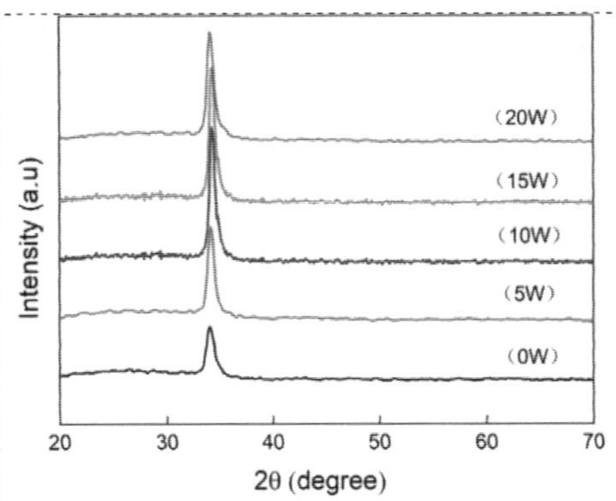

Figure 1. XRD results of SZO films.

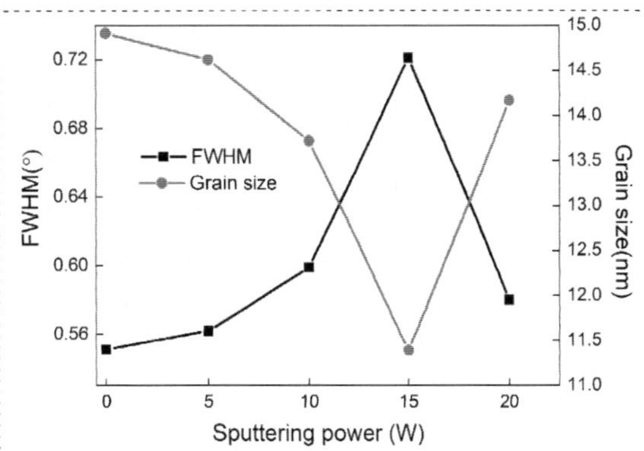

Figure 2. Grain sizes and FWHM of SZO films.

The EDS spectrum of SZO films are shown in Figure 4. Result of element composition confirms that there are O, Si, Zn, and Sn atoms in the SZO thin films, these manifest that the prepared films have high purity and no other impurities. Figure 5 shows the atomic percent of Zn and Sn. The atomic percent of the doped Sn atoms increases when DC power increasing, and this phenomenon shows that the doping concentration of Sn atoms is proportional to the DC sputtering power. The reduction of the Zn element indicates that Sn atoms have replaced Zn atoms in the film very well.

The morphology and surface characteristics of SZO films obtained by AFM are shown from Figure 5. Two-dimensional and corresponding three-dimensional images also can be obtained from Figure 5, respectively. We can observe that from the results, the size of grain decreases as the x-coordinate keeps increasing, the smallest value appear at 15 W DC power. However, as the Sn-doped DC sputtering exceeds 15 W, the grain size begins to increase again, it may be caused by impurity defects introduced by excessive doping concentration. The roughness of the Sn-doped ZnO film obtained by AFM is shown in Figure 7. The value of *Rq* (root mean square roughness) can be reduced from 8.18 nm to 1.81 nm at 15 W in Figure 6. The roughness is the smallest when the Sn-doped DC sputtering power is 15 W, suggesting that the film prepared at 15 W is the smoothest.

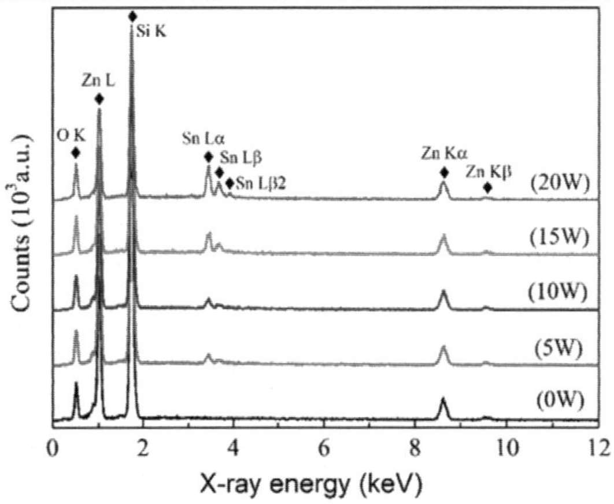

Figure 3. EDS of doped SZO films.

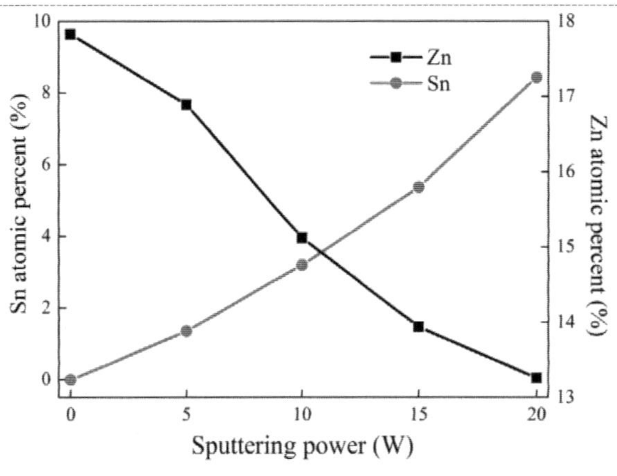

Figure 4. Atomic percent of SZO films.

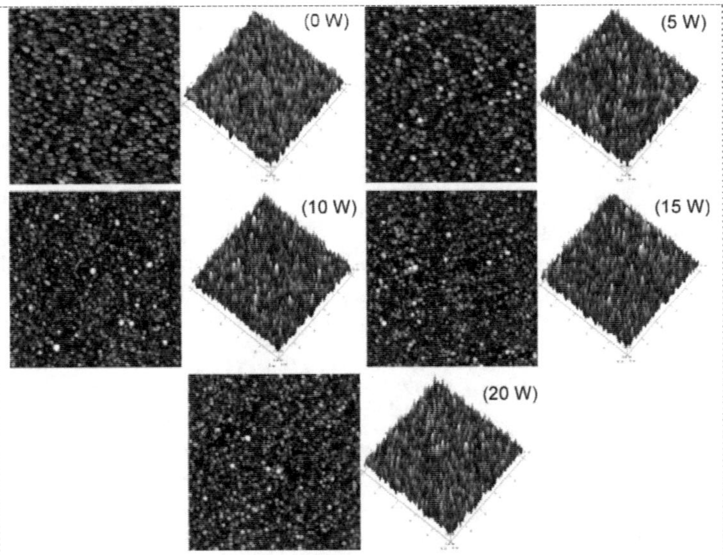

Figure 5. AFM images of SZO films.

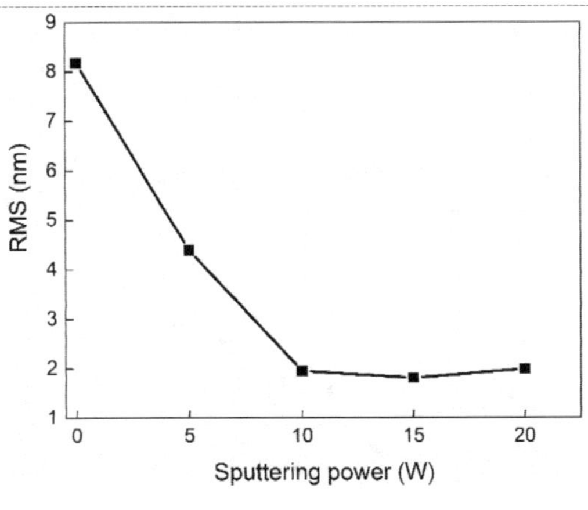

Figure 6. The surface roughness of SZO films.

3.2. *Electricity and optics of the Sn-doped ZnO*

The electricity of the SZO films are shown in Figure 7. When the sputtering power is 0 W, the resistivity is 7.53×10^{-3} Ω·cm, and it tends to get smaller. When the sputtering power is 15 W, the resistivity can reach 1.27×10^{-4} Ω·cm. When the Sn-doped DC sputtering power continues to increase, the resistivity is 3.3×10^{-3} Ω·cm at 20 W. The maximum carrier concentration is 7.99×10^{19} cm^{-3} at 15 W sputtering power, and the maximum mobility is 1047.5 cm^2/Vs, both at 15W. This is because the conduction mechanism of the ZnO film is mainly the scattering of ionized impurities, and the incompleteness of the crystal enhances the scattering, resulting in high resistivity. Since the resistivity of semiconductor is mainly determined by the majority carrier density, and the doped Sn^{4+} provides two more conductive electrons when replacing Zn^{2+}, the donor concentration is inversely proportional to the resistivity. Therefore, carrier concentration increases as the doping concentration increase when the sputtering power is increased, and the resistivity shows opposite trend of change. However, the efficiency of Sn doping decreases due to the limitation of the doping degree in the ZnO crystal when

Journal of Physics: Conference Series **2370** (2022) 012031 doi:10.1088/1742-6596/2370/1/012031

the Sn-doped sputtering power exceeds 15 W. As the doping concentration increases to a certain level, the excess Sn will occupy the interstitial sites. In addition, Sn ions will aggregate in the grain boundaries, which will strongly deteriorate the electrical properties of SZO thin films [8].

The transmittance of the SZO films is shown in Figure 8. The average transmittance of the film in the range between 390 nm and 800 nm exceeds 87.4%. The transmittance shows that the prepared SZO film has an obvious absorption edge and the position of this edge has a slight blue shift with the increase of Sn doping, which is caused by the Burstein-Moss effect. Doping concentration can change carrier concentration, when Sn^{4+} ions replace the position of Zn^{2+} ions, the Fermi level can be changed, resulting in an enlarged band gap and increased transmittance. From the results of transmittance, we can get the optical band gap (*Eg*) of SZO films from Formula (1) [9]:

$$(\alpha h\nu)^2 = A(h\nu - Eg) \quad (1)$$

According to the Tauc relationship, Figure 9 shows the plots of $(\alpha h\nu)^2$ vs $h\nu$ of SZO films. The *Eg* can be derived by extrapolating. The results reveal that from 0 W to 15 W, *Eg* gradually can reach 3.85 eV at power 15 W, it is 3.3 eV at power 0 W. However, after the power is beyond 15 W, the band gap decrease to 3.76 eV at 20 W, and the change in ionization energy of Sn-doped alters the interaction between carrier impurities, resulting in this shrinkage.

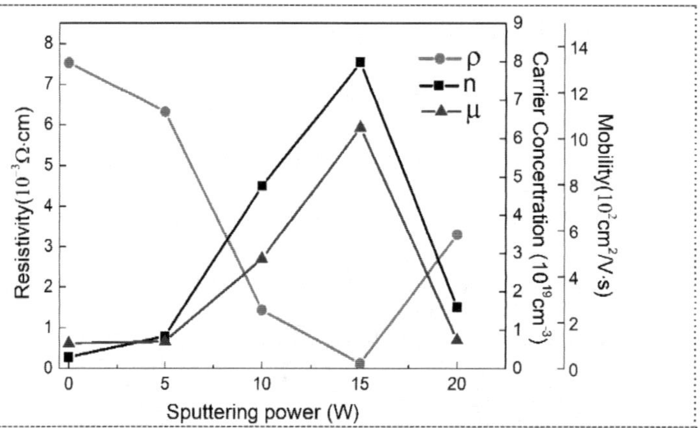

Figure 7. Resistivity, carrier concentration, mobility of SZO films.

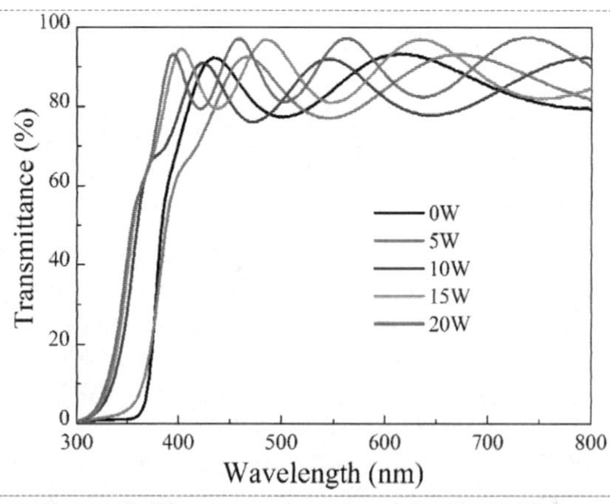

Figure 8. The transmittance of the SZO films.

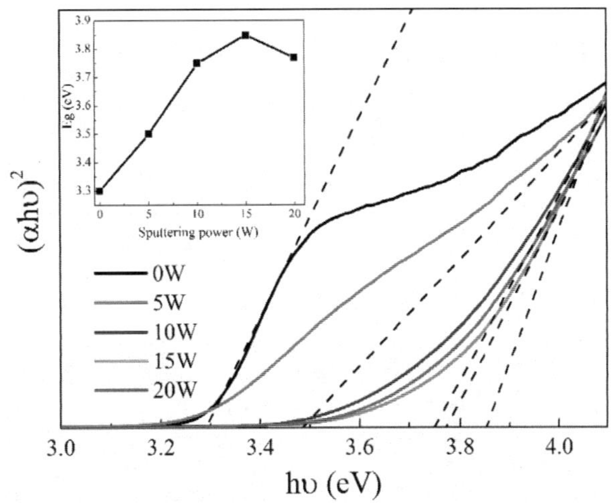

Figure 9. Plots of vs of SZO films, inset is the *Eg* of SZO films.

4. Conclusions

In conclusion, the structure, composition, surface images, and electricity and optics features of SZO thin films. The results reveal that all the SZO films had c-axis preferred orientation. EDS results indicate that Sn atoms have replaced Zn atoms in the film. Grain size is found to be decreased with increasing Sn-doped concentrations DC sputtering power, and the minimum grain size and surface roughness were obtained to be 11.39 nm and 1.81 nm at 15 W, respectively. The maximum carrier concentration and the minimum resistivity are obtained to be 7.99×10^{19} cm^{-3} and 1.27×10^{-4} Ω·cm at 15 W. *Eg* increases from 3.3 eV to 3.85 eV.

References

[1] Acharya A D, Moghe S, Panda R 2012 J. Mol. Struct. **1022 (34)** 8-15
[2] Caglar Y, Aksoy S, Ilican S 2009 *Superlattices Microstruct* **46** 469-4 75
[3] Park S, An S, Ko H, Lee S, Lee C 2013 *Sens. Actuators, B.* **188**, 1270-1276
[4] Postica V, Hoppe M, Grottrup J 2017 *Solid State Sci,* S1293255817300973
[5] Xu L, Zheng G, Xian F 2019 *Mater. Chem. Phys.* **229** 215-225
[6] Agarwal M B, Sharma A, Malaidurai M 2018 *Superlattices Microstruct.* **117** 342-350
[7] Mishra S K, Bayan S, Shankar R 2014 *Sens. Actuators, A.* **211** 8-14
[8] Mejda A, Michel, Castagné N, Kamoun, Turki 2013 *Superlattices Microstruct.* **53**
[9] Adem S, Hyemin J, Jin H K, Jonghoon Y 2016 *Ceram. Int.* **42** 3064-3071

ZnO-based Piezotronic Avalanche Devices

Ping Zhu [1], Yan Zhang [1]

[1] School of Physics, University of Electronic Science and Technology of China, Chengdu 610054, China

E-mail: zhuping_sally@163.com

Abstract. High-performance piezotronic semiconductor devices have been proven to apply in signal detection, energy converter, and other important fields. Avalanche breakdown is a classical current amplification mechanism. Combining the piezotronic polarization charges regulation of the avalanche process, we novelty proposed a piezotronic PIN diode based on ZnO and used the FEM to calculate the characteristic of the device. The gauge factor can reach 10^6. This study can guide the design of ultrahigh sensitivity piezotronic photodetectors.

1. Introduction

Ever since piezotronics and piezo-phototronics were proposed, piezoelectric nanodevices have shown high performance in many different fields. As the most extensively studied piezoelectric semiconductor material, ZnO is used to build piezotronic strain sensors [1], nano-generators [2], quantum piezotronic transistors [3], piezotronic photocells [4], and piezotronic logic devices [5] in the previous study. Combined with the regulation of the strain-induced polarization, the piezotronic devices has shown ultrafast gauge factor over 10^4 for bipolar transistor [6]. The gauge factor of piezotronic tunnelling junction-based strain sensor (Ag/HfO_2/n-ZnO) can reach 4.8×10^5 at a strain of 0.1% [7].

Due to the strong nonlinear current amplification of the avalanche devices, the most common application for avalanche devices is weak signal detection. In the field of single photon detection, compared with a traditional photomultiplier (PMT), Single Photon Avalanche Diodes (SPADs) have obvious advantages in size, cost, and non-friable features [8]. In particular, SPADs have a much faster response than PMTs. Compared with piezoelectric switches as a strain sensor, the piezoelectric avalanche diodes have obvious advantages on the performance of sensitivity and time of duration. And avalanche diodes are single-junction devices which is different with piezotronic bipolar transistor, it is simple in device preparation in practice.

In 2003, a kind of ultraviolet-enhanced n-ZnO/p-Si photodiode was proposed. Under a reverse voltage of about 30V, a photon of 380nm can be detected. The strong responsivities illustrated the heterojunction devices could be used to detect ultraviolet and visible photons simultaneously [9]. In 2005, the first ZnO p-n homojunction ultraviolet photodetector was prepared. The detection photocurrent is about 2mA when the UV light (wavelength of 325 nm) radiation irradiates on the p-n homojunction [10]. In 2021, an ultraviolet photodetector made of ZnO nanorods and polypyrrole nanoparticles was reported. The structure has higher sensitivity and a much lower detection wavelength than normal ZnO-based detectors. The device can detect wavelengths which shorter than 250 nm [11].

This study uses ZnO material to build ultrafast and ultrahigh sensitivity PIN avalanche devices. The device theoretical model is studied by the finite element method to illustrate the mechanism of strain-induced polarization. The carrier separation, transport, and recombination can be regulated by piezopotential effectively. The current amplification ability is enhanced by the piezoelectric charges in the depletion layer. It gives a new method to fabricate ultrafast devices on the photodetector. Also, this research broadens the application of device based on avalanche breakdown, like IMOS [12].

2. Theoretical analysis of piezotronic avalanche devices

The basic unit composition of avalanche devices is the p-n junction. It is very difficult to prepare stable and repeatable p-ZnO. Most ZnO-based ultraviolet photodetector is made of n-ZnO. We choose p-Si and n-ZnO to fabricate the heterojunction. The avalanche multiplication effect occurs under a large reverse bias voltage. In the depletion region, the integral breakdown condition caused by holes is given by:

$$\int_0^W \alpha_p \exp\left[-\int_0^x \left(\alpha_p - \alpha_n\right) dx'\right] dx = 1 \tag{1}$$

α_p and α_n are the hole and electron impact ionization coefficients. It is mainly determined by the electric field strength. When the n-type region is piezoelectric in piezoelectric p-n junctions, the maximum field E_m is described by [13]:

$$\left|E_m\right| = \frac{q\left(N_D W_{Dn} + \rho_{piezo} W_{piezo}\right)}{\varepsilon_s} \tag{2}$$

The change of the electric field by piezotronic polarization can influence the avalanche process. With an applied strain to the device, the built-in voltage ψ_{bi} is changed into:

$$\psi_{bi} = \frac{q}{2\varepsilon_s}\left(N_A W_{Dp}^2 + \rho_{piezo} W_{piezo}^2 + N_D W_{Dn}^2\right) \tag{3}$$

Where N_D and N_A is the donor and acceptor doping concentration, W_{Dp} and W_{Dn} is the width of depletion in each type region, ε_s and ρ_{piezo} is the dielectric constant and the polarization charges density, W_{piezo} is the piezoelectric charges distribution width. The piezoelectric p–n junction current-voltage characteristics is rewritten as:

$$J = J_{C0} \exp\left(\frac{q^2 \rho_{piezo} W_{piezo}^2}{2\varepsilon_s kT}\right)\left[\exp\left(\frac{qV}{kT}\right) - 1\right] \tag{4}$$

While J and J_{C0} is the current density in the presence and absence of piezoelectric charges, and T is the temperature. The gauge factor is always used to define the device sensitivity when the device is used as a sensor. In previous research, piezotronic devices have been proven to be much higher sensitive than traditional carbon nanotubes (1000) [1]. For our piezotronic avalanche devices, it is given by [14]:

$$G = \frac{I - I_i}{I_i \cdot \Delta\varepsilon} = \frac{M - 1}{\Delta\varepsilon} \tag{5}$$

While I_i, I is the APD output current and average initial current, and $\Delta\varepsilon$ is the applied strain change. Due to the amplification effect of the avalanche device on current, Piezoelectric avalanche PIN diodes will achieve higher gauge factors compared with typical piezoelectric devices. PIN photodetector has high quantum efficiency and frequency response because of the thick depletion region [15]. So, we use this structure to build high performance on sensitivity and response speed piezotronic devices. The experimentally possible device structure of the ZnO piezotronic avalanche PIN diodes is pictured in Fig. 1(a). We also give the equivalent circuit diagram under reverse bias in Fig.1(b).

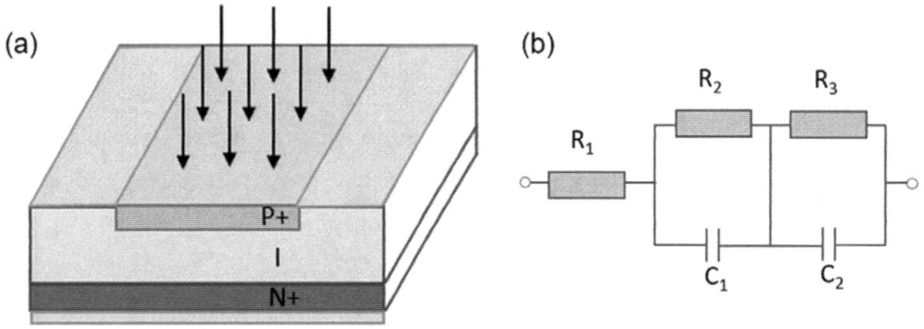

Figure 1. (a) The device structure of the ZnO piezotronic avalanche photodiode. (b) The equivalent circuit diagram of ZnO avalanche photodiode under reverse bias.

3. Results and discussion

In this study, the bandgap E_g of wurtzite ZnO is 3.4eV. ZnO relative dielectric constant is about $\varepsilon_s = 8.91$. ZnO piezoelectric constant is $e_{33} = 1.22 C/m^2$. The fixed temperature is $295\ K$. The piezo charges distribution width is $W_{piezo} = 0.25\ nm$. The doping concentration is $N_A = N_D = 1 \times 10^{18} / cm^3$, the doping concentration of the substrate is $N = 5 \times 10^{13} / cm^3$.

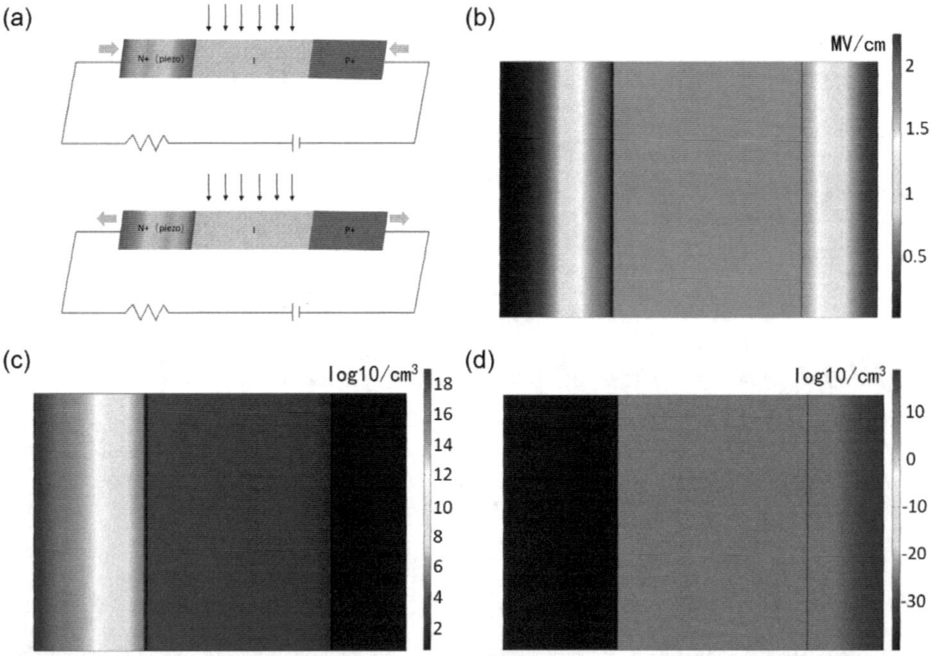

Figure 2. The FEM numerical calculation of the PIN piezotronic avalanche photodiode. (a) The structure chart of the ZnO piezotronic avalanche PIN diode. (b) Distribution of electric field in the whole device. (c, d) The plot of the electronic and hole logarithmic concentration.

We propose a PIN structure for the ZnO avalanche diode, which has a high sensitivity to photon detection. The schematic of the device used in the calculation is shown in Fig.2(a). N-type region is made of wurtzite ZnO. The piezoelectric charges are distributed in width W_{piezo}. The whole device length is 50 nm. Under reverse bias ($V_{RV} = 5V$), Fig. 2(b) is the distribution of electric field. The i-type region electric is the uniform distribution. The direction of piezoelectric electric field is decided by the applied strain direction. Fig. 2(c) and 2(d) is the electronic and hole logarithmic concentration. The applied strain tunes the carrier transport due to the piezoelectric potential. The depletion width is changed, which is one of the factors for breakdown time.

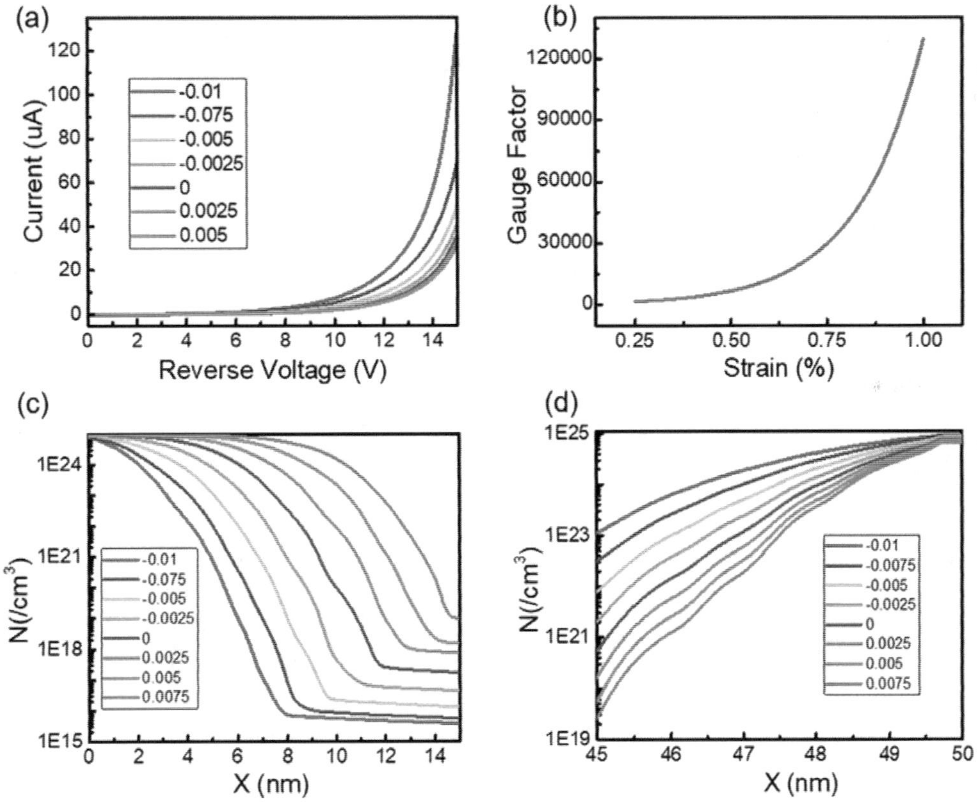

Figure 3. (a) The FEM numerical calculation of the piezotronic PIN avalanche photodiode. (b) Relationship between gauge factor and applied strain (0 %- 1 %). (c) the distribution of holes and (d) electrons concentrations across the device.

Fig. 3(a) is the current and voltage characteristics for ZnO piezotronic avalanche devices. When the reverse voltage grows gradually, the current is quite tiny before the avalanche breakdown happens. The applied strain direction decides the avalanche process direction. The tensile strain (minus) regulates the avalanche process to occur in advance and improves the performance of the piezotronic diode. The relationship between gauge factor and applied strain (0 %- 1 %) is represented in Fig. 3(b) when the reverse voltage is constant on 5V. It can reach 10^6, which is much higher than before the device. At a fixed reverse voltage of 5V, Fig. 3(c) and Fig. 3(d) are the distribution of hole and electron concentrations across the device. The piezotronic polarization charges influence the carrier distributions obviously which is in concordance with above theory analysis.

4. Conclusion

In this study, we propose a new structure for piezotronic avalanche devices based on ZnO. The working mechanism combines the semiconductor junction theory with the piezotronics theory. The strain-induced piezotronic polarization can regulate the device avalanche breakdown process efficiently. The gauge factor of ZnO avalanche devices is significantly improved and can reach 10^6. This research provides guidance for high-performance strain sensors and photon detection. It also gives immense potential to micro stress detection.

References

[1] Zhou J, Gu Y, Fei P, Mai W, Gao Y, Yang R, Bao G, and Wang Z L 2008 *Nano Lett.* **8** 3035
[2] Haroun A, Tarek M, Mosleh M, and Ismail F 2022 *Nanomaterials* **12** 17
[3] Zhou Y, Jiang Y, Dan M, Hu G, Li L, and Zhang Y 2021 *Nano Energy* **85** 106002
[4] Hu Y, Zhang Y, Chang Y, Snyder R, and Wang Z L 2010 *ACS Nano.* **4** 4220
[5] Wu W, Wei Y and Wang Z L 2010 *Adv. Mater.* **22** 4711
[6] Zhu P, Zhao Z, Nie J, Hu G, Li L, and Zhang Y 2018 *Nano Energy* **50** 744
[7] Yu Q, Ge R, Wen J, Du T, Zhai J, Liu S, Wang L and Qin Y 2022 *Nat. Commun.* **13** 778
[8] Eisaman M D, Fan J, Migdall A and Polyakov S V 2011 *Rev. Sci. Instrum.* **82** 071101
[9] Jeong I -S, Kim J H, and Im S 2003 *Appl. Phys. Lett.* **83** 2946
[10] Moon T -H, Jeong M -C, Lee W and Myoung J -M 2005 *Appl. Surf. Sci.* **240** 280
[11] Payandehdarinejad Z, Alam SN, Ghaffarinejad A, Ghafary B and Mazandarani H 2021 *Opt. Mater.* **122** 111634
[12] Zhai Y, Feng Z, Zhou Y, and Han S 2021 *Mater. Horiz.* **8** 1601
[13] Zhang Y, Liu Y, and Wang Z L 2011 *Adv. Mater.* **23** 3004
[14] Zhu P, Zhang Y, and Zhang Y 2022 *Nano Energy* **100** 107450
[15] Jang B J, Yom I B, and Lee M Q 2003 *Microwave Journal* **46** 86-+

Influence of Eu doping on the electric properties of perovskite iridate $Sr_{2-x}Eu_xIrO_4$

Hui Liu[1], Yanfei Xiong[2], Wei Shi[1] and Hao Zu[1]

[1] College of Advanced Manufacturing Engineering, Hefei University, Hefei 230601, China

[2] College of Energy Materials and Chemicals, Hefei University, Hefei 230601, China

E-mail: 32376938@qq.com (H. Liu)

Abstract. Here, we report the evolution of the physical properties of Eu-doped polycrystalline perovskite iridates $Sr_{2-x}Eu_xIrO_4$ (x=0, 0.02, 0.04, 0.06). It is found that substituting Eu^{3+} for Sr^{2+} retains the *I*41/*acd* crystal structure but causes rotation of IrO_6-octahedra decreases so that the bond angle of Ir-O1-Ir increases. The SEM micrographs indicate that the average grain size decreases, while the grain size distribution increases with doping. Compared with the resistivity of the parent material, all the doped materials show an increase, which can be further verified by the slight increase of band gap obtained from ultraviolet-visible absorption spectra. Furthermore, Eu doping makes the dielectric constant increases and thus the ability of electrical conduction decreases. All the physical properties of the compounds rely heavily on the parameters of their crystal structure, φ_{Oct} and c/a, implying spin-lattice coupling makes an important contribution to the electric behaviors.

1. Introduction

Recently, the oxide iridates have attracted lots of interest owing to the comparable strength of the spin-orbit coupling (SOC) U_{SOC}, the electron-electron correlations U_{ee}, and Crystal field effect U_{CF}, which can result in a number of exotic states such as topological insulators, Weyl semimetals, Mott insulators with spin-orbit coupling, etc. [1] This field has turned into one of the hotspots in condensed state physics. In the family of oxide iridates, there is a layered-perovskite Sr_2IrO_4 which is a $J_{eff} = 1/2$ Mott insulating antiferromagnet (AFM) from the interplay of the electronic correlations and SOC. Sr_2IrO_4 adopts weak ferromagnetism, which is due to a canting AFM state [2]. This material has a small band gap which is less than 0.62 eV [3] and a Néel temperature at around 240 K [4].

As is known that the structural and electric properties of Sr_2IrO_4 are similar to those of La_2CuO_4; thus, one may expect that novel superconductivity could occur through chemical doping. Up to now, there is no proof of superconductivity in spite of plenty of experimental efforts [5]. One possible method to induce a metal from a Mott insulator is to substitute the Sr or Ir elements via chemical doping [8]. For example, A slight doping at either the Ir^{4+} site (e.g., Rh^{3+}, Ru^{4+}, Tb^{4+}, Cu^{2+}, V^{4+}, Ga^{3+}, Sn^{4+}, and Ti^{4+}, Co^{4+}, Fe^{3+} [6-11]) or the Sr^{2+} site (e.g., La^{3+}, Pr^{3+}, Sm^{3+}, Ga^{3+}, Ca^{2+}, K^+ and Eu^{3+} [12-16]) brings about large variations in either the electric transport property or the magnetic property or both. However, the properties of Sr_2IrO_4 are extremely sensitive to chemical doping. Each substitute mentioned above gives rise to unique behavior in its own way, which indicates that the material has not been fully understood and explored. Thus, we consider that we should make further efforts to investigate more doping in Sr_2IrO_4.

In the present paper, we have carried out Sr-site substitution with Eu^{3+} in $Sr_{2-x}Eu_xIrO_4$. On the one hand, the Eu^{3+} (J=0, L = S) is nonmagnetic in the Hund's rule ground state, avoiding any magnetic interaction with Ir. On the other hand, Eu^{3+} has a different valence state compared with Sr^{2+}, which will act for electron doping in the iridate. The structure, electrical transport, spectroscopy, and dielectric properties have been investigated in $Sr_{2-x}Eu_xIrO_4$. The system still keeps its original lattice structure. The electric transport properties show an increase in resistivity and dielectric constant with doping. The spin-lattice coupling has a great influence on the physical properties.

2. Experimental details

The polycrystalline europium doped materials $Sr_{2-x}Eu_xIrO_4$ ($0 \le x \le 0.06$) were synthesized by solid-state reaction method. They were synthesized by firing $SrCO_2$, Eu_2O_3, and IrO_2 powders between 800 and 1250°C in the air for several days. The X-ray diffraction (XRD) patterns were obtained using a Rigaku-made diffractometer. The Rietveld refinement using XRD data was carried out through the FULLPROF sofrware. Data of Raman spectra were measured using a Thermo Fisher-made Laser Confocal Micro Raman Spectrometer with λ=532 nm. The grain sizes were estimated from the micrographs using Hitachi SU8010 Scanning Electron Microscope (SEM). The chemical composition of the elements was performed through a combined analysis of an SEM and an Energy Dispersive X-Ray Spectroscopy (EDX). The resistivity measurements were performed by a four-probe method (Quantum Design, PPMS). The optical absorbance spectra were collected by an ultraviolet-visible (UV-vis) absorption Spectrophotometer (V-650, Jasco). Dielectric properties were investigated by an impedance analyzer between 20 Hz and 10 MHz at room temperature (TH2839).

3. Results

3.1. Structural analysis

The XRD patterns of the $Sr_{2-x}Eu_xIrO_4$ ($0 \le x \le 0.06$) samples are given in Figure 1(a). It can be seen that all the diffraction peaks can be assigned using the $I41/acd$ space group. The parameters of refinements are plotted in Figure.1(b)-(d) and listed in Table 1. It is found that slightly substituting Eu^{3+} for Sr^{2+} (up to 6%) still keeps the tetragonal crystal structure but induces evident variations in the lattice parameters. The a and c parameters show nonmonotonic changes. Also, the c/a ratio first rises and then decreases, roughly tracking the changes in a and c. However, the unit cell volume V almost increases monotonically with the increase of the x value, indicating the change of $V = (c/a)a^3$ with doping is motivated by both a and c. In addition, it is noted that the rotation of IrO_6 octahedra in samples decreases first and then increases such that the Ir-O1-Ir bond angle rises from 156.1∘ for $x = 0$ sample to 164.1∘ for x =0.02 sample, and then declines to 158.6∘ at x = 0.06 (Figure. 1(d)). The structural distortions, such as a decrease in rotation of IrO_6 octahedra (φ_{oct}) and an increase of tetragonal distortion (c/a) as a whole, will play a significant role in the formation of electronic properties.

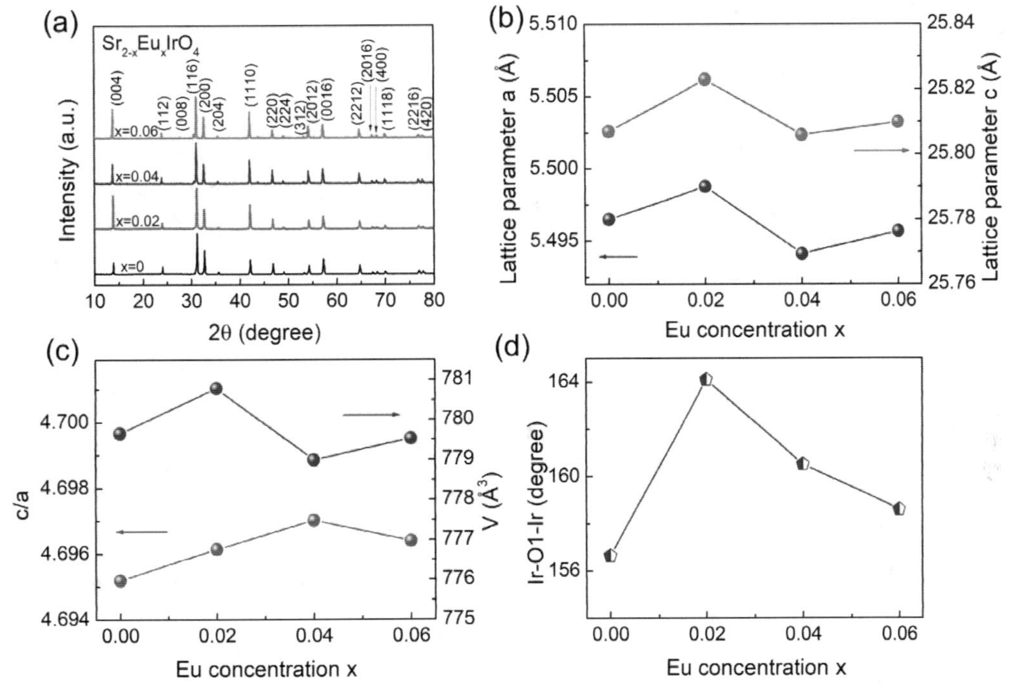

Figure 1. Powder XRD patterns of $Sr_{2-x}Eu_xIrO_4$ samples. The lattice parameters versus x value, the c/a ratio and volume V per unit cell, and the bond angle of Ir-O1-Ir are shown.

Table 1. The parameters are determined by Rietveld refinements.

	x=0	x=0.02	x=0.04	x=0.06
a (Å)	5.4965	5.4990	5.4941	5.4957
c (Å)	25.8071	25.8230	25.8060	25.8100
V (Å^3)	779.6686	780.7936	778.9632	779.5300
z(Sr)	0.0494	0.0534	0.0517	0.0531
x(O1)	0.196	0.285	0.207	0.196
z(O2)	0.9654	0.9669	0.9686	0.9667
Ir-O1 (Å)	1.987	1.963	1.973	1.993
Ir-O2 (Å)	2.33	2.37	2.41	2.37
Ir-O1-Ir (°)	156.6	164.1	160.5	158.6
φ_{oct} (°)	11.70	7.95	9.75	10.2

Figure 2 shows the SEM graphs of $Sr_{2-x}Eu_xIrO_4$, whose average particle sizes are estimated by Nano Measurer software. The powder shows almost uniform morphology, and the average crystallite sizes are 4.04μm, 3.28μm, 2.81μm, and 3.33μm for x=0 to x=0.06 samples, respectively. The average particle size slightly decreases, and the grain size distribution increases with doping. These could probably be attributed to the increasing content of Eu^{3+} with smaller ionic radii. It is also observed that the porosity in Figure 2 gradually increases with the increasing content of Eu, which is consistent with the reduced densities and increased unit-cell volume shown in Table 1. The introduction of Eu^{3+} ions causes a composition gradient between the grain bulk and grain boundaries, which will affect the electrical properties [17].

Figure 2. The SEM graphs of $Sr_{2-x}Eu_xIrO_4$ compounds.

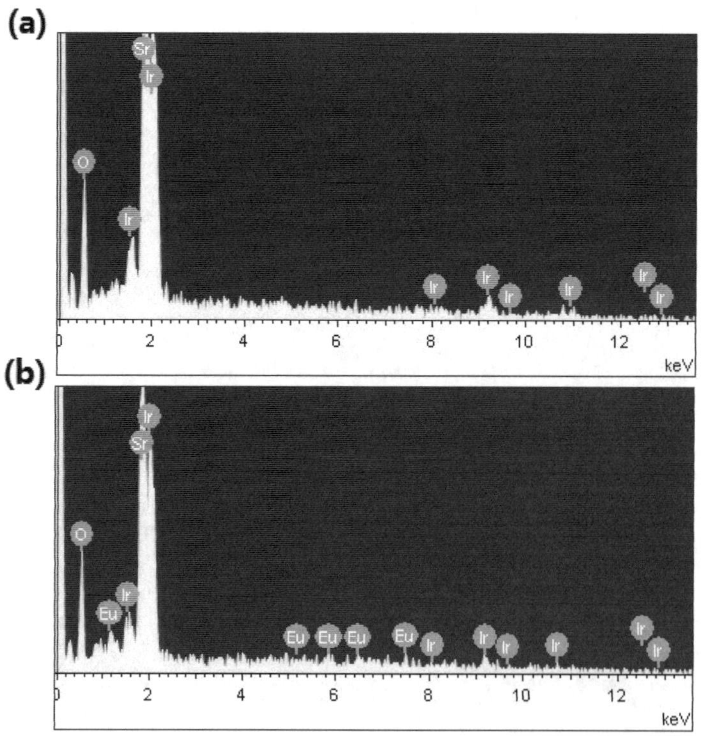

Figure 3. The EDX pictures for $x = 0$ and $x = 0.02$ compounds.

Table 2. The components of the $Sr_{2-x}Eu_xIrO_4$ series.

sample	Nominal component	Real component
0	Sr_2IrO_4	$Sr_{1.9313}Ir_{1.0685}O_{4.0683}$
0.02	$Sr_{1.98}Eu_{0.02}IrO_{4.01}$	$Sr_{1.9877}Eu_{0.0298}Ir_{0.9824}O_{3.9972}$
0.04	$Sr_{1.96}Eu_{0.04}IrO_{4.02}$	$Sr_{1.9421}Eu_{0.0517}Ir_{0.9961}O_{4.0119}$
0.06	$Sr_{1.94}Eu_{0.06}IrO_{4.03}$	$Sr_{1.9293}Eu_{0.0719}Ir_{0.9787}O_{3.9946}$

The real component of the compounds was measured by EDX spectra (see Figure 3), and the corresponding calculated result was summarized in Table 2. The data show that the real component of Eu is slightly larger than the nominal component, effectively demonstrating that the $Sr_{2-x}Eu_xIrO_4$ samples were successfully synthesized.

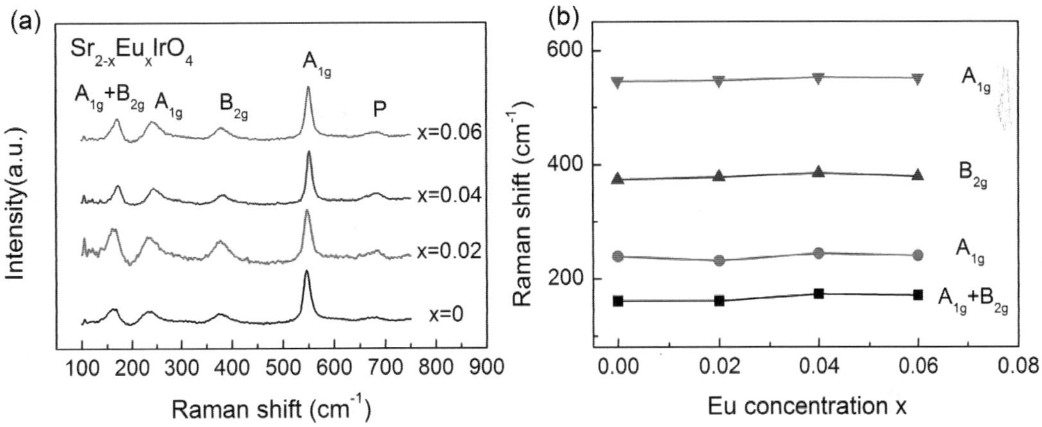

Figure 4. (a) Raman spectra of $Sr_{2-x}Eu_xIrO_4$ samples; (b) The frequencies of Raman modes versus Eu concentration x.

In Figure 4(a), we plot the Raman spectra of $Sr_{2-x}Eu_xIrO_4$ at room temperature. We are able to observe five modes. Based on previous reports, we assign the low energies mode around 161 cm^{-1} as the merging of A_{1g} and B_{2g} modes, and assign the modes around 240 cm^{-1}, and 546 cm^{-1} are A_{1g} modes [18]. The $A_{1g}+B_{2g}$ ~161 cm^{-1} expresses the IrO_6 octahedra rotation, combined with the Sr stretching against IrO_6, the A_{1g}~240 cm^{-1} is a mode of the IrO_6 rotation, and the A_{1g} ~546 cm^{-1} is the stretching mode of the top O2 atoms [1,18-20]. The 374 cm^{-1} peak is B_{2g} phonon modes, which is a bending mode of O1 squares in the plane [18-20]. The weak mode near 683 cm^{-1} whose frequency nearly does not change with doping, marked by P most likely roots in the two-phonon scattering.

All the modes undergo some degree of hardening with the increasing doping concentration (Figure 4(b)). It is known that the shift of vibrational modes relies on the mass of the cations and the bonding forces. Since the mass of the Eu^{3+} ion is larger than that of Sr^{2+}, if we consider the atomic mass only, the vibrational modes should represent redshift according to the inverse relationship between the mass of atoms and the wavenumber. Therefore, the hardening with the increasing content of Eu doping mainly depends on the decreased ionic radius (Eu^{3+}: 94.7pm; Sr^{3+}: 118 pm), which can be verified by the decreased Ir-O1 bond length in Table 1.

3.2. Resistivity study

Figure 5(a) shows the resistivity of the $Sr_{2-x}Eu_xIrO_4$ samples. Compared with the resistivity of the parent material, all the doped materials show an increase in resistivity. At the temperature range of 200–300 K, the resistivity can be fitted using the expression $\rho(T)=\rho_0\exp(-\alpha T)$ (see Figure 5(b)). The resistivity within 100–200 K follows the equation $\rho(T)=\rho_0\exp(\Delta/2k_BT)$ (see Figure 5(c)). However, at low temperatures (30–100 K), the curves can be described by a 3D VRH model written by

$\rho(T)=\rho_0\exp(T_0/T)^{1/4}$ (see Figure 5(d)). The transport behaviors divided by three temperature ranges have also been observed in previous reports [13,14]. All the fitted data are shown in Table 3. We observe that the values of Δ, which is defined as the calculated energy gap, are very small for an insulator, implying that the gap originated from electronic localization. The gap presents a nonmonotonic increase with the increase of Eu^{3+} doping, which is in conformity with the increase of resistivity.

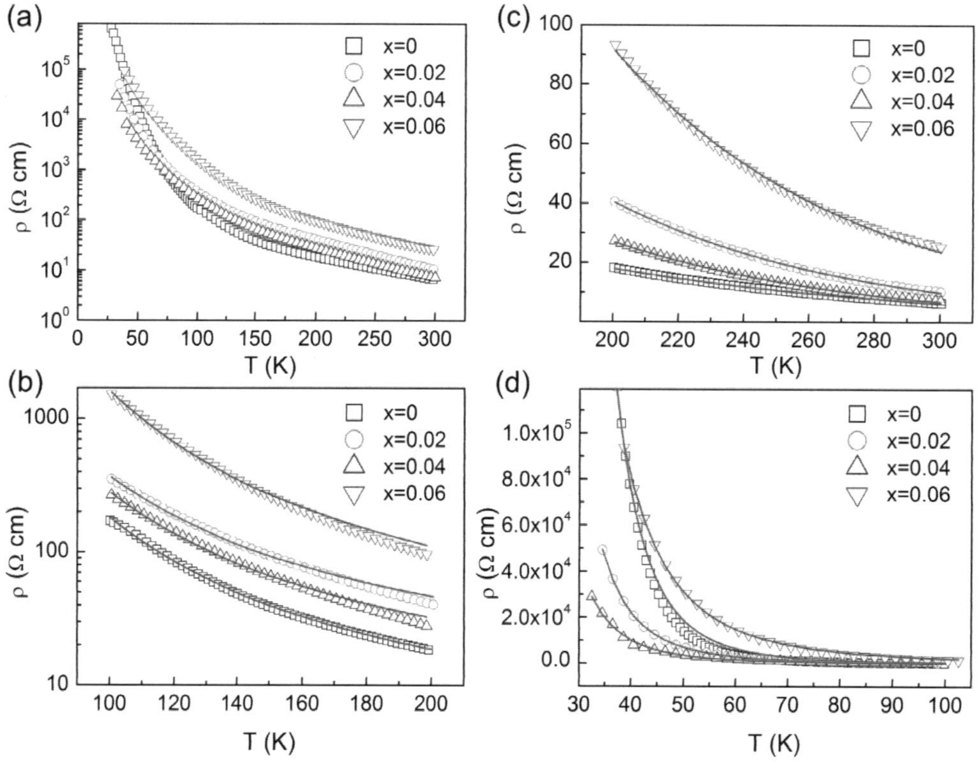

Figure 5. (a) The electric resistivity versus temperature of the $Sr_{2-x}Eu_xIrO_4$ series. The resistivity fitted by $\rho(T)=\rho_0\exp(-\alpha T)$ in the range 200-300 K (b), $\rho(T)=\rho_0\exp(\Delta/2k_BT)$ between 100 and 200 K (c), and $\rho(T)=\rho_0\exp(T_0/T)^{1/4}$ below 100 K (d), respectively. The fitting curves are marked with solid lines.

Table 3. The fitted parameters of ρ-T curves in three different temperature regions.

x	200–300 K $\rho(T)=\rho_0\exp(-\alpha T)$		100–200 K $\rho(T)=\rho_0\exp(\Delta/2k_BT)$		30–100 K $\rho(T)=\rho_0\exp(T_0/T)^{1/4}$	
	α (K^{-1})	ρ_0 (Ω cm)	Δ (meV)	ρ_0 (Ω cm)	T_0 (K)	ρ_0 (Ω cm)
0	1.04×10^{-2}	144.14	78.60	1.91	1.66×10^{7}	7.09×10^{-7}
0.02	1.41×10^{-2}	681.08	71.62	5.86	8.15×10^{6}	1.01×10^{-5}
0.04	1.38×10^{-2}	424.73	74.90	3.67	5.97×10^{6}	3.00×10^{-5}
0.06	1.36×10^{-2}	1395.45	92.75	7.51	3.86×10^{6}	1.78×10^{-3}

3.3. The ultraviolet-visible absorption spectra

UV-vis absorption spectroscopy can uncover important information between electronic bands and energy gaps in insulating materials. The optical absorbance spectra of the compounds are shown in Figure 6. For the indirect nature of the band gap in Sr_2IrO_4, the optical transition mechanism can be investigated by Taucs' formula

$$(\beta h\nu)^{1/2} = A(h\nu - E_g) \quad (1)$$

Where β denotes the absorption coefficient, the constant A depending on the electronic transition probability, and E_g is the optical energy gap. Taucs' plot and energy gap are given in the insets of Figure 6.

From the insets, it is observed that the band gap of Sr_2IrO_4 is only 0.83 eV, in agreement with the previous report [4]. The doped samples exhibit a firstly decrease and then increase band gap, which is consistent with the variation of calculated gap Δ with doping hereinbefore. The small band gap indicates that all the compounds are Mott insulators.

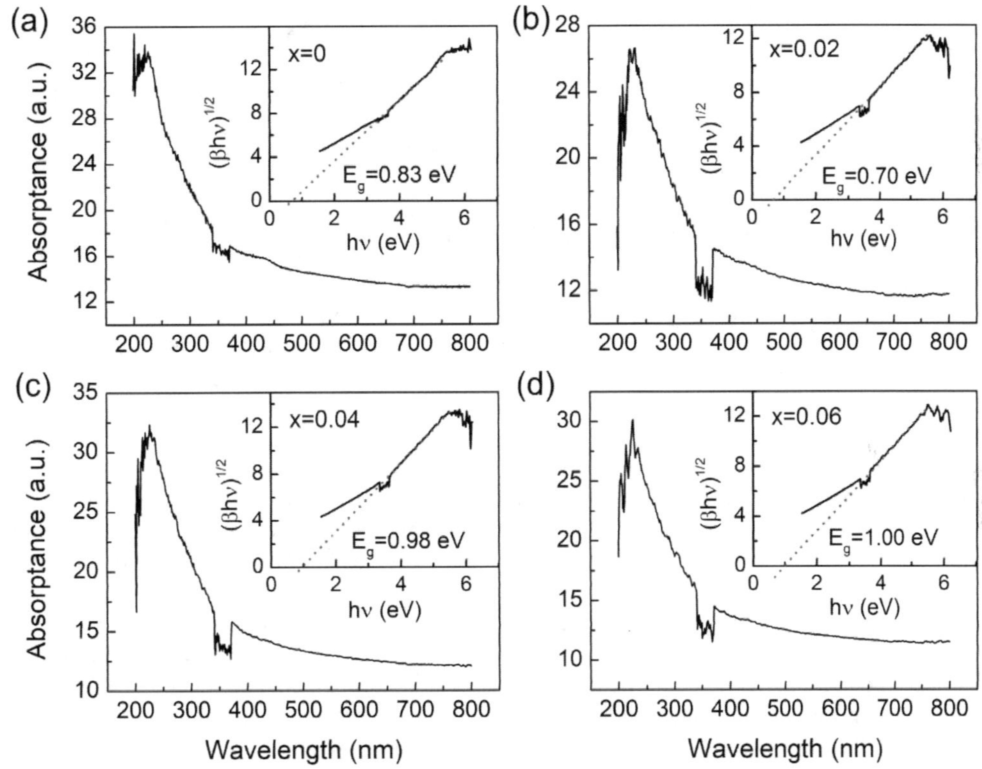

Figure 6. UV-vis absorption spectra of the $Sr_{2-x}Eu_xIrO_4$ samples. The insets plot $(\beta h\nu)^{1/2}$-$h\nu$ curves where the optical band gaps are given.

3.4. Dielectric study

Figure 7(a) and 7(b) give the dielectric constant (ε_r) and tangent loss (tanδ) versus frequency (f(Hz)) at room temperature, respectively. At low frequencies, the value of ε_r is very high, which may be due to some kinds of polarization. However, the value of ε_r decreases with frequency, suggesting the polarization decreases. This result is similar to that of the $Sr_2Ir_{0.9}V_{0.1}O_4$ compound [10]. We also find that ε_r increases with doping, which is in agreement with the increase of resistivity with doping. In addition, the tanδ initially decreases and then increases with doping. For the x=0.2 sample, it has the biggest ε_r and the smallest tanδ. At f=10^7 Hz, tanδ=0.739. It has potential application value in microelectronics at high frequency.

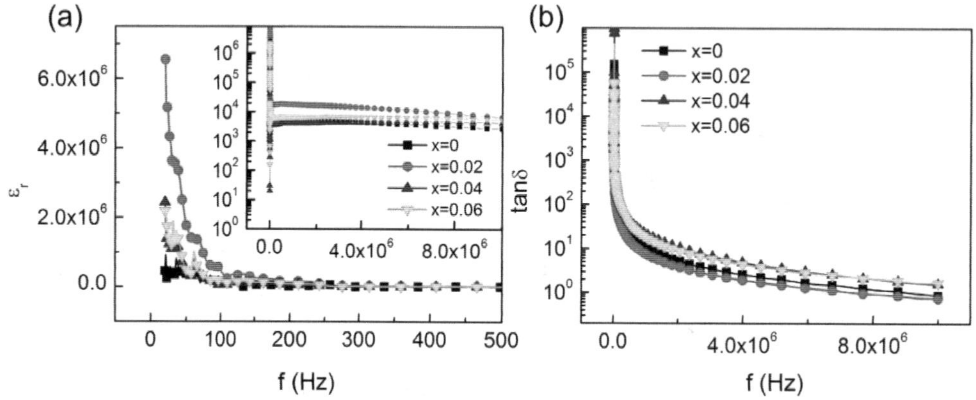

Figure 7. (a) The real part of dielectric constant ε_r vs frequency; (b) The dielectric loss tangent tanδ vs frequency.

4. Discussion

For the Sr_2IrO_4 system, the main crystal distortion is a staggered rotation of IrO_6 octahedra (φ_{oct}) and tetragonal distortion (c/a). The Ir-O-Ir bond angles show an increase with doping as a whole, which means the octahedral rotation φ_{Oct} decreases. As demonstrated in Table 1, Eu doping induces an increase in the angle of Ir-O1-Ir, which will enhance the *d*-orbital overlap or electron hopping [11]. One may expect that the electrical resistivity should decrease. However, the increment of resistivity with doping should be originated from other factors.

The increased c/a shows an increased tetragonal distortion. The enhanced tetragonal distortion originates from two factors: decreased a (Ir-O1 length) and increased c (Ir-O2 length). This means there are two competing forces at work: (1) The decreased in-plane Ir-O1 bond length and the increased Ir-O1-Ir bond angle will enhance the *d*-orbital overlap or electron hopping, favoring metallic behavior. (2) The increased out-plane Ir-O2 bond length increases the interlayer distance, which leads to localization and weakens electron hopping, resulting in insulating state. Besides, the grain boundary in polycrystalline samples will play an important role. The random distribution with doping, in general, will introduce disorder in the conducting pathway and enhance electron localization, resulting in the suppression of the electron screening effect and the increase of Coulomb interaction, which leads to the suppression of electrical conductivity. The above-competing factors lead to increased resistance to doping. The evolution trend of the resistivity ρ, the dielectric constant ε_r, and tangent loss tanδ with doping is consistent with the change of tetragonal distortion, which implies that the electric properties are mainly determined by tetragonal distortion c/a. In addition, the grain boundary in polycrystalline samples, modification of U_{ee} by electron doping through Eu^{3+}, and disorder should have contributed to the electric transport properties.

The physical properties of the compounds rely heavily on the parameters of their crystal structure, i.e., φ_{oct} and c/a. It is also noted that the structure parameters present nonmonotonic variations with doping, which may be due to energy level mismatch for the Eu and Sr sites, leading to the randomness of the Eu/Sr occupations.

5. Conclusions

We have conducted a research on the structural, electric, optics, and dielectric properties of the perovskite iridates $Sr_{2-x}Eu_xIrO_4$. All the samples belong to the *I41/acd* space group. The resistivity increases slightly with the increase of doping concentration, which can be further verified by the slight increase of band gap obtained from ultraviolet-visible absorption spectra. The mechanism of electric transport properties can be divided into three temperature ranges. In addition, Eu doping makes the dielectric constant increases; thus, the electrical conduction ability weakens. For the x=0.02 sample, it has the biggest dielectric permittivity and the smallest loss tangent, which has potential application value in microelectronics at high frequency. The electric properties are mainly determined by increased *c*/*a*. All the results indicate that the physical properties of a doped Sr_2IrO_4 system are strongly correlated with the crystal distortion.

Acknowledgments
This work was supported by the Natural Science Foundation of the Education Department of Anhui Province (Nos. KJ2020A0652, gxyq2022071, KJ2019A0840, 2021kcszsfkc360), the Natural Science Foundation of Anhui Province (No. 2108085QE198), the Talent Research Fund of Hefei University (Grant No. 18-19RC36) and the Undergraduate Training Programs (Nos. S202111059043, 202211059103, 202111059019 and 202111059071).

References

[1] Liu X, Cao Y, Pal B, Middey S, Kareev M, Choi Y, Shafer P, Haskel D, Arenholz E, and Chakhalian J 2017 *Phys. Rev. Mater.* **1** 075004
[2] Ye F, Chi S, Chakoumakos B C, Fernandez-Baca J A, Qi T and Cao G 2013 *Phys. Rev. B* **87** 140406
[3] Lane C, Zhang Y, Furness J W, Markiewicz R S, Barbiellini B, Sun J and Bansil A 2020 *Phys. Rev. B* **101** 155110
[4] Dhital C, Hogan T, Yamani Z, Cruz C de la, Chen X, Khadka S, Ren Z and Wilson S D 2013 *Phys. Rev. B* **87** 144405
[5] Chen C, Zhou Y, Chen X, Han T, An C, Zhou Y, Yuan Y, Zhang B, Wang S, Zhang R, Zhang L, Zhang C, Yang Z, DeLong L E, and Cao G 2020 *Phys. Rev. B* **101** 144102
[6] Bertinshaw J, Kim J K, Porras J, Ueda K, Sung N H, Efimenko A, Bombardi A, Kim J, Keimer B and Kim B J 2020 *Phys. Rev. B* **101** 094428
[7] Xu B, Marsik P, Sheveleva E, Lyzwa F, Louat A, Brouet V, Munzar D and Bernhard C 2020 *Phys. Rev. Lett.* **124** 027402
[8] Zwartsenberg B, Day R P, Razzoli E, Michiardi M, Xu N, Shi M, Denlinger J D, Cao G, Calder S, Ueda K, Bertinshaw J, Takagi H, Kim B J, Elfimov I S and Damascelli A 2020 *Nat. Phys.* **16** 290-294
[9] Yuan S J, Aswartham S, Terzic J, Zheng H, Zhao H D, Schlottmann P and Cao G 2015 *Phys. Rev. B* **92** 245103
[10] Gondh S and Pramanik A K 2020 AIP Conf. Proc. **2220** 040033
[11] Yuan S J, Aswartham S, Terzic J, Zheng H, Zhao H D, Schlottmann P and Cao G 2015 *Phys. Rev. B* **92** 245103
[12] Hu Y, Chen X, Peng S T, Lane C, Matzelle M, Sun Z L, Hashimoto M, Lu D H, Schwier E F, Arita M, Wu T, Markiewicz R S, Shimada K, Chen X H, Shen Z X, Bansil A, Wilson S D and He J F 2019 *Phys. Rev. Lett.* **123** 216402
[13] Liu H, Liang D, Bian J, Feng Y, Xie Y, and Fang B 2020 *Mater. Res. Express* **7** 026102
[14] Han T, Wang Y, Yang J, He L, Xu J, Liang D, Han H, Ge M, Xi C Y, Zhu W K, Zhang C, and Zhang Y 2016 *Appl. Phys. Lett.* **109** 192409
[15] Nelson J N, Parzyck C T, Faeth B D, Kawasaki J K, Schlom D G and Shen K M 2020 *Nat. Commun.* **11** 2597
[16] Gim Y, Sethi A, Zhao Q, Mitchell J F, Cao G, and Cooper S L 2016 *Phys. Rev. B* **93** 024405
[17] Guo X and Waser R 2006 *Prog. Mater. Sci.* **51** 151–210
[18] Gretarsson H, Sauceda J, Sung N H, Höppner M, Minola M, Kim B J, Keimer B and Tacon M Le 2017 *Phys. Rev. B* **96** 115138
[19] Samanta K, Ardito F M, Souza-Neto N M and Granado E 2018 *Phys. Rev. B* **98** 094101
[20] Cetin M F, Lemmens P, Gnezdilov V, Wulferding D, Menzel D, Takayama T, Ohashi K and Takagi H 2012 *Phys. Rev. B* **85** 195148

The electric transport properties of iridate $Sr_{2-x}Bi_xIrO_4$ and $Sr_{2-y}Dy_yIrO_4$

Yanfei Xiong[1], Hao Zu[2] and Hui Liu[2]

[1] College of Energy Materials and Chemicals, Hefei University, Hefei 230601, China

[2] College of Advanced Manufacturing Engineering, Hefei University, Hefei 230601, China

E-mail: liuhui@hfuu.edu.cn (H. Liu)

Abstract. Sr_2IrO_4 is predicted to be a new superconductor in theory, and its electric transport properties have attracted much attention in the field of physics. Here we prepared a series of polycrystalline samples doped with $Sr_{2-x}Bi_xIrO_4$ (x=0,0.02,0.04,0.06,0.08) and $Sr_{2-y}Dy_yIrO_4$ (y=0,0.02,0.04,0.06) respectively. The influence of the two ions on the lattice structure and electronic structure after the substitution of the Ir site is very similar, which makes the overall reduction trend of IrO_6 octahedron and Ir-O1-Ir bond angle. The resistivity shows a non-monotonic change with doping and still maintains the insulation. The low-temperature section, medium-temperature section, and high-temperature section conform to three-dimensional variable range hopping (3D-VRH), thermal activation model, two-dimensional weak localization model respectively.

1. Introduction

5d transition metal oxides with layered chalcogenide structures have been extensively researched because they express a variety of physical properties under crystal fields, spin-orbit coupling, and Coulomb interactions [1-4]. The balance between spin-orbit coupling and crystal field cleavage is easily broken, especially in 5d transition metal oxides, because the expanded d-orbital space and larger electron wave function radius, increased number of charges and wider range of electron energy bands in 5d transition metal oxides compared to 3d and 4d orbitals eventually lead to weakened Coulomb interaction forces in the oxides.5d transition metal oxides have been considered as candidates for numerous novel physical properties, such as J_{eff}=1/2 Mott state [5-7], topology Insulation, Weyl Semi-Metal [8], quantum Spin Liquids[9], and so on.

Sr_2IrO_4 with layered perovskite structure is a 5d iridium-based transition metal oxide. From the calculation of the energy band, it should be a compound with metallicity, however, after several experimental studies, it is a weakly ferromagnetic insulator [10]. This is due to the fact that the J_{eff} = 1/2 energy band is very weak and can also split an energy gap under the action of small Coulomb repulsion energy, which we call the Mott energy gap, resulting in a Mott insulator with J_{eff} = 1/2. Therefore, Sr_2IrO_4 is a Mott insulator [11]. The crystal structure and electronic structure of Sr_2IrO_4 are similar to that of La_2CuO_4, the parent of a copper oxide superconductor, both are K_2NiF_4 type and 1/2 Hubbard pseudospin with layered perovskite structure [12].

The theory predicts that Sr_2IrO_4 can achieve superconductivity through electron or hole doping, and become a new unconventional superconductor. Therefore, it has been continuously explored through a variety of experimental methods like electron and hole doping. We found that the metal insulation transition can be achieved by electron doping $(Sr_{1-x}La_x)_2IrO4$, oxygen vacancy with different concentrations in single crystal Sr_2IrO_4 samples, and substitution of Sr sites by Ga elements can achieve the metal-insulation transition by studying the results of previous studies [13-14]. Similarly, the hole doping achieved by doping Rh/Ru elements into the Ir sites can also lead to the metallic behavior of the sample [15-16].

In this work, we use Bi and Dy elements to replace Sr sites to introduce electrons and to fit the resistance temperature curves at different temperature bands by three different models to analyze the conductivity mechanism of doped samples in more detail, so as to reveal the coupling relationship between electronic structure and electrical properties.

2. Experiment

The polycrystalline doped samples $Sr_{2-x}Bi_xIrO_4$ ($0 \leq x \leq 0.08$) and $Sr_{2-y}Dy_yIrO_4$ ($0 \leq y \leq 0.06$) were produced by the traditional solid phase reaction method, where we mixed the raw materials $SrCO_3$ (99.99%), Bi_2O_3 (99.99%), Ir metal powder and Dy_2O_3 (99.99%) completely according to the stoichiometric number. The powder was calcined three times in a high-temperature muffle furnace at 800°C-1250°C, then ground twice in an agglomerate mortar and pressed to form small discs with a diameter up to 13 mm. X-ray diffraction (XRD) was performed on the prepared samples using a Rigaku diffractometer (Cu Kα radiation) to describe the crystal architecture and phase fineness of the doped specimens. The XRD data were refined by Rietveld structure using Fullprof software to obtain the lattice parameters. The particle size was estimated by Hitachi SU8010 scanning electron microscope (SEM). The chemical composition and distribution of elements were investigated by SEM and EDX. The resistivity data were collected using conventional four-probe method (Quantum Design, PPMS) at temperatures ranging from 4 to 300K.

3. Results

3.1. XRD

Figure 1(a) and 2(a) show the X-ray diffraction plots of $Sr_{2-x}Bi_xIrO_4$ (x=0,0.02,0.04,0.06,0.08) and $Sr_{2-y}Dy_yIrO_4$ (y=0,0.02,0.04,0.06) at about 300K, respectively. JCPDS), all the peaks in the $Sr_{2-x}Bi_xIrO4$ doped sample can be found and there are no spurious peaks, which indicates that the sample is a pure phase. The prepared sample of $Sr_{2-y}Dy_yIrO_4$ (y=0.04,0.06) has a small heterogeneous peak (*, the asterisk symbol) at $2\theta \sim 44.11°$, except that the doped phases of $Sr_{2-x}Bi_xIrO_4$ and $Sr_{2-y}Dy_yIrO_4$ are both with $I4_1/acd$ tetragonal structure and meet the symmetry criteria.We refined the structures of the samples using Fullprof software, and representative refined plots for x=0 and y=0.02 are given in Figures 1(b) and 2(b), respectively. Although the substitution of ions did not cause a change in the phase, the lattice parameters still differ from Sr_2IrO_4 with the change in the doping amount. a series of parameters obtained after refinement are shown in Figures 1 and 2 (c) and (d). In the doping of the two ions, we can see that the lattice parameters a and c have the same trend as doping. The overall trend is increasing, at x=0.08 and y=0.06, respectively. From Figure 1 and (d) in Figure 2, we can easily see that the angle of the Ir-O1-Ir bond tends to increase and then decrease with the increasing doping amount for the two ions, which indicates that the rotation φ ($\varphi=1/2(180°-\theta_{Ir-O1-Ir})$) of the IrO_6 octahedron is first decreasing and then increasing with the doping amount. Meanwhile, with the substitution of Bi, c/a shows a non-monotonic change, which indicates that the quadrilateral aberration shows a tendency to increase and then decrease. As the Dy is added, although a turn occurs at y=0.04, c/a still shows a gradually decreasing trend, and the quadratic aberration shows an increasing trend.

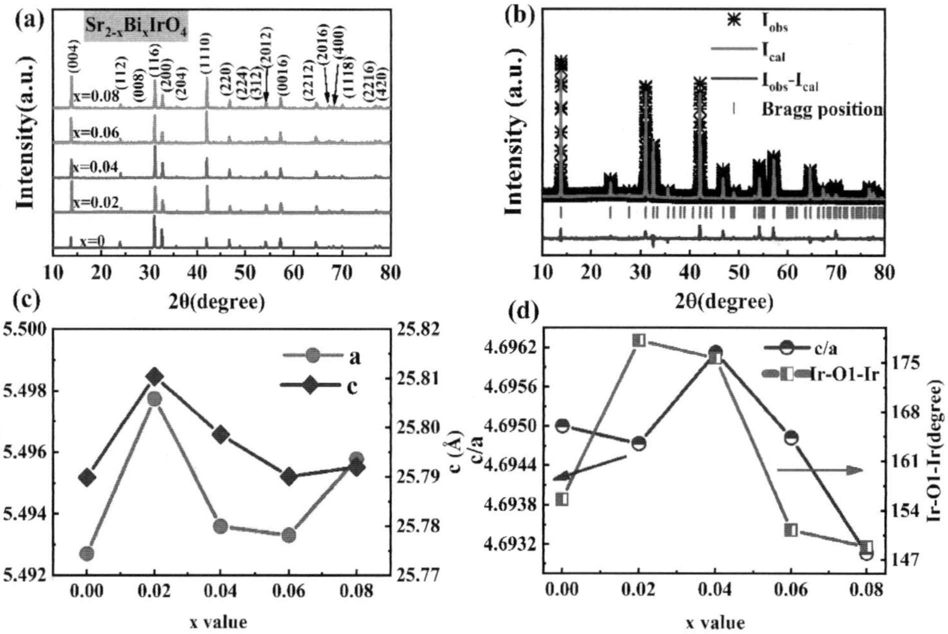

Figure 1. (a)XRD patterns of $Sr_{2-x}Bi_xIrO_4$(x=0,0.02,0.04,0.06,0.08). (b) Rietveld refinement of the x=0.02. (c) Trends in lattice parameters. (d) Curves of c/a and Ir-O1-Ir angles with x.

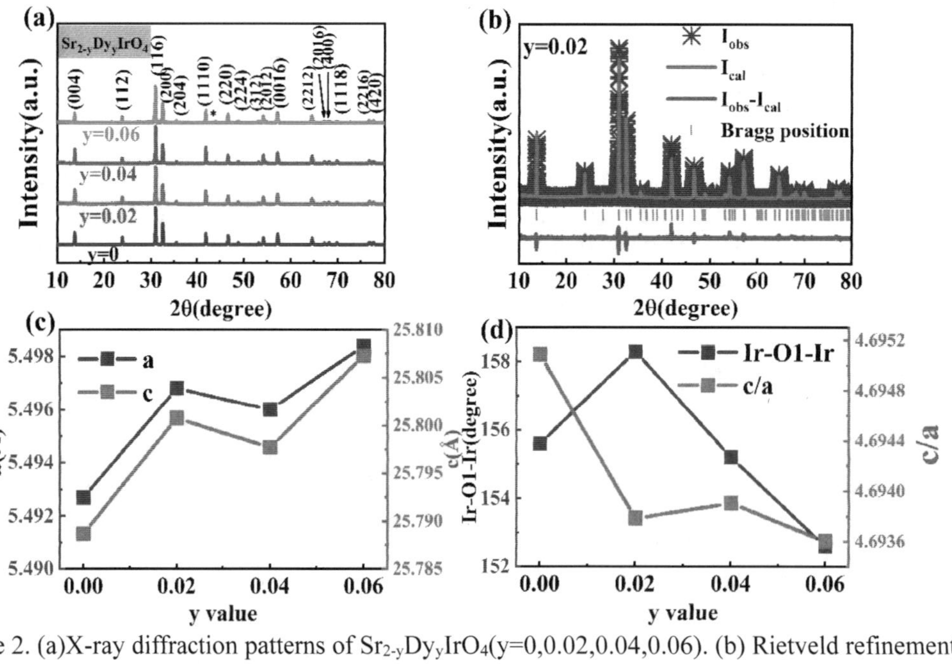

Figure 2. (a)X-ray diffraction patterns of $Sr_{2-y}Dy_yIrO_4$(y=0,0.02,0.04,0.06). (b) Rietveld refinement of the y=0.02. (c) Trends in lattice parameters. (d) Curves of c/a and Ir-O1-Ir angles with y.

3.2. EDX and SEM

The SEM images of $Sr_{2-x}Bi_xIrO_4$ and $Sr_{2-y}Dy_yIrO_4$ are given in Figure 4. It is not difficult to get from the figure that with the increase of x, the density of the sample gradually increases and the particles become smoother and fuller. However, the changing trend of $Sr_{2-y}Dy_yIrO_4$ in Figure 4 is completely opposite to that of $Sr_{2-x}Bi_xIrO_4$. The average size of the particles is 4.17um, 2.20um, 2.25um, and

Journal of Physics: Conference Series **2370** (2022) 012034 doi:10.1088/1742-6596/2370/1/012034

2.59um for y=0, 0.02, 0.04, and 0.06, respectively, and the overall size of the particles tends to decrease, indicating that the two ions are successfully doped into the Sr_2IrO_4 parent material. This is also demonstrated by the EDX spectrum in Figure 3.

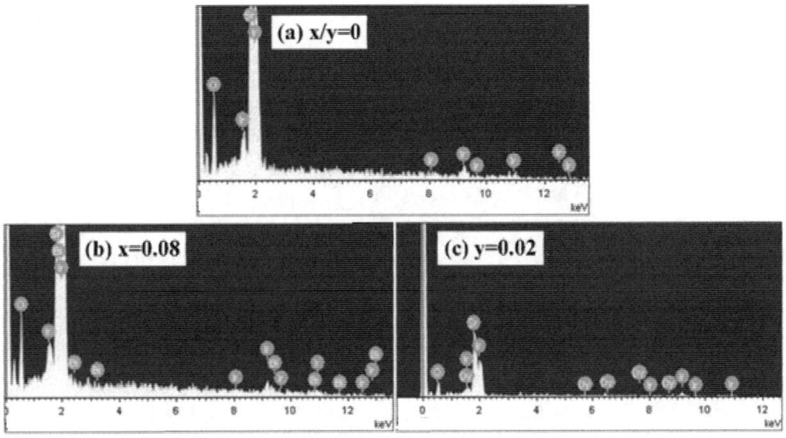

Figure 3. The EDX spectra for (a) x/y= 0, (b) x= 0.08, (c) y=0.02 compounds.

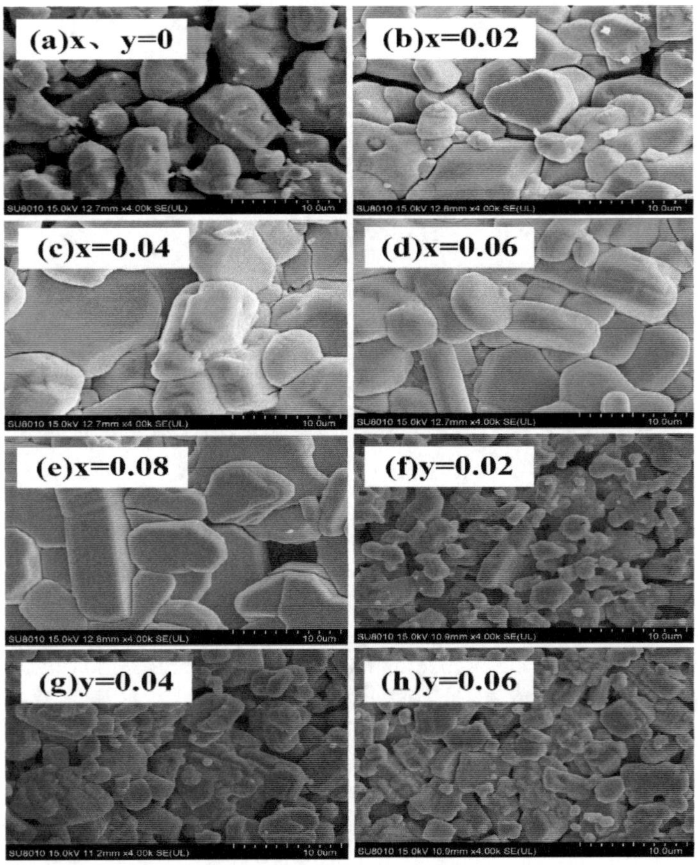

Figure 4. The SEM graphs for (a-e) $Sr_{2-x}Bi_xIrO_4$(x=0,0.02,0.04,0.06,0.08), (a) and (f-h) $Sr_{2-y}Dy_yIrO_4$(y=0,0.02,0.04,0.06).

3.3. Resistivity

The variation of resistivity with temperature for $Sr_{2-x}Bi_xIrO_4$ and $Sr_{2-y}Dy_yIrO_4$ is shown in Figures 5 and 6 (a), and we can find that all the samples remain highly insulating and show non-monotonic variation for both Bi and Dy doping. When the doping concentration x and y≤0.04, the resistivity shows a monotonically declining trend, and then gradually increases with doping. The resistivity of all doped samples except for x, y = 0.06 is lower than that of the parent material. It is clear to us that the physical properties of the material are closely related to the structural changes. We find from the above XRD data that the Ir-O1-Ir bond angle shows the non-monotonic variation which increases first and then decreases, and the larger the Ir-O1-Ir bond angle becomes, the easier the electron leap is. So by the above conclusion we have reason to speculate the decreasing resistivity when x, y ≤ 0.04 may be related to the increase of Ir-O1-Ir bond angle. Besides, an increase in the band gap from the UHB to the LHB for the J_{eff}=1/2 state is caused by the irregular distortion of IrO_6 octahedron, which can be proved by the results of DFT calculation. Therefore, the variation of resistivity with x in a nonmonotonic manner may be caused by a combination of the angle of the Ir-O1-Ir bond and the deformation by the IrO6 octahedron.

To better investigate the electrical transport properties of this series of samples, we used the following three different models to fit the segments for each temperature band.

Figures 5 and 6(b) show the variation of resistivity with the low temperature section region (40-70K), we use the three-dimensional variable-range-hopping model to fit, the 3D-VRH formula as

$$\rho(T) = \rho_0 \exp\left(\frac{T_0}{T}\right)^{1/4} \tag{1}$$

(ρ_0 is resistance coefficient and T_0 is characterization temperature). This can appear from the good linearity, the temperature band is consistent with this model. Thus, the transport properties are mainly determined by the jump conduction of local carriers, and even in the semiconductor region, the Fermi energy level may have a small DOS. The specific values of the fitting parameters ρ_0 and T_0 are listed in Table 1, the magnitude of T_0 depends on the density of states at the fermion energy scale N_{EF} and the localization length l can be described by the equation:

$$T_0 = \frac{18}{k_B N_{EF} l^3} \tag{2}$$

k_B is the Boltzmann constant. Therefore, for doped samples, density of states at fermion energy scale is the main cause for the variation of T_0 with x.

Figures 5 and 6(c) show the resistivity exhibit an Arrhenius-type behavior with temperature in the middle-temperature band (120K-190K, 110K-210K), the equation is

$$\rho(T) = \rho_0 \exp\left(\frac{\Delta}{2k_B T}\right) \tag{3}$$

From Table 1, we can see the fitted data, the undoped Δ= 82.08meV in the sample, which is approximately equal to that reported in $Sr_{2-x}Pr_xIrO_4$. Whether Bi-doped or Dy-doped, Δ decreases with increasing doping, indicating that the pseudo gap arises from the localization of the electronic states and that doping changes the structure of the conductivity.

Figures 5 and 6(d) show the variation of resistivity at temperatures close to room temperature (215K-275K, 210K-300K), which can be described by this equation:

$$\rho(T) = \rho_0 \exp(-\alpha T) \tag{4}$$

The values of ρ_0 and α are also listed in Table 1, and the perfect fit curves indicate that in quasi-two-dimensional materials, the localization of electrons is weak, and α shows an increasing tendency with doping, indicating a gradual increase in the electron interactions.

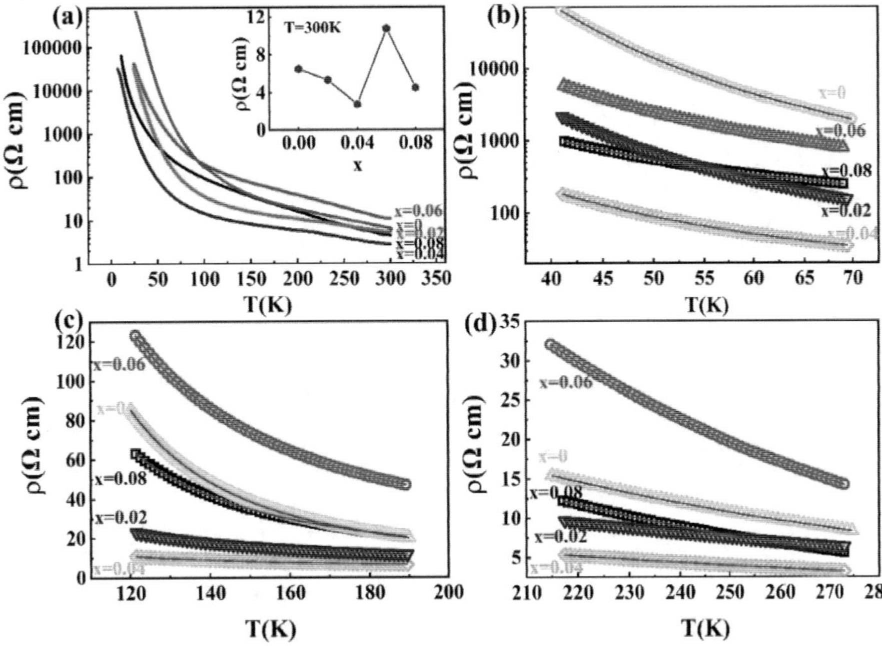

Figure 5. (a) The tendency of electrical resistivity for $Sr_{2-x}Bi_xIrO_4$ with temperature . Inset shows ρ_{300K} versus x. (b) Curves of ρ versus T between 215K and 275 K. (c) The curve of ρ versus T between 120K and 190 K. (d) The curve of ρ versus T curves between 40K and 70 K. The red solid lines are fitting data.

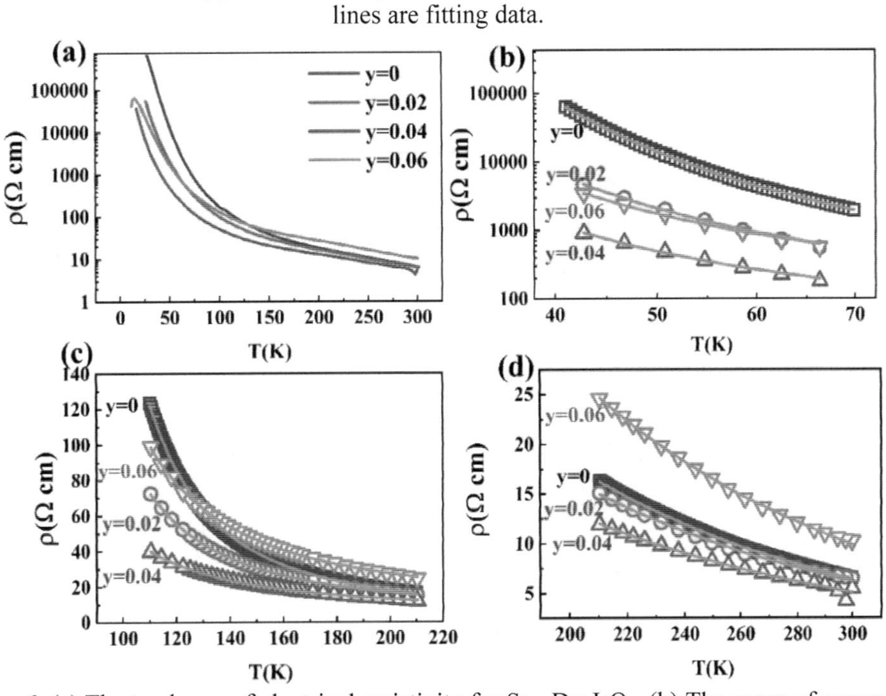

Figure 6. (a) The tendency of electrical resistivity for $Sr_{2-y}Dy_yIrO_4$. (b) The curve of ρ versus T between 40K and 70 K. (c) Curves of ρ versus T between 110K and 210 K. (d) The curve of ρ versus T between 210K and 300 K. Yellow solid lines are fitting data.

Table 1. Resistivity fitting results for $Sr_{2-x}Bi_xIrO_4$ and $Sr_{2-y}Dy_yIrO_4$ samples in the various temperature ranges.

	$\rho(T)=\rho_0\exp(T_0/T)^{1/4}$		$\rho(T)=\rho_0\exp(\frac{\Delta}{2k_B}*T^{-1})$		$\rho(T)=\rho_0\exp(-\alpha T)$	
	T_0(K)	$\rho_0(\Omega\cdot cm)$	Δ(meV)	$\rho_0(\Omega\cdot cm)$	$\alpha(K^{-1})$	$\rho_0(\Omega\cdot cm)$
0	$1.39x10^3$	13.51	81.36	1.69	$1.03x10^{-2}$	140.72
x	**40-70K**		**120-190K**		**215-275K**	
0.02	$1.07x10^3$	3.29	40.06	3.39	$7.60x10^{-3}$	50.55
0.04	$6.68x10^2$	3.13	30.98	2.49	$8.98x10^{-3}$	37.59
0.06	$8.24x10^2$	39.53	56.33	8.38	$1.39x10^{-2}$	642.87
0.08	$5.65x10^2$	31.94	65.09	2.82	$1.40x10^{-2}$	255.03
y	**40-70K**		**110-210K**		**210-300K**	
0.02	$9.68x10^2$	16.45	62.14	2.77	$9.76x10^{-3}$	116.61
0.04	$7.44x10^2$	11.86	47.16	3.36	$9.51x10^{-3}$	88.58
0.06	$8.27x10^2$	27.62	54.45	5.66	$1.01x10^{-2}$	207.98

4. Conclusions

To sum up, we specially studied the effects doping of Bi and Dy of Sr_2IrO_4 on the electronic structure as well as electronic properties. We were surprised to find that the doping results of these two ions were very close, both of which make the bond angles of IrO_6 octahedron and Ir-O1-Ir decrease as a whole, and the lattice constant changes non-monotonically, indicating that there are two kinds of interaction competition in the sample plane. Similarly, the doping of Bi and Dy makes the resistivity change non-monotonically with the increase of doping, which may be the result of the joint action of lattice disorder and octahedral regular modulation.

Acknowledgments

This work was supported by the Natural Science Foundation of the Education Department of Anhui Province (Nos. KJ2020A0652, gxyq2022071, KJ2019A0840, 2021kcszsfkc360), the Natural Science Foundation of Anhui Province (No. 2108085QE198), the Talent Research Fund of Hefei University (Grant No. 18-19RC36) and the Undergraduate Training Programs for Innovation and Entrepreneurship (Nos. S202111059043, 202211059103, 202111059019 and 202111059071).

References

[1] Okabe H, Isobe M, Takayama-Muromachi E, Takeshita N, and Akimitsu J, 2013 *Phys. Rev. B* **88** 4049-4056
[2] Cao G, Qi T F, Li L, Terzic J and Kaul R K, 2013 *Phys. Rev. B* **88** 220414
[3] Wang F, Senthil T, 2011 *Phys. Rev. Lett.* **106** 186-197
[4] Lu C, Quindeau A, Deniz H, Preziosi D, Hesse D, and Alexe M, 2014 *Appl. Phys. Lett.* **105** 076402
[5] Kim B J, Jin H, Moon S J, Kim J Y, Park B G, Leem C S, Yu J, Noh T W, Kim C and Oh S J, 2008 *Phys. Rev. Lett.* **101** 076402
[6] Moon S J, Jin H, Kim K W, Choi W S, Lee Y S, Yu J, Cao G, Sumi A, Funakubo H, Bernhard C, 2008 *Phys. Rev. Lett.* **101** 226402
[7] Kim B J, Ohsumi H, Komesu T, Sakai S, Morita T, Takagi H, Arima T, 2009 *Science* **323** 5919
[8] Yu M, Grasso S, Mckinnon R, Saunders T and Reece M J, 2017 *Adv. Appl. Ceram.* **116** 1-37
[9] Balents L, 2010 Nature. 464 199-208
[10] Kim Y K, Sung N H, Denlinger J D and Kim B J 2015 Nature. Phys. 12 37-41
[11] Kim B J, Jin H, Moon S J, Kim J Y, Park B G, Leem C S, Yu J, Noh T W, Kim C and Oh S J, 2008 *Phys. Rev. Lett.* **101** 076402
[12] Wang F and Senthil T, 2011 *Phys. Rev. Lett.* **106** 186-197
[13] Xiang C, Hogan T, Walkup D, Zhou W, Pokharel M, Yao M, Wei T, Ward T Z, Zhao Y and Parshall D, 2015 *Phys. Rev. B* **92** 075125

[14] Qi T F, Korneta O B, Chikara S, Ge M, Parkin S and Long L D, 2011 *J. of Appl. Phys.* **109** 07D906
[15] Qi T F, Korneta O B, Li L, Butrouna K, Cao V S, Wan X, Schlottmann P, Kaul R K and Cao G, 2012 *Phys. Rev. B* **86** 125105
[16] Yuan S J, Aswartham S, Terzic J, Zheng H, Zhao H D, Schlottmann P and Cao G, 2015 *Phys. Rev. B* **92** 245103

AUTHOR INDEX